Reversible Deactivation Radical Polymerization

Also of interest

Twin Polymerization.
New Strategy for Hybrid Materials Synthesis
Spange, Mehring (Eds.), 2018
ISBN 978-3-11-050067-7, e-ISBN 978-3-11-049936-0

Polymer Engineering.
Tylkowski, Wieszczycka, Jastrzab, 2017
ISBN 978-3-11-046828-1, e-ISBN 978-3-11-046974-5

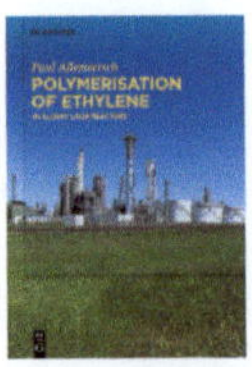

Polymerisation of Ethylene.
In Slurry Loop Reactors
Allemeersch, 2015
ISBN 978-3-11-029214-5, e-ISBN 978-3-11-029219-0

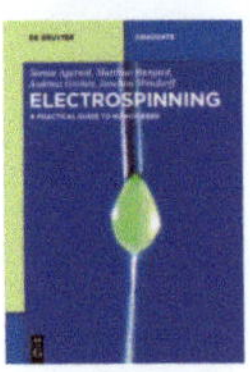

Electrospinning.
A Practical Guide to Nanofibers
Agarwal, Burgard, Greiner, Wendorff, 2016
ISBN 978-3-11-033180-6, e-ISBN 978-3-11-033351-0

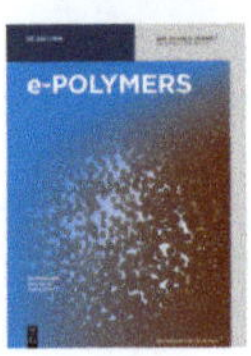

e-Polymers.
Editor-in-Chief: Seema Agarwal
ISSN 2197-4586, e-ISSN 1618-7229

Reversible Deactivation Radical Polymerization

Synthesis and Applications of Functional Polymers

Edited by
Nikhil K. Singha, Jimmy Mays

DE GRUYTER

Editors
Prof. Dr. Nikhil K. Singha, FRSC
Former Chair Rubber Technology Centre
Indian Institute of Technology
Kharagpur-721302
India
nks@rtc.iitkgp.ac.in

Prof. Dr. Jimmy Mays
University of Tennessee
Department of Chemistry
552 Buehler Hall
Knoxville 37996-1600
USA
jimmymays@utk.edu

ISBN 978-3-11-063999-5
e-ISBN (PDF) 978-3-11-064369-5
e-ISBN (EPUB) 978-3-11-064017-5

Library of Congress Control Number: 2019944952

Bibliographic information published by the Deutsche Nationalbibliothek
The Deutsche Nationalbibliothek lists this publication in the Deutsche Nationalbibliografie; detailed bibliographic data are available on the Internet at http://dnb.dnb.de.

Typesetting: Integra Software Services Pvt. Ltd.
Printing and binding: CPI books GmbH, Leck
Cover image: Science Photo Library / Laguna Design

www.degruyter.com

Dedicated to my Beloved Father

Preface

Free radical polymerization (FRP) is one of the most widely used methods to prepare polymers. The major advantages of this polymerization method are its utility with a large number of vinyl monomers, its robustness, its tolerance toward adventitious impurities and even water, leading to its use in emulsion and suspension polymerizations. However, there are several disadvantages of FRP. Via FRP it's difficult to prepare polymers with precisely controlled molecular weights and narrow dispersity (*Đ*). It's also difficult to prepare polymers with well-defined architectures, such as block copolymers, graft copolymers with controlled backbone and graft lengths, star polymers, core-shell polymers, and so on. Pioneering work on FRP using thio-iniferters in the 1980s is the path-breaking work in the field of controlled radical polymerizations (CRP), widely used since the mid-1990s. CRP is now "renamed" as reversible deactivation radical polymerization (RDRP), as recommended by IUPAC. In this book, almost in all cases, the term RDRP has been used to indicate the controlled nature of the radical polymerization. Among the different RDRP methods, atom transfer radical polymerization (ATRP), reversible addition fragmentation chain transfer (RAFT), and nitroxide mediated polymerization (NMP) are widely used. The main objective of this book is to assemble information on the preparation of tailor-made functional polymers and their composites using different RDRP methods.

"Click" reactions are an important class of organic reactions that are used to prepare different advanced polymeric materials and their nanocomposites via post-polymerization modification of polymers. These materials have a variety of specialty applications. Among the different click reactions, Diels–Alder reactions and thiol-ene reactions are widely used. A further objective of this book is to review the preparation of functional polymers via different RDRP methods and various "click" reactions. This book also covers the potential application of these materials as specialty polymer composites, for example in self-healing applications and in different biomaterials applications. Chapter 1 delineates the different methods of RDRP. Chapter 2 describes the historical development of RDRP and use of different RDRP techniques in the preparation of polymer nanohybrid materials via in situ and surface-initiated polymerizations. Chapter 3 covers design and synthesis of tailor-made end-functional polymers via ATRP. Chapters 4 and 5 elucidate strategies for preparing functional polymers via the combination of RDRP and click reactions, such as Diels–Alder, as well as thiol-ene reactions, respectively. Chapter 6 depicts the preparation of fluorinated polymers via different RDRP processes. Finally, Chapter 7 reports biomedical applications of functional polymers prepared via different RDRP methods. This book will be of great interest to polymer chemists and materials scientists who are engaged in research and design of new polymeric materials for various applications. Students engaged in the above disciplines will be also benefitted by studying the book.

https://doi.org/10.1515/9783110643695-202

We are thankful to all the authors for the contribution of their chapters. Thanks are due to Dr. Arindam Chakrabarty, an SRF with CSIR, New Delhi, as well as a Fulbright Fellow for his assistance in editing this book.

List of contributors

Prof. Bruno Ameduri
Engineering and Macromolecular Architectures
Institut Charles Gerhardt UMR (CNRS) 5253
Ecole Nationale Supérieure de Chimie de Montpellier
8, Rue Ecole Normale
34296 Montpellier Cedex 5
France
e-mail: bruno.ameduri@enscm.fr

Prof. Andrew B. Lowe
Nanochemistry Research Institute
Department of Chemistry
School of Science
Faculty of Science & Engineering
Curtin University
Kent Street
Bentley WA 6102
Australia
e-mail: andrew.b.lowe@curtin.edu.au

Dr. Nabendu B. Pramanik
Deparment of Chemistry
Lehigh University
6 E Packer Ave
Bethlehem, PA 18015
USA
e-mail: nabenduchem@gmail.com

Dr. Sanjib Banerjee
Department of Chemistry
Indian Institute of Technology Bhilai
GEC Campus, Sejbahar
Raipur 492015
Chhattisgarh
India
e-mail: sanjib.banerjee@iitbhilai.ac.in

Dr. Arindam Chakrabarty
JSPS Post-doctoral Fellow
Division of Forest and Biomaterials Science
Graduate School of Agriculture
Kyoto University, Kyoto 606-8502
Japan
e-mail: arindamchakrabarty.vu@gmail.com

Prof. Priyadarsi De
Polymer Research Centre
Department of Chemical Sciences
Indian Institute of Science Education and Research Kolkata
Mohanpur - 741246
Nadia, West Bengal
India
e-mail: p_de@iiserkol.ac.in

Prof. Raghavachari Dhamodharan
Department of Chemistry
Indian Institute of Technology Madras
Chennai 600 036
India
e-mail: damo@iitm.ac.in

Dr. Raghuraman G. Karunakaran
Evonik Corporation
7201 Hamilton Boulevard
Building R3 / Room D252
Trexlertown, PA 18195
USA
e-mail: raghuraman.govindan-karunakaran@evonik.com

Dr. Saswati Ghosh Roy
Polymer Research Centre
Department of Chemical Sciences
Indian Institute of Science Education and Research Kolkata
Mohanpur - 741246
Nadia
West Bengal
India
e-mail: saswati.ghoshroy9@gmail.com

Dr. Dhruba J. Haloi
Department of Chemistry
Bodoland University
Kokrajhar 783370
Assam
India
e-mail: dhruba2k3@gmail.com

https://doi.org/10.1515/9783110643695-203

Prof. Nikhil K. Singha
Rubber Technology Centre
Indian Institute of Technology Kharagpur
Kharagpur- 721302
West Bengal
India
e-mail: nks@rtc.iitkgp.ac.in

Prof. Bert Klumperman
South African Research Chair on Advanced Macromolecular Architectures
Department of Chemistry and Polymer Science
Stellenbosch University
Private Bag X1
Matieland 7602
South Africa
e-mail: bklump@sun.ac.za

Prof. Dr. Jimmy Mays
University of Tennessee
Department of Chemistry
552 Buehler Hall
Knoxville 37996-1600
e-mail: jimmymays@utk.edu

Mr. Prantik Mondal
Rubber Technology Centre
Indian Institute of Technology Kharagpur
Kharagpur- 721302
West Bengal
India
e-mail: mprantik3@gmail.com

Dr. Prakash P. Wadgaonkar
Polymers and Advanced Materials Laboratory
Polymer Science and Engineering Division
CSIR-National Chemical Laboratory
Dr. Homi Bhabha Road
Pune 411008
India
e-mail: pp.wadgaonkar@ncl.res.in

Mr. Sachin S. Patil
Reliance corporate park
Reliance Industries Limited, Ghansoli
Navi Mumbai-400709
Maharashtra
India
e-mail: patilss9011@gmail.com

Dr. Prakash S. Sane
Asian Paints Limited
Research & Technology Center
TTC Industrial area, Pawne MIDC
TURBHE
Navi Mumbai-400703
India
e-mail: pssane@gmail.com

Dr. Dnyaneshwar V. Palaskar
ELANTAS Beck India Limited
Mumbai-Pune Road
Pimpri
Pune- 411018
Maharashtra
India
e-mail: dvpalaskar@gmail.com

Contents

Bert Klumperman

1 Introduction to reversible deactivation radical polymerization

1.1 Introduction

Radical polymerization is among the oldest techniques used for polymer synthesis. It is a typical example of a chain growth reaction in which the active center at a growing chain end is an unpaired electron or radical. A wide variety of monomers that carry a C = C unsaturated functionality can be polymerized via radical polymerization. These monomers include styrene derivatives, acrylates, methacrylates, vinyl esters, *N*-vinyllactams, and many others. Despite its long history in polymer synthesis, what is now called conventional radical polymerization has always suffered from inherent shortcomings. The two most important of those shortcomings are the lack of stereoselectivity and the inability to control chain topology (broad molar mass distributions, no block copolymers, or other advanced architectures). Since the 1980s, the development of radical polymerization techniques with characteristics that resemble living chain growth polymerization methods (e.g., anionic polymerization) has received considerable attention. Through the 1990s, three main techniques were published that collectively are now known under the IUPAC recommended name reversible deactivation radical polymerization (RDRP). In earlier studies, controlled radical polymerization (CRP) and living radical polymerization (LRP) have extensively been used as names to describe the general concept of RDRP. The three main techniques within the field of RDRP are, in chronological order: nitroxide-mediated polymerization (NMP) [1], atom transfer radical polymerization (ATRP) [2, 3], and reversible addition-fragmentation chain transfer (RAFT) mediated polymerization [4]. In this introductory chapter, these three techniques will be described and discussed in relation to the synthesis of functional polymers.

1.2 Nitroxide-mediated polymerization (NMP)

The lack of control over the molar mass distribution in conventional radical polymerization stems from the continuous initiation and termination of polymer chains throughout the course of the reaction. The statistical nature of these processes leads to a dispersity (*Đ*) of 1.5 or 2, depending on the mode of termination (combination or disproportionation respectively). To confer a living character to radical

Bert Klumperman, Department of Chemistry and Polymer Science, Stellenbosch University, Matieland, South Africa

https://doi.org/10.1515/9783110643695-001

polymerization, two main criteria need to be fulfilled. First, initiation should take place early in the polymerization to ensure that all chains have the opportunity to grow throughout the course of the reaction. However, in radical polymerization this would normally lead to an enormous radical concentration with an inevitably high rate of termination. Therefore, a second criterion is needed, which entails the reversible deactivation of radical chain ends to lower the instantaneous radical concentration and consequently minimize the occurrence of bimolecular termination. In NMP, the most common way to initiate chains is via the homolytic dissociation of an alkoxyamine into a transient carbon-centered radical and a persistent nitroxide radical as is shown in Scheme 1.1 for the case of styrene polymerization with 2,2,6,6-tetramethylpiperidine-1-oxyl (TEMPO) as the mediating nitroxide.

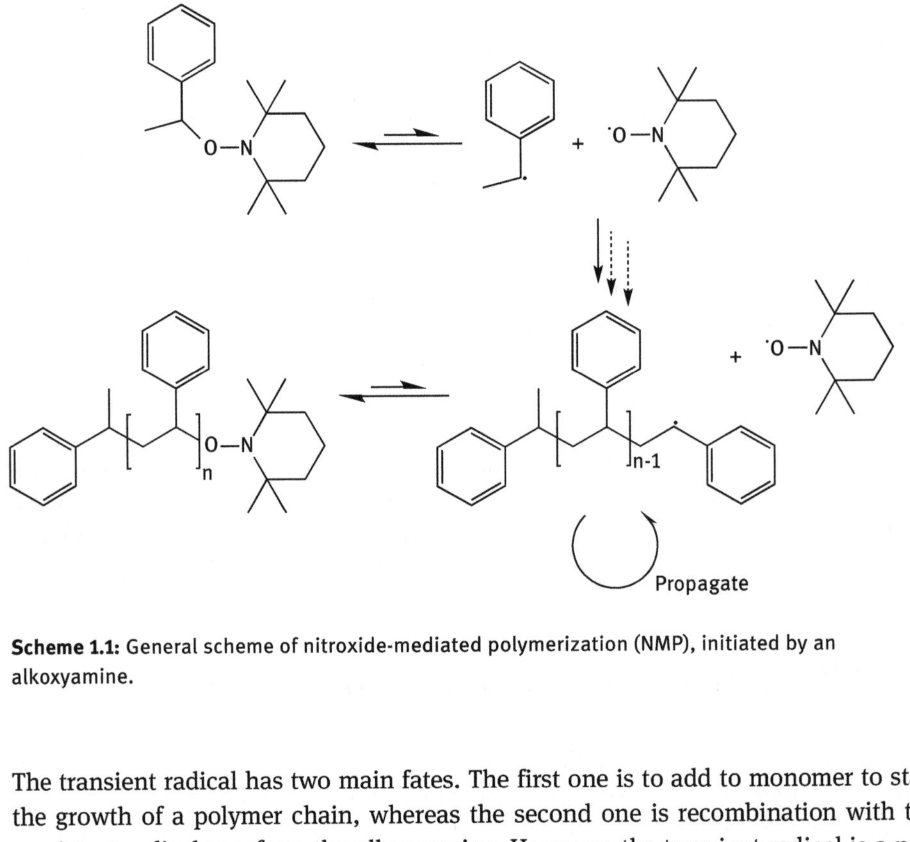

Scheme 1.1: General scheme of nitroxide-mediated polymerization (NMP), initiated by an alkoxyamine.

The transient radical has two main fates. The first one is to add to monomer to start the growth of a polymer chain, whereas the second one is recombination with the persistent radical to reform the alkoxyamine. However, the transient radical is a normal carbon-centered radical that can undergo all the usual reactions in radical polymerization including propagation, chain transfer, and termination. In a conventional radical polymerization, the time it takes for a complete chain to grow from initiation to termination is in the order of a second. The reversible deactivation process in NMP cuts this one-second into short intervals with a duration in the order of a millisecond. In between two consecutive intervals of chain growth, a dormant period is inserted that may last in the order of 10 milliseconds. During each activation–deactivation cycle, only one or very few monomer units are added to the polymer chain. As a consequence, all chains grow with a very similar rate, resulting in a narrow molar mass distribution. In addition, the reversible deactivation leads to a situation where only a very small fraction of all chains is in its active form at any point in time. This leads to a strong reduction in irreversible bimolecular termination, which also contributes to a reduction in the width of the molar mass distribution. To some extent, NMP has a self-regulating mechanism that reduces the transient radical concentration, namely the persistent radical effect (PRE) [5]. As soon as irreversible bimolecular termination takes place, an excess of persistent radicals is created. As in any equilibrium reaction, the build-up of concentration of a species leads to a shift in equilibrium. In this specific case, the increased concentration of persistent radicals leads to a shift of the activation–deactivation equilibrium toward the dormant side. This effectively means that the transient radical concentration is reduced, which leads to a reduction in rate of polymerization and a strong reduction in rate of termination. Ultimately, this leads to an enhanced living character of the polymerization, which is shown by a narrow molecular weight distribution (MWD) and the majority of polymer chains carrying well-defined α- and ω-chain end functionalities. Georges and coworkers contributed a lot to the optimization of NMP in the 1990s. They found that the addition of additives like camphor sulfonic acid has a rate-enhancing effect on NMP [6, 7]. The mechanism of rate enhancement is not yet fully elucidated [8].

In the case of NMP, the α-chain end is defined by the choice of the initiating primary radical that forms part of the initial alkoxyamine. Typical examples are esters that resemble acrylate- or methacrylate-propagating radicals as depicted in Figure 1.1. The ω-chain end is an alkoxyamine and there are two ways to make use of its functionality in, for example, conjugation reactions. The first option is to use an alkoxyamine that carries a suitable functionality on the nitroxide fragment, such

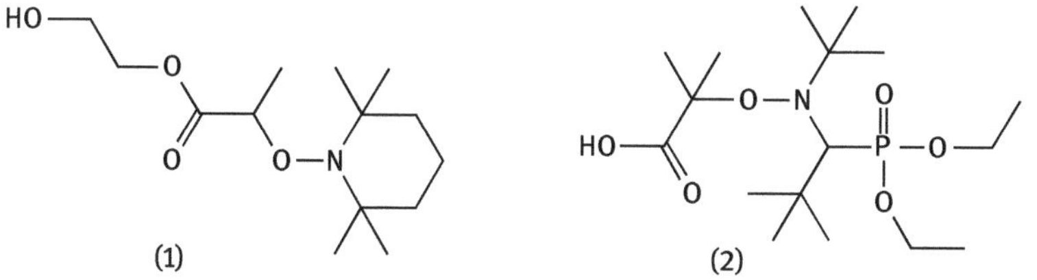

(1) (2)

Figure 1.1: Examples of alkoxyamines, **(1)** 2-hydroxyethyl 2-((2,2,6,6-tetramethylpiperidin-1-yl)oxy) propanoate and **(2)** 2-((*tert*-butyl(1-diethoxyphosphoryl)-2,2-dimethylpropyl)amino)oxy)-2-methylpropanoic acid, also known as BlocBuilder or MAMA-SG1.

as, for example, 4-hydroxy-TEMPO. The second option is to end cap the ω-chain end in a bimolecular termination reaction. An example of this strategy is to decompose a radical initiator in the presence of the alkoxyamine terminal polymer chains. When a large excess of radicals is generated of an initiator like 4,4′-azobis(4-cyanopentanoic acid)(ACPA) in the presence of the suitable macro-alkoxyamine, hetero-bimolecular termination will lead to the introduction of an ω-chain end carboxylic acid functionality as shown in Scheme 1.2.

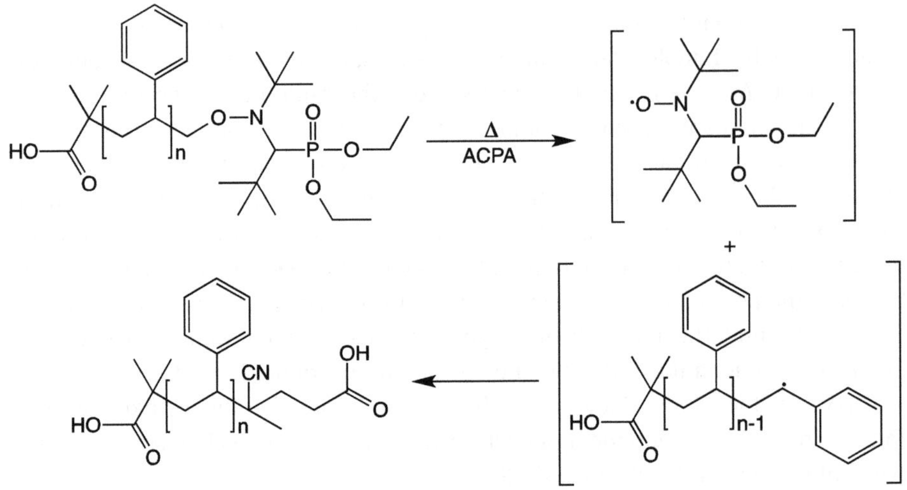

Scheme 1.2: End-group modification of SG1-terminal PSTY by radical addition with ACPA.

1.3 Transition metal-mediated radical polymerization

In the mid-1990s, Matyjaszewski and Sawamoto first published atom transfer radical polymerization (ATRP) independently [2, 3]. A suitably stabilized alkyl halide is typically used as an initiator where a transition metal-based catalyst in its lower oxidation state (Cu[I], Ru[II], etc) acts as the activator. Cu[I] complexes are most frequently used as activator/catalyst and in a one-electron redox process, the halide gets abstracted from the initiator to create a carbon-centered primary radical and oxidize the copper complex to its Cu[II] state. The carbon-centered transient radical can add to monomer to initiate the growth of a polymer chain, and any carbon-centered radical (primary radical or growing chain) can reversibly be deactivated by the Cu[II] complex. Similar to NMP, a dynamic equilibrium between carbon-centered transient radical and dormant species, in this case halides, is established. The large difference between NMP and ATRP is that in the case of NMP, the dynamic equilibrium is thermally induced, whereas in the case of ATRP the transition metal catalyst is required to mediate the equilibrium. In the early days of ATRP, equimolar amounts of the copper catalyst would typically be used relative to the initiator. This led to serious concerns about the applicability of the technique for practical applications, especially, in the biomedical field. Significant research efforts were spent on methods to prevent the contamination of the final product with the copper catalyst. These efforts included the use of immobilized catalysts [9], biphasic reaction mixtures [10, 11], and advanced purification techniques [12–14]. However, the real breakthrough in the lowering of residual copper only came with the development of techniques that rely on much lower catalyst concentrations. The kinetics of ATRP are controlled by the ratio of Cu[I] over Cu[II] [15]. The consequence is that for the rate of polymerization, the absolute concentration of copper in the reaction mixture is not important, but only the indicated ratio. However, the dispersity of the resulting molar mass distribution is dependent on the Cu[II] concentration [15]. It was also experimentally confirmed that when the Cu[II] concentration drops below a certain value, the dispersity starts to increase to undesired values. On the basis of these considerations, Matyjaszewski and coworkers developed a technique for which they coined the name activators regenerated by electron transfer (ARGET) ATRP [16–18]. In this method, a low overall copper concentration is used and the ratio between Cu[I] and Cu[II] is controlled by the addition of reducing agents such as glucose, ascorbic acid, or tin[II] 2-ethylhexanoate ($Sn(EH)_2$) that counteract the build-up of Cu[II] caused by the PRE.

In recent years, a different approach toward transition metal-mediated polymerization has received significant interest for which the name single electron transfer-living radical polymerization (SET-LRP) has been coined [19–21]. From a practical point of view, SET-LRP strongly resembles ATRP, but there are some characteristic differences. In their first publications on SET-LRP, Percec and coworkers pointed out that the activator in their case is Cu[0]. The activation step leads to the formation of Cu[I] that quickly disproportionates into Cu[0] and Cu[II], where the latter acts as deactivator similar to ATRP. There has been quite some debate in literature on the mechanistic aspects of SET-LRP versus ATRP, but there seems to be relatively strong evidence that both mechanisms can occur. In a recent publication, supplemental activator and reducing agent (SARA) ATRP has been compared with SET-LRP. The mechanistic differences between the two systems are depicted in Scheme 1.3 [22]. A complex combination of factors, including type of monomer, solvent, ligand, and more, seems to determine which one of the two mechanisms takes place for a specific system. Haddleton and coworkers have recently published some very elegant work in which SET-LRP is used for the precision control of water-soluble homopolymers and block copolymers [23]. In their work, they show that Cu[I]Br in the presence of Me_6TREN as a ligand in aqueous media disproportionates very rapidly as evidenced by the appearance of a blue color in solution (Cu[II]) and

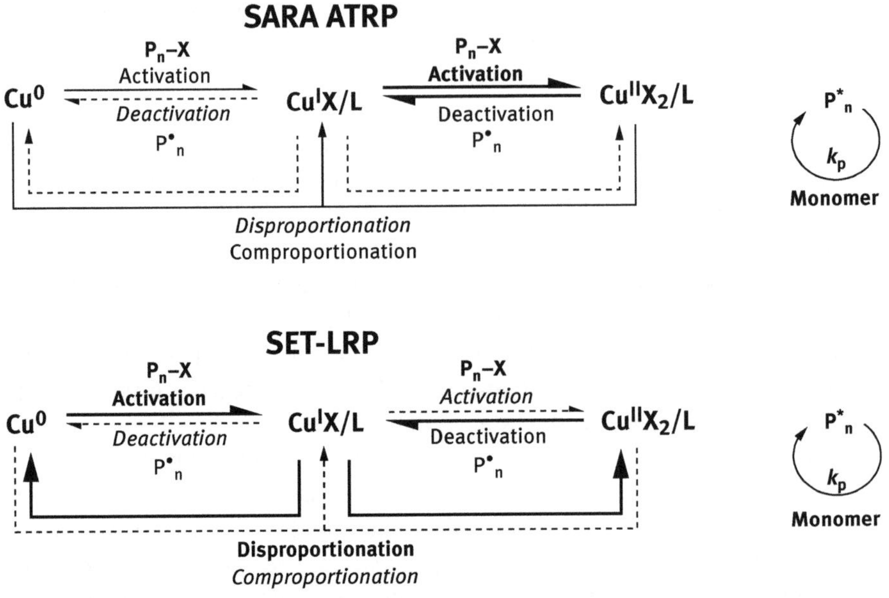

Scheme 1.3: Proposed SARA ATRP and SET-LRP mechanisms. Line thicknesses denote the contribution of the reaction, with bold reactions being dominant, thin solid lines contributing, and thin dashed lines being negligible. For simplicity, the products of activation and deactivation, and stoichiometric balance in comproportionation and disproportionation reactions are omitted. Reproduced with permission from D. Konkolewicz, P. Krys and K. Matyjaszewski, Accounts of Chemical Research, 2014, 47, 10, 3028. ©2014, ACS [22].

brownish precipitate in the vessel (Cu[0]). The use of this pre-disproportionated catalyst in the polymerization of *N*-isopropylacrylamide (NIPAM) leads to polymers with dispersities in the order of $Đ = 1.10$. Multiple chain extensions of such polymers with NIPAM under the same reaction conditions (H_2O, 0 °C) prove the excellent chain end fidelity, even when the monomer conversion after each individual chain extension is at least >96%. In a similar fashion, block copolymers of NIPAM, 2-hydroxy ethyl acrylate (HEA), and polyethyleneglycol acrylate (PEGA) can be prepared with excellent control.

1.4 Reversible addition-fragmentation chain transfer (RAFT)-mediated polymerization

While NMP, ATRP, and SET-LRP rely on reversible deactivation through the reaction between a polymer radical and a low molar mass species (nitroxide radical or transition metal complex in its higher oxidation state, e.g., Cu[II]), this is not the only option to achieve RDRP. In the late 1990s, a system based on degenerative chain transfer was published [4]. Reversible addition-fragmentation chain transfer (RAFT)-mediated polymerization relies on the chain transfer activity of thiocarbonyl thio compounds. At the start of a typical RAFT-mediated polymerization, a low molar mass dithioester, trithiocarbonate, dithiocarbamate, or xanthate is present. Radicals are generated through a conventional thermal radical initiator, and after the addition of one or very few monomers, reaction with the S = C bond of the thiocarbonyl thio compound takes place. This leads to the formation of an intermediate radical, which can largely undergo one of two fates. The first one is splitting off the radical that just added to the thiocarbonyl thio compound to regenerate the original compounds, and the other is the fragmentation of the so-called leaving (R) group radical. In the latter case, the leaving group radical can reinitiate a new chain that can in turn react with a RAFT agent to form the intermediate radical, after which the process repeats itself. In this sequence of events, the low molar mass RAFT agent effectively undergoes the insertion of one or very few monomers. The selective insertion of one monomer unit to form the single-monomer adduct is called initialization. After initialization has been completed, subsequent formation of intermediate radicals creates symmetrical versions of those intermediate radicals, with essentially identical leaving groups, albeit of possibly slightly different degrees of polymerization. The general reaction scheme of RAFT-mediated polymerization is shown in Scheme 1.4.

RAFT-mediated polymerization is compatible with virtually all monomers that are susceptible to radical polymerization. However, the nature of the R group and especially the stabilizing (Z) group needs to be tuned to the reactivity of the specific monomer [25]. Figure 1.2 provides an overview of typical RAFT agent leaving groups and stabilizing groups and an indication of the monomers that are compatible with them. The monomers can roughly be divided in two classes, denoted as more-activated monomers (MAMs) and less-activated monomers (LAMs). Typical examples of MAMs are styrene, acrylates, and methacrylates, whereas typical examples of LAMs are vinyl esters and vinyl lactams. The polymerization of MAMs is best controlled by dithioesters and trithiocarbonates, whereas LAMs are best controlled by dithiocarbamates and xanthates.

The need to use different RAFT agents for different classes of monomers puts some restrictions to the ability to synthesize block copolymers of arbitrarily chosen comonomer pairs.In order to circumvent this intrinsic problem of RAFT-mediated polymerization, a RAFT agent with a tunable reactivity was developed by Rizzardo and co-workers [26]. They synthesized *N*-(4-pyridinyl)-*N*-methyldithiocarbamates as a new class of so-called switchable RAFT agents that made it possible to control the polymerization of MAMs as well as LAMs. The switching ability in these RAFT agents is induced by the protonation/deprotonation of the pyridine substituent on the Z-group. Very recently it was shown that the use of semi-batch operation can also be used to mitigate the effect of a relatively poor RAFT agent on the width of

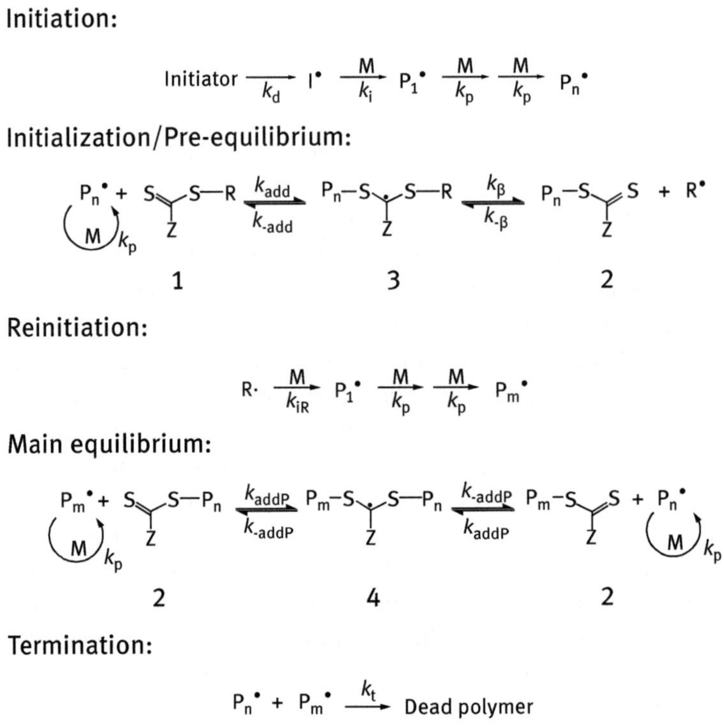

Scheme 1.4: General reaction scheme of RAFT-mediated polymerization. Reproduced with permission from D.J. Keddie, Chemical Society Reviews, 2014, 43, 2, 496. ©2014, RSC [24].

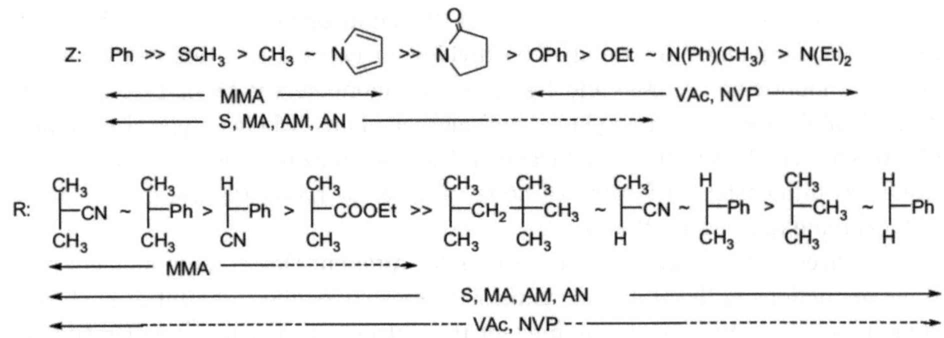

Figure 1.2: Guideline for the selection of RAFT agents for a variety of monomer types. Reproduced with permission from G. Moad, E. Rizzardo and S.H. Thang, Polymer, 2008, 49, 5, 1079. ©2008, Elsevier [25].

the MMD [27]. This technique has potential applicability in the synthesis of more complex polymer architectures.

1.5 Functional polymers

Functional polymers can be defined in many different ways. For the purpose of this chapter, we will stick to the definition for functional polymers of macromolecules carrying functional end-groups or functional side-groups. Access to polymers with functional side-groups via RDRP is relatively straightforward. The general rule-of-thumb is that those polymers can be synthesized if the monomer carrying the desired functionality is compatible with radical polymerization. There are a few exceptions of monomers that are compatible with conventional radical polymerization, but where the specific functional group interferes with the mediating agent of the RDRP system. An example that has long been quoted in this respect is the polymerization of acid-functional monomers such as acrylic acid, methacrylic acid, and styrene sulfonic acid via ATRP. In more recent years, it has become clear that the deprotonated acids are fully compatible with ATRP [28]. Similarly, a monomer containing an unprotected primary amine would not be compatible with RAFT-mediated polymerization, since aminolysis of the thiocarbonyl thio end-group would occur.

Interesting from a synthetic point of view are the options for end-functional polymers via RDRP. For ATRP as well as RAFT-mediated polymerization, there are many examples of chain end functionalities on the α- and ω-chain ends. Generally speaking, introduction of an α-chain end functionality requires the introduction of the desired group into the initiating radical (ATRP initiator or RAFT leaving group). A strategy that has been used very frequently is the esterification of a suitable alcohol to create the corresponding bromopropionate or bromo-isobutyrate ATRP initiator [29]. This initiator can also be used to include the same fragment as a leaving group in a RAFT agent. An example for the synthesis of an ATRP initiator is given in Scheme 1.5.

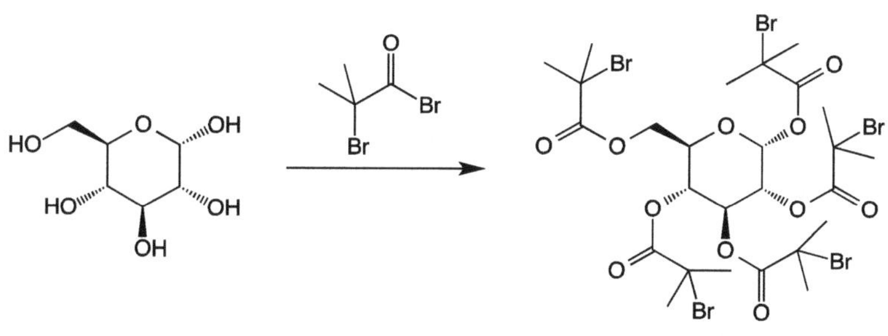

Scheme 1.5: Example of a multifunctional ATRP initiator based on glucose. Reproduced with permission from D. M. Haddleton, R. Edmonds, A.M. Heming, E.J. Kelly and D. Kukulj, New Journal of Chemistry, 1999, 23, 5, 477. ©1999, RSC [29].

A less frequently used general approach to introduce a leaving group in a RAFT agent is via click chemistry. In that case, an alkyne-functional RAFT-agent precursor is initially synthesized after which the alkyne is converted into a triazole. The latter is pseudo-aromatic and acts as a stabilizing group for the leaving group radical in pretty much the same way as a phenyl substituent does. The advantage of this approach is that there is no hydrolyzable link between polymer and functional group, which obviously enhances the stability of the end-group under various pH conditions [30]. The synthetic procedure for the synthesis of such RAFT agents is depicted in Scheme 1.6.

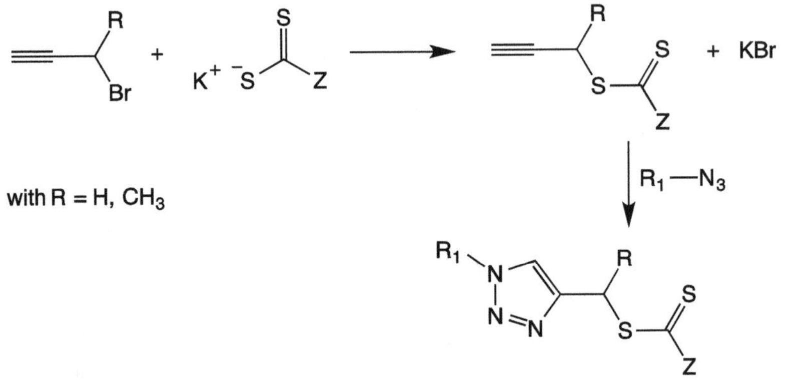

Scheme 1.6: Synthetic scheme for a triazole-based RAFT agent. Reproduced with permission from N. Akeroyd, R. Pfukwa and B. Klumperman, Macromolecules, 2009, 42, 8, 3014. ©2009, ACS [30].

The ω-chain end functionality is usually introduced via post-polymerization modification of a polymer synthesized via ATRP or RAFT-mediated polymerization. One strategy that has been implemented in several studies on polymers synthesized via ATRP is the introduction of an azide functionality, which is simply achieved by stirring a solution of the halide end-functional polymer with sodium azide or *p*-toluenesulfonyl azide (tosyl azide) [31]. The introduced azide is then in subsequent reactions suitable for coupling with an alkyne-functional compound via the Huisgen copper-catalyzed 1,3-dipolar cycloaddition (CuAAC) or stress-promoted azide-alkyne coupling (SPAAC) [32, 33]. Another strategy is to induce a bimolecular radical coupling reaction. To achieve this, the halide end-functional polymer is brought under ATRP condition, while simultaneously generating a relatively large concentration of highly mobile small molecule radicals [34]. A polymer radical will under those conditions have a high probability to undergo bimolecular termination with one of the small radicals and thus introduce the chain-end functionality of that radical, provided that the termination reaction occurs via coupling and not via disproportionation. For polymers synthesized via RAFT-mediated polymerization, there are quite a range of reactions that can be performed on the thiocarbonyl thio chain end. Arguably the most common one is aminolysis or hydrolysis, which leads to the introduction of a thiol chain end [35]. This thiol can subsequently be used for various coupling reactions, such as the formation of disulfides, Michael addition with a suitable C = C double bond, from, e.g., acrylate, maleimide [36], vinyl sulfone [37], and so on. An alternative chain end modification is the introduction of an unsaturated chain end via thermolysis of the thiocarbonyl thio end group [38]. Finally, there are very specific reactions that are only suitable under certain circumstances. A prime example is the introduction of an aldehyde chain end in xanthate-mediated synthesis of poly(*N*-vinylpyrrolidone) via consecutive hydrolysis and thermolysis of the xanthate under neutral or acidic conditions [39]. Many of the end group transformations have been carried out with high specificity and high yield.

1.6 Concluding remarks

RDRP techniques have only been around for approximately two decades, but they currently play a major role in research toward functional polymers. The general compatibility of radical polymerization with functional monomers is a major advantage over other techniques for polymer synthesis such as anionic polymerization and metallocene-catalyzed polymerization. A further advantage is that ATRP and RAFT-mediated polymerization have enormous potential regarding the introduction of α-chain end functionalities. In addition, both these polymerization techniques provide several possibilities to perform post-polymerization chain end modifications on the ω-chain end. In further chapters of this book, many examples will be highlighted of applications of the chemistry that has been briefly introduced in this chapter.

References

[1] D.H. Solomon, E. Rizzardo and P. Cacioli, inventors, New polymerization process and polymers produced thereby; EP135280A2, 1985.

[2] J.S. Wang and K. Matyjaszewski, Controlled/"living" radical polymerization. atom transfer radical polymerization in the presence of transition-metal complexes, Journal of the American Chemical Society, 1995, 117(20), 5614.

[3] M. Kato, M. Kamigaito, M. Sawamoto and T. Higashimura, Polymerization of methyl methacrylate with the carbon tetrachloride/dichlorotris- (triphenylphosphine)ruthenium(II)/ methylaluminum bis(2,6-di-tert-butylphenoxide) initiating system: possibility of living radical polymerization, Macromolecules, 1995, 28(5), 1721.

[4] J. Chiefari, Y.K. Chong; F. Ercole, J. Krstina, J. Jeffery, T.P.T. Le, R.T.A. Mayadunne, G.F. Meijs, C.L. Moad, G. Moad, E. Rizzardo and S.H. Thang, Living free-radical polymerization by reversible addition-fragmentation chain transfer: the RAFT process, Macromolecules, 1998, 31(16), 5559.

[5] H. Fischer, The persistent radical effect in "living" radical polymerization, Macromolecules, 1997, 30(19), 5666.

[6] P.G. Odell, R.P.N. Veregin, L.M. Michalak, D. Brousmiche and M.K. Georges, Rate enhancement of living free-radical polymerizations by an organic acid salt, Macromolecules, 1995, 28(24), 8453.
[7] M.K. Georges, R.P.N. Veregin, P.M. Kazmaier, G.K. Hamer and M. Saban, Narrow polydispersity polystyrene by a free-radical polymerization process-rate enhancement, Macromolecules, 1994, 27(24), 7228.
[8] A. Goto and T. Fukuda, Kinetics of living radical polymerization, Progress in Polymer Science, 2004, 29(4), 329.
[9] D.M. Haddleton, D. Kukulj and A.P. Radigue, Atom transfer polymerisation of methyl methacrylate mediated by solid supported copper catalysts, Chemical Communications, 1999, 1, 99.
[10] D.M Haddleton, S.G. Jackson and S.A.F. Bon, Copper(I)-mediated living radical polymerization under fluorous biphasic conditions, Journal of the American Chemical Society, 2000, 122(7), 1542.
[11] B. Zhang, X. Jiang, L. Zhang, Z. Cheng and X. Zhu, Fe(III)-mediated ICAR ATRP in a p-xylene/PEG-200 biphasic system: facile and highly efficient separation and recycling of an iron catalyst, Polymer Chemistry, 2015, 6(37), 6616.
[12] K. Matyjaszewski, T. Pintauer and S. Gaynor, Removal of copper-based catalyst in atom transfer radical polymerization using ion exchange resins, Macromolecules, 2000, 33(4), 1476.
[13] M.E. Honigfort, W.J. Brittain, T. Bosanac and C.S. Wilcox, Use of precipitons for copper removal in atom transfer radical polymerization, Macromolecules, 2002, 35(13), 4849.
[14] I. Ydens, S. Moins, F. Botteman, P. Degee and P. Dubois, Removal of copper-based catalyst in atom transfer radical polymerization using different extraction techniques, E-Polymers, 2004, 039.
[15] W. Tang and K. Matyjaszewski, Kinetic modeling of normal ATRP, normal ATRP with [CuII]0, reverse ATRP and SR&NI ATRP, Macromolecular Theory and Simulations, 2008, 17(7–8), 359.
[16] K. Matyjaszewski, H. Dong, W. Jakubowski, J. Pietrasik and A. Kusumo, Grafting from surfaces for "everyone": ARGET ATRP in the presence of air, Langmuir, 2007, 23(8), 4528.
[17] L. Mueller, W. Jakubowski and W. Tang, Successful chain extension of polyacrylate and polystyrene macroinitiators with methacrylates in an ARGET and ICAR ATRP, Macromolecules, 2007, 40(18), 6464.
[18] W. Jakubowski, B. Kirci-Denizli, R.R. Gil and K. Matyjaszewski, Polystyrene with improved chain-end functionality and higher molecular weight by ARGET ATRP, Macromolecular Chemistry and Physics, 2008, 209(1), 32.
[19] V. Percec, A.V. Popov, E. Ramirez-Castillo, M. Monteiro, B. Barboiu, O. Weichold, A.D. Asandei and C.M. Mitchell, Aqueous room temperature metal-catalyzed living radical polymerization of vinyl chloride, Journal of the American Chemical Society, 2002, 124(18), 4940.
[20] V. Percec, T. Guliashvili, J.S. Ladislaw, A. Wistrand, A. Stjerndahl, M.J. Sienkowska, M J. Monteiro and S. Sahoo, Ultrafast synthesis of ultrahigh molar mass polymers by metal-catalyzed living radical polymerization of acrylates, methacrylates, and vinyl chloride mediated by SET at 25 °C, Journal of the American Chemical Society, 2006, 128(43), 14156.
[21] M.J. Monteiro, T. Guliashvili and V. Percec, Kinetic simulation of single electron transfer–living radical polymerization of methyl acrylate at 25 °C, Journal of Polymer Science Part A: Polymer Chemistry, 2007, 45(10), 1835.
[22] D. Konkolewicz, P. Krys and K. Matyjaszewski, Explaining unexpected data via competitive equilibria and processes in radical reactions with reversible deactivation, Accounts of Chemical Research, 2014, 47(10), 3028.

[23] Q. Zhang, P. Wilson, Z. Li, R. McHale, J. Godfrey, A. Anastasaki, C. Waldron and D.M. Haddleton, Aqueous copper-mediated living polymerization: exploiting rapid disproportionation of CuBr with Me6TREN, Journal of the American Chemical Society, 2013, 135(19), 7355.
[24] D.J. Keddie, A guide to the synthesis of block copolymers using reversible-addition fragmentation chain transfer (RAFT) polymerization, Chemical Society Reviews, 2014, 43(2), 496.
[25] G. Moad, E. Rizzardo and S.H. Thang, Radical addition–fragmentation chemistry in polymer synthesis, Polymer, 2008, 49(5), 1079.
[26] M. Benaglia, J. Chiefari, Y.K. Chong, G. Moad, E. Rizzardo and S.H. Thang, Universal (switchable) RAFT agents, Journal of the American Chemical Society, 2009, 131(20), 6914.
[27] A. Ilchev, R. Pfukwa, L. Hlalele, M. Smit and B. Klumperman, Improved control through a semi-batch process in RAFT-mediated polymerization utilizing relatively poor leaving groups, Polymer Chemistry, 2015, 6(46), 7945.
[28] P.D. Iddon, K.L Robinson and S.P. Armes, Polymerization of sodium 4-styrenesulfonate via atom transfer radical polymerization in protic media, Polymer, 2004, 45(3), 759.
[29] D. M. Haddleton, R. Edmonds, A.M. Heming, E.J. Kelly and D. Kukulj, Atom transfer polymerisation with glucose and cholesterol derived initiators, New Journal of Chemistry, 1999, 23(5), 477.
[30] N. Akeroyd, R. Pfukwa and B. Klumperman, Triazole-based leaving group for RAFT-mediated polymerization synthesized via the Cu-mediated Huisgen 1,3-dipolar cycloaddition reaction, Macromolecules, 2009, 42(8), 3014.
[31] V. Coessens, Y. Nakagawa and K. Matyjaszewski, Synthesis of azido end-functionalized polyacrylates via atom transfer radical polymerization, Polymer Bulletin, 1998, 40(2–3), 135.
[32] J.A. Opsteen and J.C.M. van Hest, Modular synthesis of block copolymers via cycloaddition of terminal azide and alkyne functionalized polymers, Chemical Communications, 2005, 1, 57.
[33] P. Sun, Q. Tang, Z. Wang, Y. Zhao and K. Zhang, Cyclic polymers based on UV-induced strain promoted azide–alkyne cycloaddition reaction, Polymer Chemistry, 2015, 6(22), 4096.
[34] T. Sarbu, K.Y. Lin, J. Spanswick, R.R. Gil, D.J. Siegwart and K. Matyjaszewski, Synthesis of hydroxy-telechelic poly(methyl acrylate) and polystyrene by atom transfer radical coupling, Macromolecules, 2004, 37(26), 9694.
[35] M.R. Whittaker, Y.-K. Goh, H. Gemici, T.M. Legge, S. Perrier and M.J. Monteiro, Synthesis of monocyclic and linear polystyrene using the reversible coupling/cleavage of thiol/disulfide groups, Macromolecules, 2006, 39(26), 9028.
[36] M. Li, P. De, H. Li and B.S. Sumerlin, Conjugation of RAFT-generated polymers to proteins by two consecutive thiol–ene reactions, Polymer Chemistry, 2010, 1(6), 854.
[37] G.N. Grover, S.N.S. Alconcel, N.M. Matsumoto and H.D. Maynard, Trapping of thiol-terminated acrylate polymers with divinyl sulfone To generate well-defined semitelechelic Michael acceptor polymers, Macromolecules, 2009, 42(20), 7657.
[38] A. Postma, T.P. Davis, G. Moad and M.S. O'Shea, Thermolysis of RAFT-synthesized polymers. A convenient method for trithiocarbonate group elimination, Macromolecules, 2005, 38(13), 5371.
[39] G. Pound, J.M. McKenzie, R.F.M. Lange and B. Klumperman, Polymer–protein conjugates from ω-aldehyde endfunctional poly(N-vinylpyrrolidone) synthesised via xanthate-mediated living radical polymerization, Chemical Communications, 2008, 27, 3193.

Raghuraman G. Karunakaran and Raghavachari Dhamodharan

2 Tailor-made polymer–nanohybrid materials via reversible deactivation radical polymerization (RDRP)

2.1 Introduction

Organic–inorganic hybrid materials offer a wide range of applications in the field of material science and technology. In these materials, two different components each with its own advantages and drawbacks are combined at the molecular level to produce composite materials with different structural, physical, and chemical properties. This in turn offers unusual combinations of materials properties such as weight, strength, stiffness, permeability, electrical, biodegradability, and optical properties that are difficult to attain separately from the individual components [1, 2]. By exploring the properties of organic or inorganic components such as biochemical activity, electronic, optical (luminescence for example), and magnetic properties, it is possible to design materials that can be applied in the fields of sensors, membrane devices, electrochemical devices, catalysis, and so on. The most important aspect of hybrid materials is the capacity to synthesize combinations that can enhance the superior properties of each component.

A wide range of hybrid materials have been made with different combinations, starting from inorganic clusters, fullerenes, and metal or metal oxide nanoparticles dispersed in a polymer matrix, to organic, organometallic compounds, biomolecules, macrocycles, or polyethylene chains intercalated into silicate or clay materials. Literature reveals that a wide variety of silicates and polysiloxanes have been modified with different organic groups to enhance their stability and mechanical properties [3, 4]. By introducing suitable polymer layers onto the surface of metal or metal oxide nanoparticles, it is possible to design novel hybrid materials [3, 5, 6].

2.2 Nanoparticles

In the last two decades, metal and metal oxide nanoparticles have been extensively studied due to their potential applications in fields including catalysis,

Raghuraman G. Karunakaran, Evonik Corporation, Trexlertown, USA
Raghavachari Dhamodharan, Currently working in Momentive Performance Materials Pvt Ltd, Bangalore, India

https://doi.org/10.1515/9783110643695-002

nanoelectronics, optoelectronics, and biotechnology. Nanoparticles are isolable particles between 1 and 100 nm in size that are prevented from agglomeration by protective shells. Nanoparticles are of great scientific interest as they are effectively a bridge between bulk materials and atomic or molecular structures. A bulk material should have constant physical properties regardless of its size, but at the nanoscale this is often not the case. Size-dependent properties are observed such as quantum confinement in semiconductor particles, surface plasmon resonance in metal particles (e.g., gold, silver, etc.) and superparamagnetism in magnetic materials (e.g., iron oxide nanoparticles). The properties of materials change as their size approaches the nanoscale. The interesting properties of the nanoparticles are partly due to the aspect of the surface of the material dominating the properties as compared to the bulk properties. The percentage of atoms at the surface of a material becomes significant as the size of that material approaches the nanoscale, that is, they have large surface-to-volume ratios. The surface area of the nanoparticles is high. In other words, the number of particles exposed at the surface is higher when the size of the aggregate decreases, as represented in Figure 2.1.

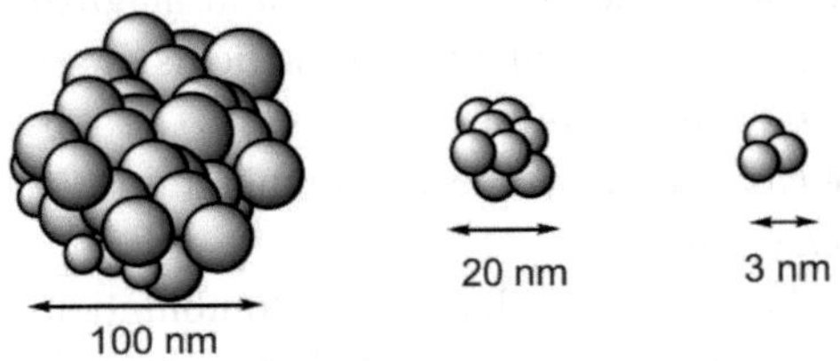

Figure 2.1: A comparison of surface-to-volume ratio of nanoparticles. (Source: G.K. Raghuraman and R. Dhamodharan, Ph.D. thesis entitled "Attachment of polymer monolayers-A convenient approach to modify nanoparticulate and flat surfaces" submitted to Indian Institute of Technology, Madras, India, 2007) [6].

2.3 Synthesis of nanoparticles

Two basic approaches have been employed to synthesize colloidal metal nanoparticles. They are i) the top-down process and ii) the bottom-up process. In the top-down process, the bulk metal is broken down while simultaneously stabilizing the resulting nanopowders with protecting agents. On the other hand, the bottom-up process involves the reduction of the corresponding metal salts (by electrochemical pathway among others) and subsequent stabilization of the particles by protective agents to obtain metal nanoparticles. In the case of metal oxide nanoparticles, the metal salts are hydrolyzed followed by oxidation. The chemical reduction of transition metal salts to produce zero-valent metal colloids in the presence of protecting agents was first reported by Faraday in 1857 [7]. Later, Turkevich reported the preparation of gold sols by the reduction of chloroaurate ion $[AuCl_4]^-$ with sodium citrate. The mechanism involved in the formation of nanocolloids involves three steps, namely, nucleation, growth, and agglomeration of particles [8–10]. In the first step, the metal salt is reduced to give metal atoms in the zero-valent state, which can collide further with metal ions or metal atoms to form clusters or irreversible seeds of stable nuclei. The protective/stabilizing agent added in the medium, namely, the citrate counterion in this case, stabilizes the irreversible seed nuclei. A schematic representation of the process of nucleation and growth is depicted in Scheme 2.1. Generally, the particles obtained by the Turkevich method are monodisperse and are highly stable. However, it is not possible to vary the size of the resulting metal colloids by varying the reaction conditions [11].

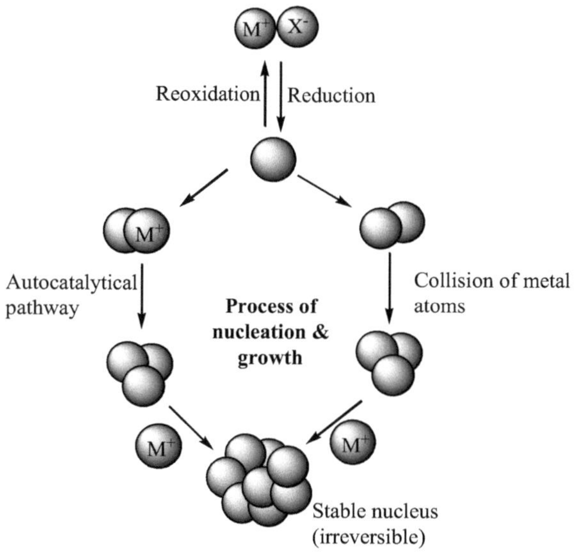

Scheme 2.1: Illustration depicting the process of nucleation and growth. Reproduced with permission from H. Bönnemann and R.M. Richards, European Journal of Inorganic Chemistry, 2001, 2001, 10, 2455. ©2001, John Wiley & Sons [11].

Numerous techniques have been employed for the synthesis of metal oxide nanoparticles that include hydrolysis of metal salts followed by precipitation [12, 13], hydrothermal processes [14, 15], microemulsion processes [16, 17], electrochemical processes [18, 19], sol–gel processes, and so on [20–22]. The controlled synthesis of metal oxide nanoparticles is an important criterion for its successful application and it can be achieved by solution-phase methods in a precise manner. The most commonly employed method to prepare metal oxide nanoparticles is the sol–gel process, where the reaction is stopped before gelation occurs, similar to precipitation methods. The synthetic procedure can be divided into several stages, including precursor formation, followed by nucleation, growth, and aging. The formation of precursor molecules usually takes place via hydroxylation and hydrolysis reactions, resulting in a zero-charge molecule or precursor. In order to have condensation of the precursor molecules, specific water/hydroxyl content is essential, and once the precursor concentration is above the nucleation threshold, nucleation can occur until the precursor concentration is again below the nucleation threshold. After the nucleation and growth processes, the average particle size and size distribution may vary upon aging. The detailed interplay between the rates of precursor formation, nucleation, growth, and aging determines the final particle size and size distribution [11].

In many applications, a narrow particle size distribution is expected. This can, in principle, be achieved by controlling the relative rates of the precursor formation, nucleation, growth, and aging. In general, a single, instantaneous nucleation step, forming uniform sized nanoparticles is preferred, and subsequent growth should occur at the same rate for all particles. This is possible only when the rate of formation of precursor molecules is much smaller than the rate of nucleation and also the growth rate should be even smaller than the nucleation process to prevent significant growth during nucleation step. Finally, aging should maintain the particle distribution the same, or may even narrow the distribution. The relative rates of this process, solvent, solute, solid properties, and the temperature play an important role in determining the size and size distribution of the nanoparticles obtained. Apart from these factors, the surface chemistry plays an important role to control the particle size and distribution, for example, by using protective groups such as surfactants.

2.4 Application of nanoparticles

When the particle size is reduced from the microscale to the nanoscale, size-dependent properties are observed as exemplified by new applications in the fields of catalysis, in semiconducting devices (size in the order of 1–10 nm), and also in biomedical applications. Nanoparticles such as gold and silver exhibit a size-specific property termed as *surface plasmon resonance* in the visible region, and with change in the size of the nanoparticles this property of the material changes [23, 24]. When the metal nanoparticles are irradiated by an electromagnetic radiation, the oscillating electric field causes the electrons in the conduction band to oscillate. When the frequency of oscillation becomes equal to that of frequency of incident light, then resonance takes place and the absorption peak appears in the visible region known as *surface plasmon absorption* or *surface plasmon resonance*. With the variation in the size of the nanoparticles (for Ag and Au colloids), there is a variation in the surface plasmon absorption maximum. The variation of this optical property of nanoscale metal colloids with size was employed in a wide variety of applications, which includes chemical sensing, biomedical imaging, high-throughput screening analysis, and nanoscale photonics. Metal oxide nanoparticles such as titanium dioxide, silica, magnetite, cerium oxide, zinc oxide are also widely used in many applications. Magnetite nanoparticles (MNPs) dispersed in a solvent, also referred to as *ferrofluids*, find applications in drug delivery systems for magnetic resonance imaging [25–27]. Titanium dioxide nanoparticles are a well-known photocatalyst, and it is also used as a filler material and pigment in the paint industry [28, 29]. Silica nanoparticles (SiNPs) are used as fillers in sealants, gaskets, and so on. Zinc oxide nanoparticles and cerium oxide nanoparticles are used as inorganic UV absorbers [30–32].

2.4.1 Semiconductor nanoparticles (quantum dots)

A semiconductor is a material that can be characterized by low band gap values of the order of >0–4 eV. Their electrical conductivity is intermediate between that of an insulator and a conductor. The most commonly used semiconductors in electronic devices are silicon, germanium, gallium arsenide (Ga-As), indium phosphide (InP), and so on. The energy difference between the valence band and the conduction band (i.e., band gap) is small such that the electrons can jump from valence band to the conduction band upon thermal agitation or by a photonic impact. In the ground state, the valence band is completely filled, and when sufficient energy is applied to a semiconductor, it becomes conducting by excitation of electrons from the valence band into the conduction band. When the electron jumps from the valence band to conduction band, it creates a hole (or positive charge) in the valence band and thus creates "electron–hole pairs" as shown in Figure 2.2. They are also referred to as "excitons" if they have not dissociated. Semiconducting materials are widely used in

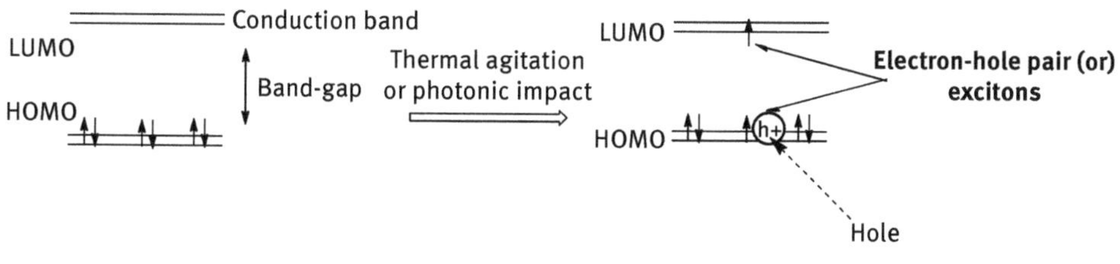

Figure 2.2: Formation of an "electron–hole pair" induced by external energy source. (Source: G.K. Raghuraman and R. Dhamodharan, Ph.D. thesis entitled "Attachment of polymer monolayers-A convenient approach to modify nanoparticulate and flat surfaces" submitted to Indian Institute of Technology, Madras, India, 2007) [6].

electronics industry and find applications as light-emitting diodes and also in personal computers, in sensor fields, and so on.

When the size of the semiconductors is reduced to nanometer regime, it leads to a fascinating class of novel materials (quantum dots) that possess characteristics between the bulk and molecular descriptions and exhibits properties of quantum confinement [33]. Quantum confinement is observed in nanoparticles with radii smaller than the average distance between the electron and the hole, known as *bulk exciton Bohr radius*. The band gap energies of the nanoscopic semiconducting materials are strongly dependent on their size, and this size-dependent property attracts technological significance. Semiconductor nanocrystals or quantum dots have been reported having a wide variety of compositions, including CdSe, CdTe, cadmium sulfide (CdS), Si, GaAs, PbSe, and InP [33–37].

2.4.2 Nanoparticles in catalysis

Catalysts are usually nanoparticles composed of clusters of atoms, often metal colloids (mono-, bimetallic colloids) with particle size varying between 1 and 20 nm. Enzymes are nature's catalysts and they are composed of inorganic nanoclusters surrounded by high molecular weight proteins. They are responsible for the human body to function and for the growth of plants. Synthetic catalysts (either homogeneous or heterogeneous) are often metallic colloids that are used to catalyze reactions with higher selectivity. Rhodium colloids dispersed in aqueous solution was used as effective hydrogenation catalysts for olefins [38]. Palladium hydrosols were used to catalyze a number of reactions, namely, the oxidative acetoxylation reactions, oxidative carbonylation of phenol to diphenyl carobonate, hydrogen-transfer reduction of multiple bonds by formic acid, the reduction of nitriles and nitroarenes, and so on [39, 40]. Tetraalkyl ammonium salts stabilized Pd nanoparticles and Pd/Ni colloids in dimethylacetamide were used to catalyze Heck and Suzuki carbon–carbon bond coupling reactions [41, 42]. Nanosized Pd colloids generated in situ by the reduction of Pd (II) to Pd (0) are involved in the catalysis of phosphane-free Heck and Suzuki reactions [43].

2.5 Agglomeration and stabilization of nanoparticles

When the nanoparticles are dispersed in a solvent, the two important forces that act on it are van der Waals attraction (and occasionally it can be electrostatic as well) and Brownian motion, whereas the influence of gravity becomes negligible.

van der Waals attraction is a weak force and becomes significant only at a very short distance. van der Waals attraction arises from sympathetic fluctuations in the particles and its electron distributions. However, van der Waals attraction takes place only at separations smaller than a few nanometers. More widely separated particles are prevented from the attraction by thermal collisions with surrounding solvent molecules, whereas the Brownian motion results in the collision of the nanoparticles with each other and with the solvent molecules. The combination of van der Waals attraction force and Brownian motion can and does result in the agglomeration or aggregation of the nanoparticles. The two common modes of preventing the agglomeration and stabilizing nanoparticles are: i) electrostatic stabilization and ii) steric stabilization. The size and size distribution of the nanoparticles are dependent on the protective/stabilizing agents that are used during the synthetic step. In general, if the protective agents interact strongly with the nanoparticles, the resulting materials have smaller size and uniform size distribution.

2.5.1 Electrostatic stabilization

The electrostatic stabilization of nanoparticles in a suspension was successfully described by the Derjaguin, Landau, Verwey, and Overbeek theory [44]. According to Derjaguin, Landau, Verwey, and Overbeek theory, the Coulombic repulsion between the particles caused by the electrical double layer formed by ions adsorbed at the particle surface and the corresponding counter ions results in the electrostatic stabilization. According to this theory, the interaction between two particles in a suspension is considered as a combination of van der Waals attraction potential and the electrostatic repulsion potential. A plot of the combination of these two opposite potentials as a function of distance from the surface of spherical nanoparticles is shown in Figure 2.3. At a distance far from the surface, both van der Waals attraction potential and electrostatic repulsion potential are close to zero. A maximum is observed when the electric repulsion potential dominates the van der Waals attraction potential. The maximum is also known as repulsive barrier. If the barrier is greater than ~10 kT, where *k* is the Boltzmann constant, the collisions of two particles produced by Brownian motion will not overcome the barrier and agglomeration will not occur. The stabilization of gold nanoparticles (AUNPs) synthesized by the Turkevich method serves as an example for the application of this model. The citrate ion functions the dual role of reducing agent as well as an electrostatic stabilizer [8–10]. The limitations of the electrostatic stabilization method are: i) it is a kinetic stabilization method and is applicable only to dilute systems; ii) it is not applicable to electrolyte sensitive systems; and iii) it is not possible to redisperse the agglomerated particles.

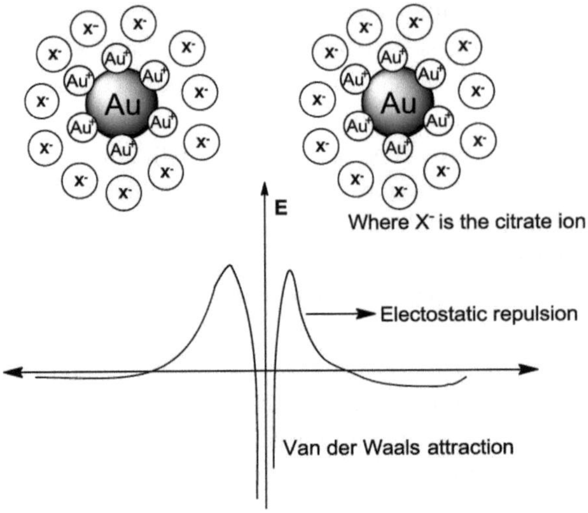

Figure 2.3: Electrostatic stabilization of gold nanoparticles by citrate ion. Reproduced with permission from H. Bönnemann and R.M. Richards, European Journal of Inorganic Chemistry, 2001, 2001, 10, 2455. ©2001, John Wiley & Sons [11].

2.5.2 Steric stabilization

The coordination of sterically demanding organic molecules (e.g., long-chain alcohols, carboxylic acids, phosphines, amines, thiols, various surfactants molecules, etc.), polymers, silanes, and siloxane networks can act as a protecting shield for the steric stabilization of metal colloids, as depicted in Figure 2.4. It is a thermodynamic

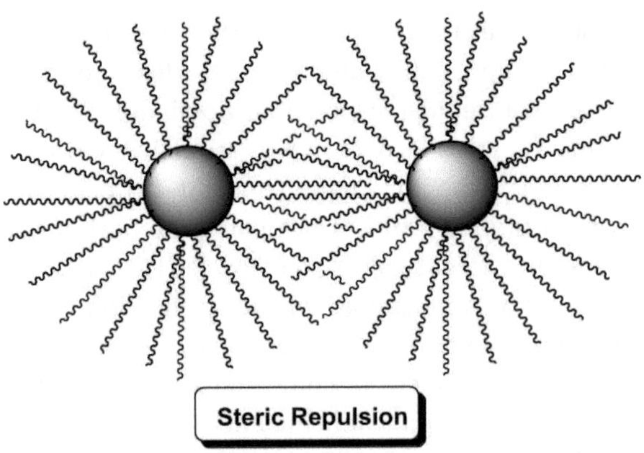

Figure 2.4: Steric stabilization of nanostructured metal colloids. Reproduced with permission from H. Bönnemann and R.M. Richards, European Journal of Inorganic Chemistry, 2001, 2001, 10, 2455. ©2001, John Wiley & Sons [11].

model and is applicable to electrolyte sensitive systems. This method of stabilization is widely used for colloidal dispersions.

2.5.3 Polymers as steric stabilizing agents

Nanoparticle stabilization through polymer layers (or) networks (polymer as a steric stabilizer) is a widely used method that has several advantages over electrostatic stabilization of colloidal materials [45–47]. Using polymeric stabilizer, it is possible to synthesize nanoparticles of narrow size distribution for which electrostatic stabilizers fail. Polymer layer adsorbed/attached on the surface of nanoparticles serve as a diffusion barrier for further growth of the species, resulting in a diffusion-limited growth. The diffusion-limited growth would reduce the size distribution of the initial nuclei, leading to monodisperse nanoparticles. Hence, this mode of stabilization is widely used for the synthesis of nanoparticles. When a polymer chain is subjected to solvent interaction, it tends to expand to reduce the overall Gibbs free energy of the system; such a solvent is called a "good solvent" for that polymer chain. If the polymer in a solvent tends to coil up and form a more compact structure, then the solvent is considered to be a "poor solvent" for that polymer. For a given system, that is, a given polymer in a given solvent, whether the solvent is a "good" or "poor" solvent is dependent on the temperature. In general, at higher temperatures, the polymer coil expands, whereas at lower temperatures it tends to collapse. The temperature, at which a poor solvent transfers to a good solvent, is the Flory theta temperature at which the polymer exhibits unperturbed dimensions [48]. In a good solvent, in which polymer expands, if the coverage of polymer on the nanoparticulate surface is not complete (less than 50% coverage), then the two polymer layers tend to interpenetrate so as to reduce the available space between polymers. Such an interpenetration of two polymer layers of the two approaching particles would result in a reduction of the free movement of polymer chain, which leads to a reduction of entropy. This in turn increases the overall Gibbs free energy of the system, assuming that the change of enthalpy due to the interpenetration of two polymer layers negligible. As a result of this, the two particles repel one another and the distance between two particles must be equal to or larger than twice the thickness of polymer layers. When the coverage of polymer is high, there would be no interpenetration. As a result, the two polymer layers will be compressed, leading to the coiling up of the polymer chains and hence increases in the overall Gibbs free energy of the system. This results in the repulsion of the two particles. Regardless of the difference in the coverage, if the two particles are covered with polymer layers, then they are prevented from agglomeration by the space exclusion or steric stabilization. When the polymer adsorbs strongly and full surface coverage is obtained, then the interaction between two polymer layers produces a purely repulsive force and results in an increased free energy. This is the same as that of anchored polymer at full coverage. When only a partial coverage is achieved, then the nature of solvent has a significant influence on the interaction between two nanoparticles.

2.6 Synthesis of polymer layers

The synthesis of polymers on the surface of nanoparticles in brush form with well-defined compositions, architectures, and functionalities is quite interesting in the area of polymeric nanomaterials. Polymerization is the process that is based on the repetitive reaction of monomers resulting in higher molecular weight molecule. Synthetic polymers of high molecular weights are produced by step-growth polymerization or chain-growth polymerization techniques. In the step-growth polymerization, condensation of bifunctional molecules leads to form a linear polymer with high molecular weight only at >99% functional group conversion. In contrast, in the chain growth polymerization, high molecular weight polymers are obtained at low conversion. The four fundamental steps involved in chain polymerization are: i) initiation, ii) propagation, iii) chain transfer (CT), and iv) termination. The scope of the properties obtained from the chain and step growth formation was inadequate; thus, copolymers were synthesized. However, the properties of naturally occurring copolymers with well-defined chain sequence are far superior when compared to that of synthetic polymeric materials. Thus, the synthesis of well-defined polymer chains with controlled architecture is the key parameter in designing the polymeric nanomaterials.

2.6.1 Reversible deactivation radical polymerization (RDRP)

The term *living polymerization* is used to describe a chain growth polymerization, which proceeds without the kinetic steps of termination and CT. It has been demonstrated that in living polymerization all chains grow at the same rate (e.g., the living polymerization of ethylene oxide initiated with alkoxide) [49]. If the initiation is faster than propagation, it allows the control of molecular weight through the variation of the relative concentrations of the initiator versus monomer. The first example of living polymerization was the polymerization of styrene initiated by sodium–naphthalene complex, in which the chain growth continued with further addition of pure monomer until destroyed or terminated by the addition of terminating agent [50, 51]. Although conventional free radical polymerization processes have several advantages over living polymerization such as applicability to wide range of monomers (including functional monomers), relatively mild reaction conditions (such as the exclusion of oxygen from the reaction medium), their main disadvantage is that the resulting polymers are polydisperse. In order to have control over the dispersity, various CT agents and catalysts have been employed.

Before the invention of RDRP techniques, ionic polymerizations (anionic or cationic) were the only "living" techniques available that efficiently controlled the structure and architecture of vinyl polymers. Although these techniques resulted in controlled molecular weight, low-dispersity, and well-defined chain ends, they were not applicable for the polymerization and copolymerization of a wide range of functionalized vinyl monomers. This limitation is due to the incompatibility of the growing polymer chain end (anion or cation) with numerous functional groups and certain monomer families. In addition, these polymerization techniques require stringent reaction conditions, including the use of ultrapure reagents and the near-total exclusion of water and oxygen. The necessity to overcome all these limitations emboldened synthetic polymer chemists to develop new concepts, which would allow for a more controlled free-radical polymerization process. RDRP processes are now the most common methodology used currently in academic laboratories for the synthesis of macromolecules with predictable and well-defined primary structures and precise molecular architectures. In these methods, the radical generated by conventional method establishes a rapid dynamic equilibrium between the propagating radical and dormant species throughout polymerization. If the radical undergoes bimolecular coupling or termination, it results in dead polymer. RDRP can be defined as a methodology by which an otherwise short-lived free radical (life time of the order of 10^{-8} s) is stabilized in such a manner that there is an equilibration between the growing free radicals and various dormant species (e.g., a persistent radical), which allows the addition of monomer throughout the polymerization period. The basic requirements for RDRP are: i) there should be fast exchange between the dormant species and growing free radical and ii) there must be fast and quantitative initiation. The historical developments in RDRP are listed in Table 2.1. The controlled/living nature of RDRP makes it possible to obtain polymers with predetermined molecular weights and narrow distributions, block copolymers with well-defined structures, end-functionalized polymers, as well as specially shaped polymers such as star-branched, comb-shaped, and macrocyclic ones by combining living polymers with the appropriate reagents [52, 53].

2.6.2 Iniferter

Iniferters are species that induce radical polymerization that proceed via initiation, propagation, primary radical termination, and transfer to initiator. Otsu studied the initiating ability of these compounds in radical polymerization of styrene and methylmethacrylate (MMA) and found that various sulfides and disulfides (e.g., phenyl, benzoyl, benzothiazoyl, thiuram, and dithiocarbamate derivatives) could serve as efficient photoinitiators [61, 62]. Among thiuram disulfides, tetraethylthiuram

Table 2.1: Summary of RDRP techniques inventory.

RDRP techniques	Year invented	Inventors	Reagents and monomers used
Iniferter	1982	Otsu et al. [54]	Dithiocarbamate/MMA
NMP or TEMPO	1993	Georges et al. [55]	TEMPO/styrene
Atom transfer radical polymerization (ATRP)	1995	Sawamoto [56] and Matyjaszewski [57]	CCl_4/$RuCl(PPh_3)$/ MMA 1-PECl/CuCl/bpy/ styrene
Reversible addition-fragmentation chain transfer (RAFT)	1998	Rizzardo et al. [58]	Dithioesters/AIBN/ acrylate, MMA
Single-electron transfer-living radical polymerization (SET-LRP)	2006	Percec et al. [59]	Alkyl halide/acrylates/ vinyl chloride/Cu (I) X
Single-electron transfer – reversible addition fragmentation chain transfer (SET-RAFT)	2008	Dhamodharan et al. [60]	Transition metal/alkyl halides/ligands/dithio compounds

disulfide (Scheme 2.2) was observed to be the most desired photo- and thermal initiator. From kinetic studies, Otsu found that tetraethylthiuram disulfide not only acted as an initiator but also as a retarder, terminator, and transfer agent. Otsu further observed that the polymerization of the monomer (M) proceeded via the dissociation of the initiating species, namely, thiuram disulfide, which is followed by the initiation, propagation, primary radical termination, and CT to initiating species. The polymerization reaction was reversible only under photochemical conditions. The thiuram disulfide groups were found to be stable under thermal condition. These experiments provided the concept that the radical mechanism can be used to synthesize well-defined polymers by manipulating the reaction conditions appropriately. The use of well-designed iniferters would give polymers or oligomers bearing controlled end groups and these polymerizations proceed by a living radical polymerization mechanism in a homogeneous system [63].

2.6.3 Nitroxide-mediated polymerization

The development of RDRP processes has resulted in broadening the other areas of research that were hitherto unavailable or severely restricted. A major example is the synthesis of block copolymers by mechanistic transformations or sequential polymerizations. This resulted in extensive research on the RDRP process [52]. It is

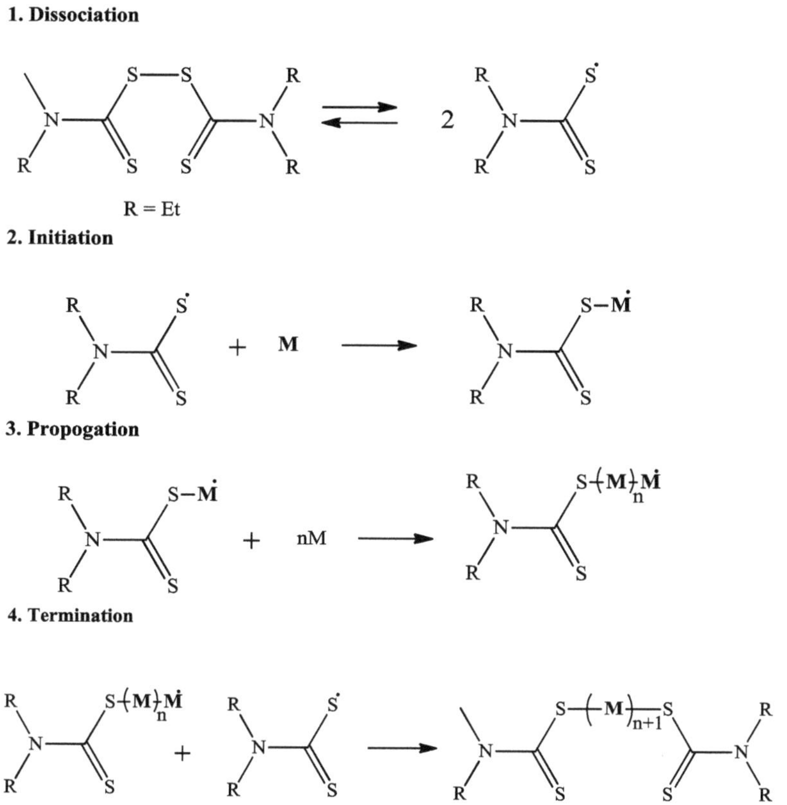

Scheme 2.2: Mechanism of Iniferter process. (Source: T. Otsu, Journal of Polymer Science Part A: Polymer Chemistry, 2000, 38, 12, 2121 ©2000, John Wiley & Sons) [63].

interesting to note the similarity between the iniferter mechanism outlined in Scheme 2.2 and the general outline of RDRP mechanism (Scheme 2.3).

The reversible capping of the growing polymeric chain is the key step for reducing the overall concentration of the propagating radical chain end. In the absence of other side reactions, namely, the reaction between the mediating radical and vinylic monomer, the concentration of reactive chain ends is extremely low, minimizing irreversible termination reactions such as combination or disproportionation. The growing polymer chain would be initiated only from the desired initiating species and growth should occur in a pseudo-living manner, allowing better control over the polymerization process and resulting in well-defined polymeric architectures. In 1986, Fischer described the method to control the fast radical reaction through persistent radical effect (PRE) in which the stationary concentration of radicals is reduced and therefore the contribution of termination was minimized or suppressed [64, 65]. The "persistent radical effect" occurs when concentration of transient or propagating radicals and persistent radicals are formed at equal rate in

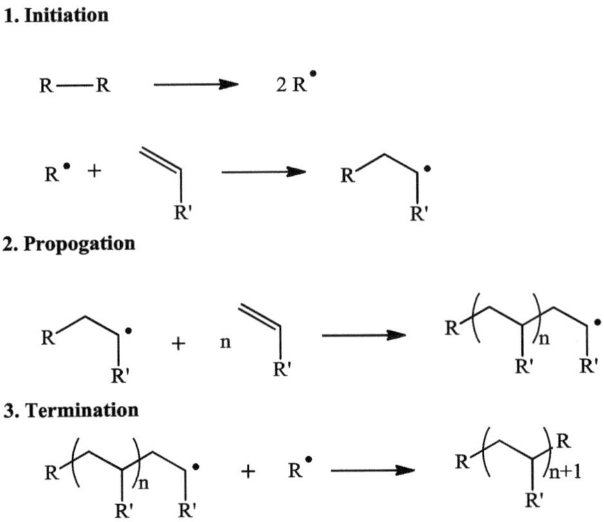

Scheme 2.3: General outline of free radical polymerization of vinyl monomer.

a single step. Since the transient radicals can undergo fast termination via coupling and disproportionation, their concentration decreases, and therefore the concentration of persistent radicals build up. When the concentration of persistent radical reaches an optimum value, then the rate at which the propagating radicals reacts with the persistent radical in a deactivation step is much faster than the rate at which the propagating radicals react with each other in an irreversible termination step. Thus, highly selective addition chemistry involving free radical intermediates can be performed over radical coupling and disproportionation.

In the case of a nitroxide-mediated polymerization reaction, the persistent radical is the nitroxide species and the transient radical is always the carbon radical. This leads to repeated coupling of the nitroxide to the growing end of the polymer chain, which would ordinarily be considered a termination step, but in this case it is reversible. Because of the high rate of coupling of the nitroxide to the growing chain end, there is little coupling of two active growing chains, which would be an irreversible terminating step, limiting the chain length. The nitroxide binds and unbinds to the growing chain, protecting it from termination steps. This ensures that any available monomer can be easily scavenged by active chains. Because this polymerization process does not naturally self-terminate, this polymerization process is described as "living," as the chains continue to grow under suitable reaction conditions whenever there is reactive monomer to "feed" them. Because of the PRE, it can be assumed that at any given time, almost all of the growing chains are "capped" by a mediating nitroxide, meaning that they dissociate and grow at very similar rates, creating a largely uniform chain length and structure.

The nature of alkoxyamines is critical to the success of NMP techniques and a variety of different persistent or stabilized radicals have been employed. The carbon-centered radicals were trapped by the stable nitroxy radical, namely, 2,2,6,6-tetramethylpiperidine-1-oxyl radical (TEMPO), which was demonstrated by Solomon and coworkers [66]. TEMPO-mediated controlled radical polymerization of styrene was carried out at 130 °C as shown in Scheme 2.4 [55]. The polymerization was carried out in bulk and benzoyl peroxide was used as the initiator and TEMPO was used as a reversible capping agent. The dispersities of the resulting polymer were 1.2, which is significantly lower for a conventional free radical initiated polymerization.

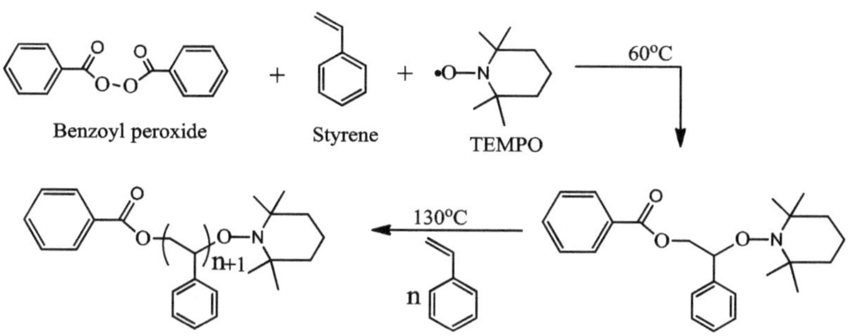

Scheme 2.4: TEMPO-mediated polymerization of styrene. (Source: M.K. Georges, R.P.N. Veregin, P.M. Kazmaier and G.K. Hamer, Macromolecules, 1993, 26, 11, 2987. ©1993, ACS) [55].

The carbon–oxygen bond of the dormant alkoxyamine (TEMPO) becomes labile at higher temperature (130 °C) and results in nitroxide and a carbon-centered radical as shown in Scheme 2.4. The radical thus produced can propagate or undergo termination or transfer reaction until it is trapped again by a nitroxide. The control in the NMP depends on different parameters including the rate of activation and deactivation, the concentration of the nitroxide, the temperature, and the monomer type. The main disadvantages of NMP are that it has to be performed at elevated temperatures (often requires above 120 °C) and is also limited to certain monomers [67].

2.6.4 Atom transfer radical polymerization (ATRP)

One of the most successful synthetic procedures utilizing the PRE concept is the metal-catalyzed RDRP that is the atom transfer radical polymerization (ATRP), which was derived from the well-known metal-catalyzed Kharasch addition or atom transfer radical addition (ATRA) [68, 69]. The mechanism for Kharasch addition reaction and ATRP is described in Scheme 2.5. In this addition reaction, a metal catalyst such as copper (I) halide complexed by suitable ligands undergoes an inner-sphere

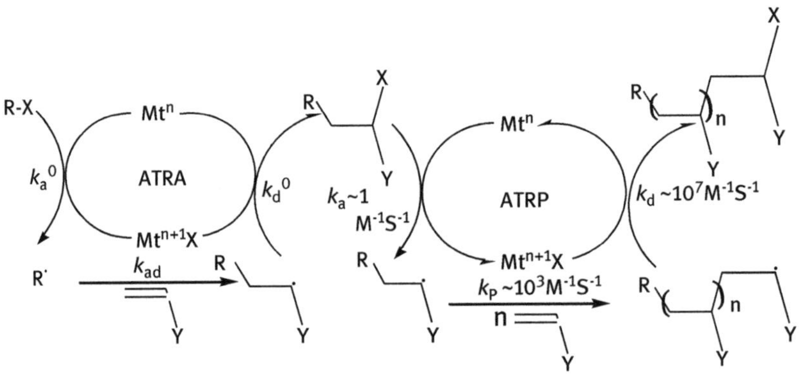

Scheme 2.5: Mechanism of Kharasch addition reaction. (Source: J.S. Wang and K. Matyjaszewski, Journal of the American Chemical Society, 1995, 117, 20, 5614. ©1995, ACS) [57].

oxidation with concomitant abstraction of a halogen atom from a starting compound, which generates a copper (II) complex and an organic radical (k_a). The organic radical can then add (k_{ad}) to a double bond in an inter- or intramolecular manner and then reabstract a halogen atom from the copper (II) complex (k_d) to complete the Kharasch addition. Alternatively, it can abstract the halogen atom from the copper (II) complex and return back to the original dormant organic–halide species. The copper (I) complex is reformed in both the cases, completing the catalytic cycle (activation–addition–deactivation or activation–deactivation). The radicals may also react with another radical, but the copper (II) complex acts as a "persistent radical" and controls the steady-state concentration of radicals and also reduces the contribution of termination. ATRA can become chain growth polymerization under suitable experimental conditions, if the radical species generated before and after the addition of the unsaturated substrate possess comparable reactivity. Under these conditions, the activation–addition–deactivation cycle will repeat until the consumption of the monomers and produces polymer with narrow dispersity [56, 57].

The basic components for ATRP are initiator (alkyl halides), monomer, catalyst, ligand, solvent, and temperature. A wide range of monomers have been successfully polymerized using ATRP, which includes styrene, (meth)acrylates, (meth) acrylamides, and their substituted derivatives that can stabilize the propagating radicals [52].

2.6.5 Reversible addition-fragmentation chain transfer (RAFT)

Reversible addition-fragmentation chain transfer (RAFT) polymerization is a reversible deactivation radical polymerization and is one of the more versatile methods available for the preparation of vinyl polymers with well-defined characteristics. It was first demonstrated by Rizzardo et al. at the Commonwealth Scientific and Industrial Research Organisation (CSIRO) of Australia in 1998 [58]. RAFT makes use of a chain transfer agent (CTA; termed a *RAFT agent*). The most common RAFT agent used is a dithiocarbonyl compound, which controls the molecular weight and polydispersity. The mechanism involves the insertion of monomer between the carbon–sulfur bonds as shown in Scheme 2.6.

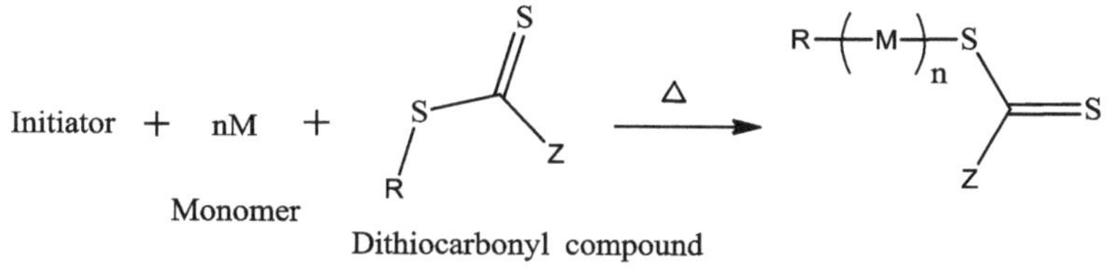

Scheme 2.6: General mechanism of RAFT process. (Source: J. Chiefari, Y.K. Chong, F. Ercole, J. Krstina, J. Jeffery, T. P. T. Le, R.T.A. Mayadunne, G.F. Meijs, C.L. Moad, G. Moad, E. Rizzardo and S.H. Thang, Macromolecules, 1998, 31, 16, 5559. ©1998, ACS) [58].

In RAFT polymerizations, the structure of CTA (RAFT agent) plays an important role in controlling the radical polymerization. Due to thermal energy (or photon flux if the initiation is photochemical), the labile bond between the thiol group and R [which can be alkyl (dithioesters), alkoxy (xanthates), NR_2 (dithiocarbomates), or SR (trithiocarbomates)] in the RAFT agent breaks during the course of propagation step. RAFT polymerization can be considered as a typical conventional free radical polymerization in the presence of a CTA. In RAFT polymerization, the initiation can be accomplished with traditional initiators such as azo compounds, peroxides, redox initiating systems, photoinitiators, and any other radiative means of initiation. The actual initiation happens when the initiator radical reacts with the dithiocarbonyl compound and then the insertion of monomer occurs in the labile S–R bond. The retention of the dithiocarbonyl end groups in the polymeric product renders the process suitable for synthesizing block copolymers and end functional polymers. With appropriate choice of reagents and polymerization conditions, RAFT polymerization can be used in the synthesis of well-defined homo, gradient, diblock, triblock, and star polymers, as well as more complex architectures including microgels and polymer brushes.

2.6.6 Single electron transfer living radical polymerization (SET-LRP)

The disproportionation of Cu(I) into Cu (0) and Cu(II) in water has been known for over 100 years and Percec et al. discovered that Cu(I)X species disproportionate into extremely reactive atomic Cu(0) and Cu(II)X_2 species in protic, dipolar aprotic, ethylene, and propylene carbonate, in ionic liquids and other solvents in the presence of nitrogen containing ligands. They utilized this rapid disproportionation reaction to control the radical polymerization [59]. They have termed this process as single-electron transfer living radical polymerization (SET-LRP). The mechanism for SET-LRP is outlined in Scheme 2.7. The Cu(I) halide in the presence of suitable ligand disproportionates to produce Cu(0) and Cu(II), which is stabilized by the ligand and functions as a persistent radical. When monomer is introduced, equilibrium between monomer addition to the radical and reversible atom transfer is established. This in turn results in the formation of polymer with a halogen end group.

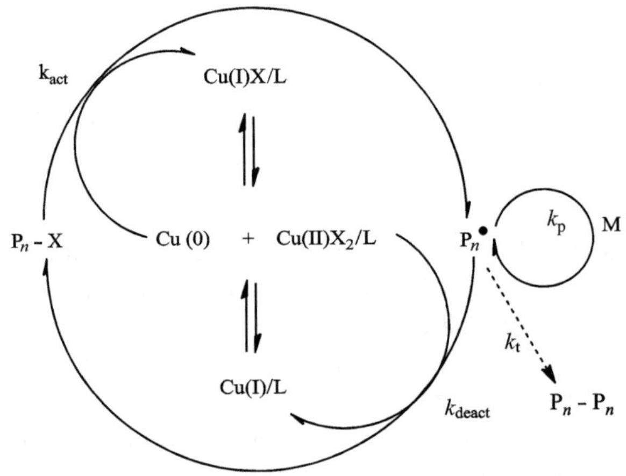

Scheme 2.7: General mechanism of SET-LRP Process. Reproduced with permission from V. Percec, T. Guliashvili, J.S. Ladislaw, A. Wistrand, A. Stjerndahl, M.J. Sienkowska, M.J. Monteiro and S. Sahoo, Journal of the American Chemical Society, 2006, 128, 43, 14156. ©2006, ACS [59].

In SET-LRP, dissociation of P–X and P_n–X is achieved through a heterolytic outer sphere single-electron transfer process wherein Cu(0) donates an electron to P_n/P_n–X resulting in a radical anion $[P_n/P{-}X]^-$, which degrades via a stepwise or a concerted pathway to P_n. and X^-. The Cu(I) species, during the single-electron transfer process or thereafter, becomes associated with the N-ligand. Cu(I)X is rapidly disproportionated in coordinating solvents, such as water, protic, dipolar aprotic, ionic liquids, and other polar solvents in the presence of N-ligands to regenerate Cu(0) and generate Cu(II)X_2/L. Cu(II)X_2/L, either from the initial reaction mixture or generated via disproportionation of Cu(I)X/L, is responsible for the reverse outer sphere oxidation of P/P_n. to P/P_n–X. Since the disproportionation Cu(I)X is the driving force of SET-LRP methodology, the nature of the solvent is expected to exert a strong influence on this polymerization. For solvents, such as water, ethanol, methanol, and acetone, K_{dis} could be correlated with the solvent polarity and was found to decrease systematically with the dielectric constant of the solvent in the presence of a perchlorate anion [59]. However, this correlation could not be extended to solvents that coordinate strongly to copper ions. In pure DMSO, Cu(I) is markedly stabilized relative to Cu (II) and metallic Cu(0) due to the strong solvation of this oxidation state. However, this disproportionation of Cu(I) can be significantly altered by other complexing agents, such as N-ligands, that stabilize Cu(I)and Cu(II) in a different manner. This is supported by a density functional theory computational study [70] that demonstrates that certain N-ligands, such as Me_6-TREN and TREN, form stronger complexes with Cu(II) rather than Cu(I) and mediate an effective SET-LRP. Apart from these two factors, namely, solvent and ligand, the particle size of the catalyst [71–73], monomer, and initiator structures [74] also influence the single-electron transfer living radical polymerization process.

2.6.7 Single-electron transfer reversible addition-fragmentation chain transfer (SET-RAFT) process

In single-electron transfer living radical polymerization, the Cu(I)X disproportionates into Cu(0) and Cu(II), which is responsible for the polymerization of various monomers. In another work, the single-electron transfer process was combined with a CTA (RAFT agent) to successfully demonstrate the polymerization of styrene, at ambient temperature [60]. The polymerization was initiated by the disproportionation reaction of Cu(I)X and controlled through the propagation regulated by the CTA via the RAFT mechanism as shown in Scheme 2.8.

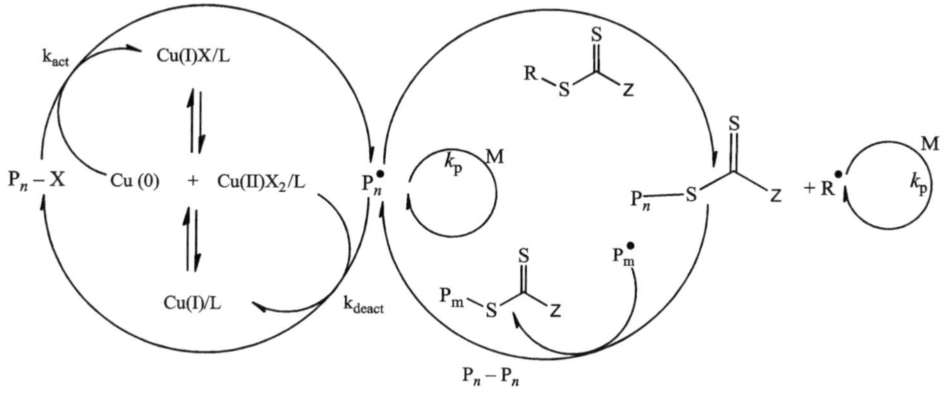

Scheme 2.8: The mechanism of SET-RAFT Process. Reproduced with permission from S. H. Subramanian, R.P. Babu and R. Dhamodharan, Macromolecules, 2008, 41, 1, 262. ©2008, ACS [60].

The factors that influence the SET-RAFT polymerization have not been documented adequately. Since it involves the concept of single-electron process, the factors that influence the disproportionation reaction in SET-LRP might also play a key role in this mechanism. The SET-RAFT mechanism was utilized for the preparation of fluorescent polymers [75], and polymers of few aliphatic cyclic methacrylates such as cyclohexyl methacrylate and isobornyl methacrylate [76]. SET-RAFT has also been utilized for the preparation of side-chain-functionalized polymer [77] and also for the polymerization of MMA using ascorbic acid and copper oxide [78].

2.7 Grafting of polymer brushes from various nanoparticulate surfaces

Nanoparticles that are stabilized by polymer layer or networks form a very good dispersion in solvents, and also in that process, the hybrid materials thus obtained exhibit combinations of different properties that include weight, strength, stiffness, permeability, electrical, biodegradability, and optical properties. Polymer–nanoparticle networks have been formed by introducing polymer layers anchored covalently or noncovalently (through ionic interactions) to particulate substrates. The covalent attachment of polymer layer onto a surface can be obtained either by the reacting the active site of polymer layers with that of nanoparticles; polymer layers were also introduced by initiating the polymerization from the surface, thereby forming polymer brushes onto the surface of nanoparticles. A polymer brush is defined as "polymers attached by one end to an interface at relatively high surface coverage and stretching away from the interface to avoid overlap" [79]. Polymer brushes attracted attention in the 1950s when it was found that grafting polymer molecules to colloidal particles was a very effective way to prevent flocculation [80]. In this manner, the brushes present on the surface of two approaching particles resist overlapping (due to the repulsive force between brushes arises ultimately from the high osmotic pressure inside the brushes) and thus stable colloidal dispersion is achieved.

The three important synthetic routes that can be employed to graft polymer brushes from various nanoparticulate surfaces are "grafting from," "grafting to," and "grafting-in-between" [81]. In the case of "grafting to" method, polymer chains carrying reactive groups at the end or the side chains are covalently coupled with the surface by physisorption, chemisorption, or photochemical grafting. In the "grafting from" method, polymer chains are grown from surface-attached initiator by in situ polymerization via thermal or photochemical means. The growth of polymer chains at a surface via the "grafting from" method provides a polymer in the brush form. In this case, the tethering is sufficiently dense that the polymer chains are crowded and are forced to stretch away from the surface or interface to avoid overlapping, sometimes much further than the typical unstretched form of a chain.

The "grafting from" process can be accomplished by immobilizing initiators on to the surface of nanoparticles, followed by polymerization. The growth of polymer brushes using "grafting from" process is more attractive due to a high density of initiators on the surface and a well-defined initiation mechanism. In addition, it is possible to control the chain length of the growing polymer chains. Polymerization methods that have been used to synthesize polymer brushes include cationic, anionic, ring-opening polymerization, conventional free radical polymerization, group transfer, and RDRP techniques like NMP, ATRP, and RAFT. The surface of the nanoparticles was modified with silane-based anchoring molecules in the case of metal oxide nanoparticles and thiol-based anchor molecules for metal nanoparticles such as gold and silver. A wide variety of reports are available in the literature for the growth of polymer brushes/layers from the surface of nanoparticles and are summarized in Table 2.2.

Table 2.2: A summary of literature reports for the growth of polymer layers/brushes onto the surface of nanoparticles.

Nanoparticles	Year	Invented by	Polymer layer/method Of growth/polymerization temperature
Silica	1990	Boven et al. [82]	PMMA/conventional free radical polymerization/70 °C
Magnetite	2003	Takahara et al. [83]	Polystyrene (PS)/NMP/125 °C
Silica	2002	Hawker et al., [84]	PS/NMP/120 °C
Silica	2000	Mandal et al. [85]	Poly(benzyl methacrylate)/ATRP/110 °C
Gold	2002	Fukuda et al. [86]	PMMA/ATRP/40 °C
Magnetite	2004	Fukuda et al. [87]	PMMA/ATRP/70 °C
Cadmium sulfide/silica	2001	Patten et al. [88]	PMMA/ATRP/100 °C
Titanium dioxide	2006	Matsuno et al. [89]	PS/NMP/125 °C
Silica	2005	Li et al. [90]	PS/RAFT/65 °C
Gold	2001	Jordan et al. [91]	Poly(2-oxazolines)/cationic polymerization/reflux conditions

2.7.1 Immobilization of initiators/cross-linkable groups onto the surface of nanoparticles

In the self-assembled monolayer process, an important group of coupling agents is silanes [92–94]. Specific end groups such as chloro, hydrido, methoxy, and ethoxy functional groups present at one end of a silane coupling agent reacts with the surface hydroxyl (–OH) group present on many nanoparticulate surfaces resulting in the covalent attachment to the surface. A schematic depiction of silane coupling agents is shown in Figure 2.5. The reactive group "R" is either an initiating moiety (to initiate the polymerization) or crosslinkable group, such as epoxy, benzophenone, azide, sulfonyl azide, vinyl groups, and so on, that crosslinks with the polymeric chain to attain covalent attachment. The initiating moiety "R" can be a free-radical initiating group such as azo-initiator (or) other RDRP initiating group. A large amount of work has been done with thiol end-functionalized self-assembled monolayer (or) multilayers for the protection of metals [95, 96] such as gold and silver. Dense, homogeneous, and monodisperse polymer monolayer were formed on various nanoparticulate surfaces such as SiNPs, AUNPs, CdS/SiO_2, titanium dioxide nanoparticles, and MNPs. Different silane initiators have been used to synthesize polymer brushes through RDRP techniques namely NMP, ATRP, RAFT, and so on [84, 85, 88, 90].

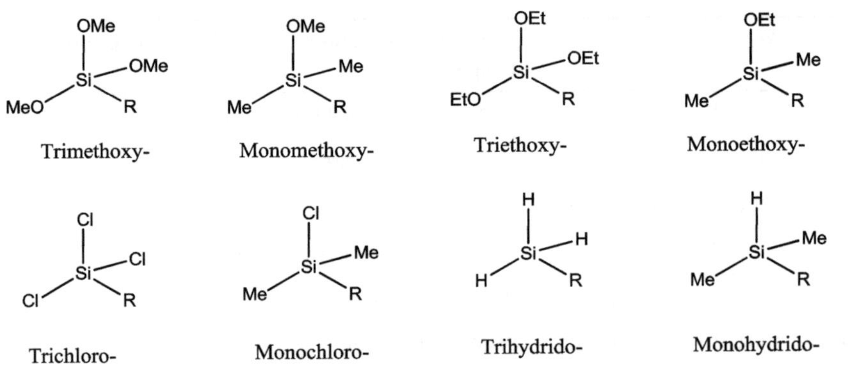

Figure 2.5: Representation of various mono- and triend groups of silanes with reactive group "R.".

2.7.2 Silica nanoparticles

SiNPs have attracted much attention due to their potential applications in various fields and ease of preparation [97]. The significance and advantages of monodispersed nanometer-sized SiO_2 particles were shown not only in the scientific field, but also in various industrial applications, for example, catalysts, pigments, pharmacy, and so on [98]. SiNPs are used to make electronic substrates, thin film substrates, electrical insulators, thermal insulators, humidity sensors, and so on, and play a different role in each of these products. The quality of some of these products is highly dependent on the size and size distribution of the silica particles. The need for well-defined SiNPs has increased, as high-tech industries (e.g., computer and biotechnology/ pharmaceuticals) provide an elevated demand for such materials. Surface functionalization of SiNPs with polymer layers permits their use in many fields including ceramics, chemical mechanical polishing, coatings, glazes, emulsifiers, strengtheners, and binders [98, 99]. Various polymerization techniques have been employed to grow polymer chains on silica surface, which include surface-initiated (SI) anionic polymerization, cationic polymerization, ring-opening polymerization, radical polymerization, and RDRP techniques such as NMP, ATRP, RAFT, and so on [100–111]. In 1990, Boven et al. grafted polymethyl methacrylate (PMMA) brushes onto the surface of SiO_2 nanoparticles by introducing azo initiator onto the surface and subsequently carried out SI free radical polymerization at 70 °C [82]. In 1998, monochlorosilane-based azo initiator was synthesized and thick polystyrene (PS) brush was grafted as shown in Scheme 2.9 [81]. In this methodology, the chlorosilane head group helps to immobilize the initiator moiety onto the surface of the SiNPs and the azo initiator moiety initiates the polymerization of various monomers.

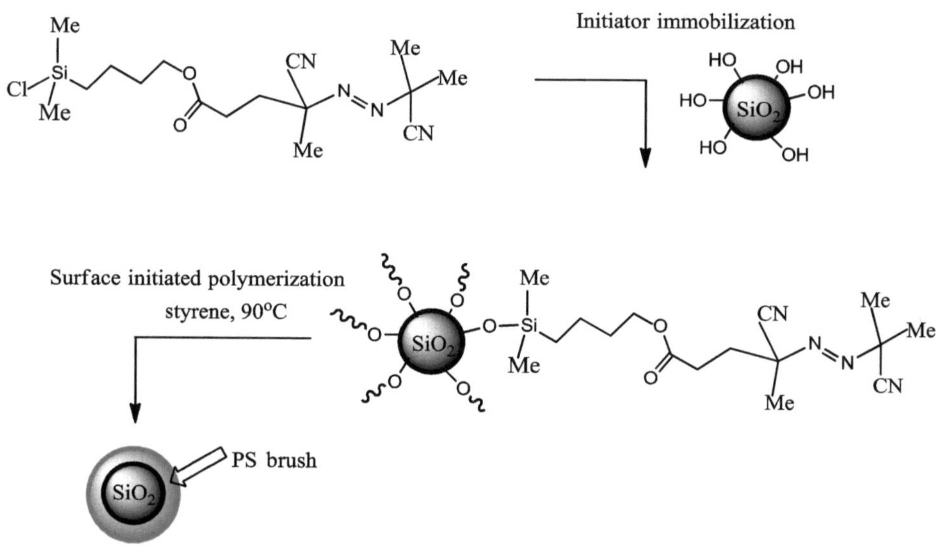

Scheme 2.9: Surface-initiated free radical polymerization of styrene on SiO_2 particles. (Source: O. Prucker and J. Rühe, Macromolecules, 1998, 31, 3, 592. ©1998, ACS) [81].

To have better control over the polymerization and molecular weight of the polymer brush, SI-RDRP was employed. Hawker et al. synthesized trichlorosilane-based nitroxide and immobilized onto the surface of SiNPs. Polymerization of styrene at 120 °C was then carried out from the surface-immobilized nitroxide molecules as shown in Scheme 2.10 [84].

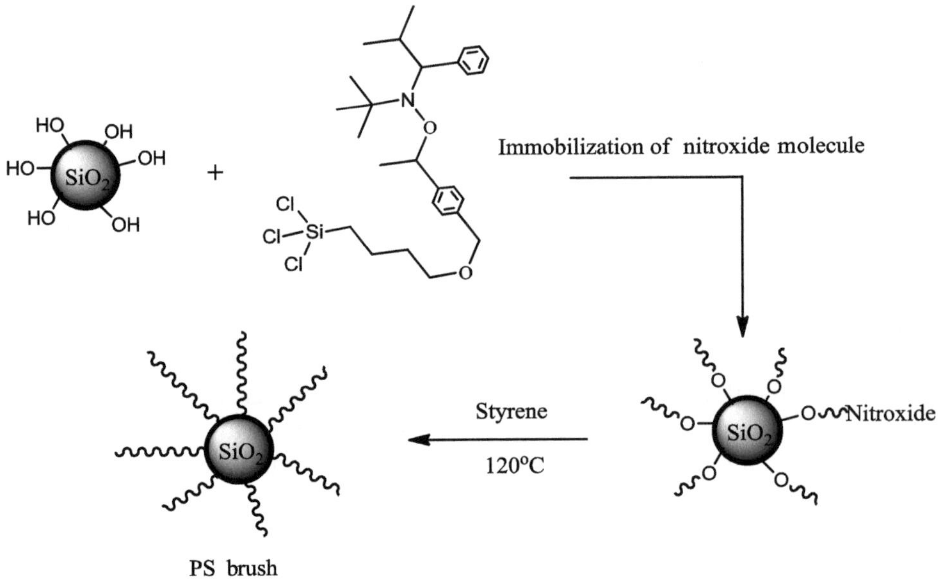

Scheme 2.10: Surface-initiated nitroxide-mediated polymerization of styrene on SiO_2 particles. (Source: S. Blomberg, S. Ostberg, E. Harth, A.W. Bosman, B.V. Horn and C.J. Hawker, Journal of Polymer Science Part A: Polymer Chemistry, 2002, 40, 9, 1309. ©2002, John Wiley & Sons) [84].

Although NMP allows control over the polymerization, it was carried out at higher temperature and polymerization at lower temperature avoids spontaneous thermal polymerization and other side reactions. Ambient polymerization temperatures could lead to better control of the polymerization reaction and accordingly improve the structural homogeneity of the grafted brushes [112, 113]. In order to carry out the SI polymerization at ambient conditions, SI-ATRP is normally employed. In our work, we have synthesized chlorosilane-based ATRP initiator, namely, (3-(2-bromoisobutyryl)propyl)dimethylchlorosilane, **2**, which was prepared from allyl alcohol and 2-bromoisobutyrlbromide in the presence of triethylamine to give allyl bromoester, **1**, which was then hydrosilylated with dimethylchlorosilane in the presence of Pt (C) as shown in Scheme 2.11 [113]. This initiator was immobilized onto the surface of SiNPs (particle size 55 ± 2 nm), which was synthesized by Stöber's process [114]. The ATRP initiator, **2**, was then immobilized onto the surface of SiNPs through chlorosilane moieties.

The ATRP of benzyl methacrylate (BnMA) was carried out, at 30 °C, from the initiator immobilized St ber's silica, **3,** using CuBr/PMDETA as the catalyst in anisole. Ethyl-2-bromoisobutyrate was added as the free initiator to control the polymerization. The addition of the free initiator creates the necessary concentration of the [Cu(II)] complex, which in turn controls the polymerization from the substrate as well as in solution [115, 116]. Control experiments were carried out under

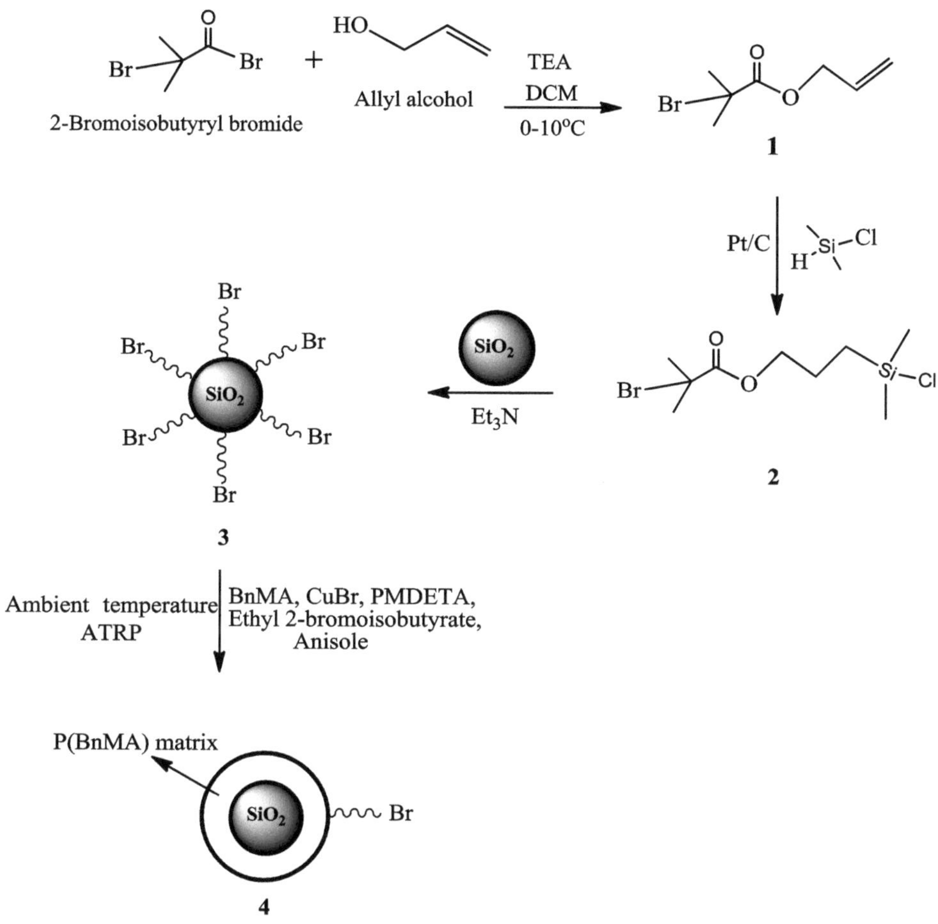

Scheme 2.11: SI-ATRP of benzyl methacrylate (BnMA) on SiO_2 particles. Reproduced with permission from S. Munirasu, G.K. Raghuraman, J. Rühe and R. Dhamodharan, Langmuir, 2011, 27, 21, 13284. ©2011, ACS [113].

identical polymerization conditions but with no initiator on the silica surface. From these experiments, it could be inferred that the entire free polymer formed is essentially from the free initiator and that the polymer that had adsorbed onto the unmodified surface could be removed by extraction procedures. In order to determine the molecular weight of the grafted polymer layers, the PBnMA layers were degrafted from the surface of silica particles using HF and then subjected to GPC measurements. The M_n and M_w/M_n (*dispersity*) of the free polymer formed in solution from the free initiator as well as the degrafted polymer are plotted as a function of polymerization time as shown in Figure 2.6. A linear increase in the molecular weight (M_n) of the polymer is observed with polymerization time and

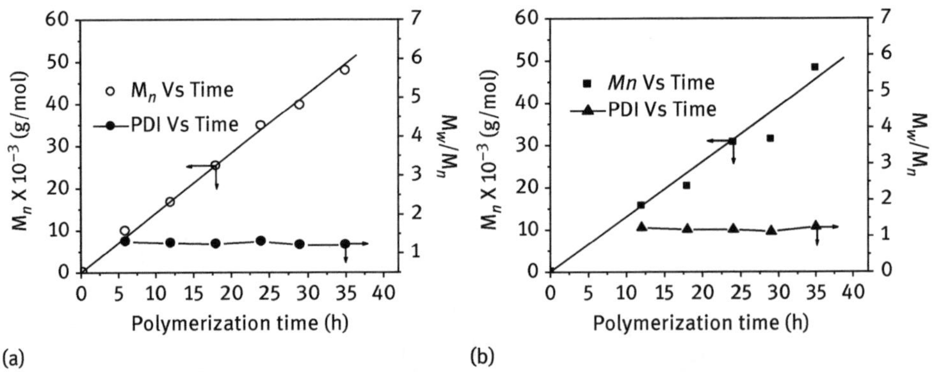

Figure 2.6: Plot of M_n and M_w/M_n with polymerization time. Reproduced with permission from S. Munirasu, G.K. Raghuraman, J. Rühe and R. Dhamodharan, Langmuir, 2011, 27, 21, 13284. ©2011, ACS [113].

dispersity values remained fairly low and constant, which confirms that the polymerization is well controlled [113].

SI-RAFT polymerization was carried out by introducing the RAFT agent onto the surface of SiNPs as shown in Scheme 2.12. The particles were first modified with 3-aminopropyl dimethylethoxysilane and subsequently reacted with activated 4-cyanopentanoic acid dithiobenzoate [111]. The first polymerization was conducted using the SiNPs with the RAFT agent. Following the deactivation of the active RAFT agent from the chain ends, a second RAFT agent was attached to the surface of the nanoparticles grafted with the first polymer chain population. The second population of chains can be made with the same monomer or monomers different from the first polymerization. Using this SI-RAFT process, mixed/binary/bimodal polymer brushes were grafted onto the surface of SiNPs.

2.7.3 Magnetite nanoparticles

Magnetite particles in the size range from nanometer to micrometer are attractive materials mainly due to their diverse properties, which are related to their size and composition when compared to the bulk material. Great interest in the use of MNPs are in areas such as bio-nanotechnology and biomedicine applications and these areas need MNPs that form stable colloidal suspensions, in addition to being compatible, nontoxic, and nonimmunogenic [117, 118]. Among the derivatives of iron oxide NPs are magnetite (Fe_3O_4), maghemite (γ-Fe_2O_3), and hematite (α-Fe_2O_3), where MNPs are the most studied materials due to their response to magnetic field through the super-paramagnetic behavior, at room temperature, with high saturation magnetization. In addition, their nontoxicity and high biocompatibility are also suitable for

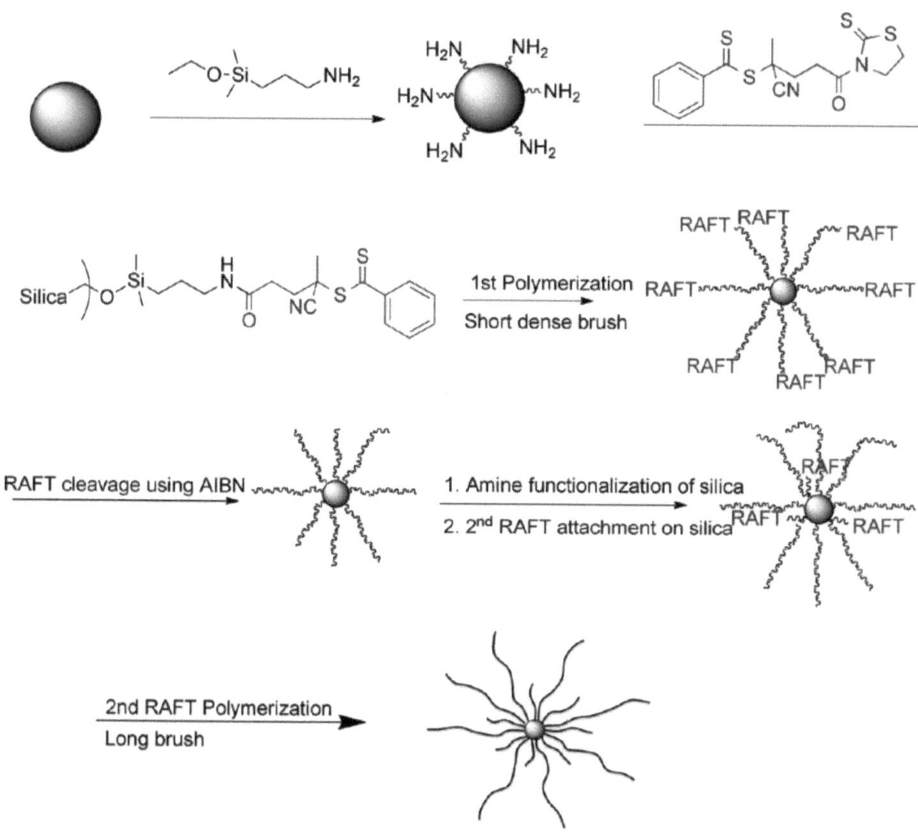

Scheme 2.12: Grafting of polymer brushes onto SiNPs by SI-RAFT technique. Reproduced with permission from A. Rungta, B. Natarajan, T. Neely, D. Duked, L.S. Schadler and B.C. Benicewicz, Macromolecules, 2012, 45, 23, 9303. ©2012, ACS [111].

biotechnology areas [119]. Magnetic nanoparticles that display high saturation magnetization and high magnetic susceptibility are of great interest for medical applications, which require the nanoparticles to be in uniform size, shape, and well dispersed in a solvent. For example, a particle of size in the nanometer range can pass through a capillary vessel to bind with a protein or DNA, whereas a micrometer particle will react only with cells [120, 121]. The synthesis of MNPs has been performed through methods such as sol–gel, coprecipitation, hydrothermal synthesis, thermal decomposition, microemulsion, and colloidal chemistry method [122–126]. Among all these synthesis routes, the coprecipitation method is easy and it is inexpensive to prepare aqueous dispersions of MNPs because the synthesis is conducted in water. In this process, aqueous solutions of Fe^{2+} and Fe^{3+} salts were prepared and they were coprecipitated by the addition of a base as shown in Scheme 2.13 [127, 128].

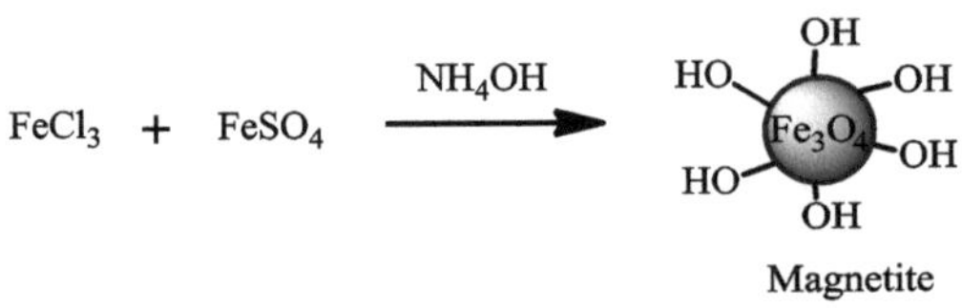

Scheme 2.13: Synthesis of magnetite nanoparticles by coprecipitation. (Source: G.K. Raghuraman and R. Dhamodharan, Ph.D. thesis entitled "Attachment of polymer monolayers-A convenient approach to modify nanoparticulate and flat surfaces" submitted to Indian Institute of Technology, Madras, India, 2007) [6].

The control of size, shape, and composition depends on the salts used (chlorides, sulfates, nitrates, etc.), Fe^{2+} and Fe^{3+} ratio, pH, and ionic strength of the medium.

There has been increased interest in grafting/attaching polymeric materials from these MNP surfaces for greater dispersion of these nanoparticles in various solvents. In principle, these polymer-grafted MNPs find applications in many fields by a simple change in the polymer matrix. These polymer attached MNPs can be used as a filler material in polymer fibers, which can be used as a sensor [129]. It is preferable to have polymer layers with controlled architecture such as chain length and its distribution, which can be achieved by SI-RDRP. Matsuno et al. reported on the synthesis of PS-grafted MNPs by NMP using phosphonic acid anchor groups as shown in Scheme 2.14 [130, 131]. Zhijun et al. grafted poly-4-vinylpyridine brushes by NMP technique. They have modified the surface of MNPs with 3-methacryloxypropyl trimethoxysilane and subsequently reacted with the methacryloxy group with nitroxide. SI-NMP of 4-vinylpyridine was then carried out [132].

As discussed earlier in Section 2.6.4, the initiating species for ATRP is alkyl halide. Various silane-based alkyl halide groups capable of initiating SI-ATRP are listed in Figure 2.7 and the monomers, catalysts, that were employed in SI-ATRP from magnetite surface are summarized in Table 2.3.

The dispersion of MNPs in various solvents is an important factor for commercial applications. The PMMA-grafted MNPs forms stable dispersion in organic solvents such as toluene as shown in Figure 2.8. The unmodified MNPs settle down completely after 6 h, whereas the PMMA-grafted magnetite forms stable dispersion [6]. The large difference in the dispersibility between MNPs and PMMA-grafted MNPs suggests that the PMMA layer acts as a steric stabilizing layer to prevent agglomeration. Apart from dispersion stability, by combining the magnetic properties of NPs and suitable polymer layers, it is possible to design nanomaterials for specific applications. Polydimethylaminoethyl methacrylate (PDMAEMA) brushes-grafted MNPs were prepared via SI-ATRP of DMAEMA, using 2-(2-bromoisobutyryloxy)ethyl)phosphonic acid (BiBEP) anchoring groups. The amino groups in PDMAEMA brushes were subsequently quarternized resulting in highly efficient antibacterial magnetic nanoparticles, which exhibited a response to external magnetic fields and were easily removed from

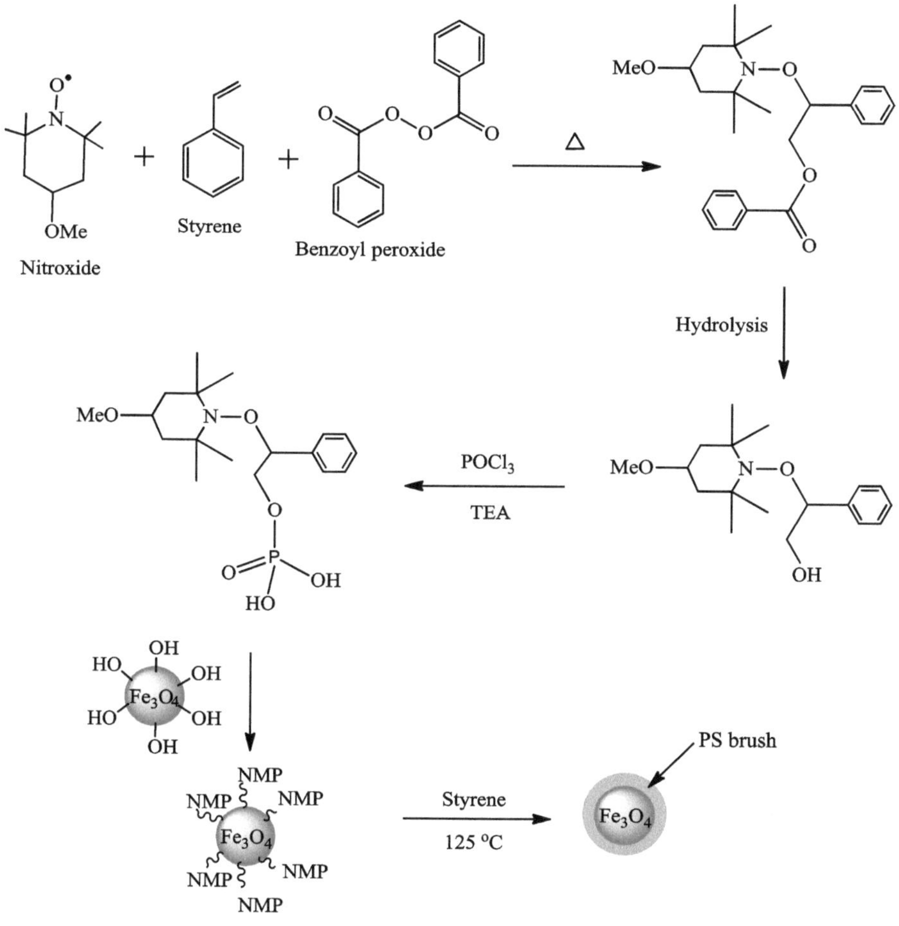

Scheme 2.14: PS grafting on magnetite nanoparticles by NMP. Reproduced with permission from R. Matsuno, K. Yamamoto., H. Otsuka and A. Takahara, Chemistry of Materials, 2003, 15, 1, 3. ©2003, ACS [130].

water after antibacterial tests. The PQA-modified MNPs retained 100% biocidal efficiency against *Escherichia coli* (105–106 *E. coli*/mg nanoparticles) without washing with any solvents or water, throughout eight cycles [136].

Another strategy to graft polymer layers onto the surface of nanoparticles is to combine the synthetic techniques of RAFT polymerization and click chemistry [141, 142]. Zhou et al. described the synthesis of azide-functionalized RAFT CTA, which was used to mediate RAFT polymerization of poly(ethylene glycol) monomethacrylate (PEGMA) on the alkyne-functionalized Fe_3O_4 NPs surface via click coupling reactions. RAFT and click coupling were combined to graft PEGMA onto the surface of Fe_3O_4 NPs [143]. SI-RAFT polymerization was also performed on MNPs to obtain

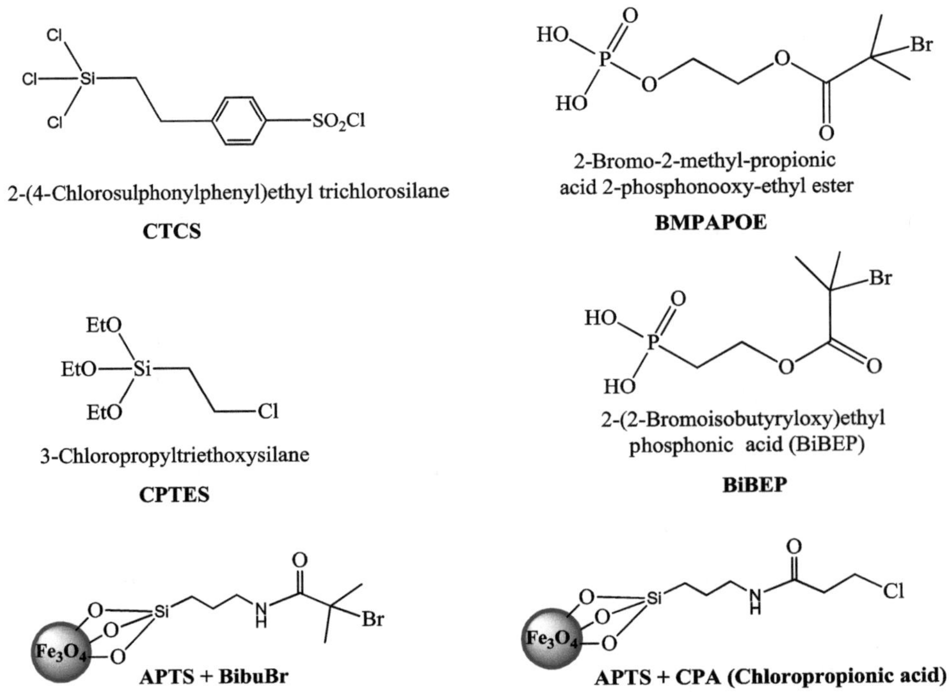

Figure 2.7: Various surface anchoring groups with ATRP initiators immobilized on magnetite surface.

Table 2.3: A summary of polymer brushes grafted onto the surface of magnetite nanoparticles through SI-ATRP.

Anchoring group	Catalyst/initiator	Polymer layer/Method of growth/polymerization temperature	Reference
CTCS	Cu(I)Br/spartine/TsCl	PMMA/70 °C	Fukuda et al., 2004 [87]
	Cu(I)Br/bipyridyl/TsCl	PS/100 °C	Garcia et al., 2007 [133]
BMPAPOE	Cu(I)Br/PMDETA	PMMA/30 °C	Dhamodharan et al., 2008 [134]
	Cu(I)Br/PMDETA	PBnMA/30 °C	Dhamodharan et al., 2009 [135]
BiBEP	Cu(I)Cl/$CuCl_2$/HMTETA	Polydimethylaminoethyl methacrylate/40 °C	Matyjaszewski et al., 2011 [136]

Table 2.3 (continued)

Anchoring group	Catalyst/initiator	Polymer layer/Method of growth/polymerization temperature	Reference
APTS + BibuBr	Cu(I)Br/PMDETA	PMMA/30 °C	Dhamodharan et al., 2006 [137]
	Cu(I)Br/PMDETA	PS/110 °C	Peng et al., 2007 [138]
CPTES	Cu(I)Br/bipyridyl/ TsCl	PMMA/90 °C	Galeotti et al., 2011 [139]
APTS + CPA	Cu(I)Br/bipyridyl	PEGMA/70 °C	Zhou et al., 2008 [140]

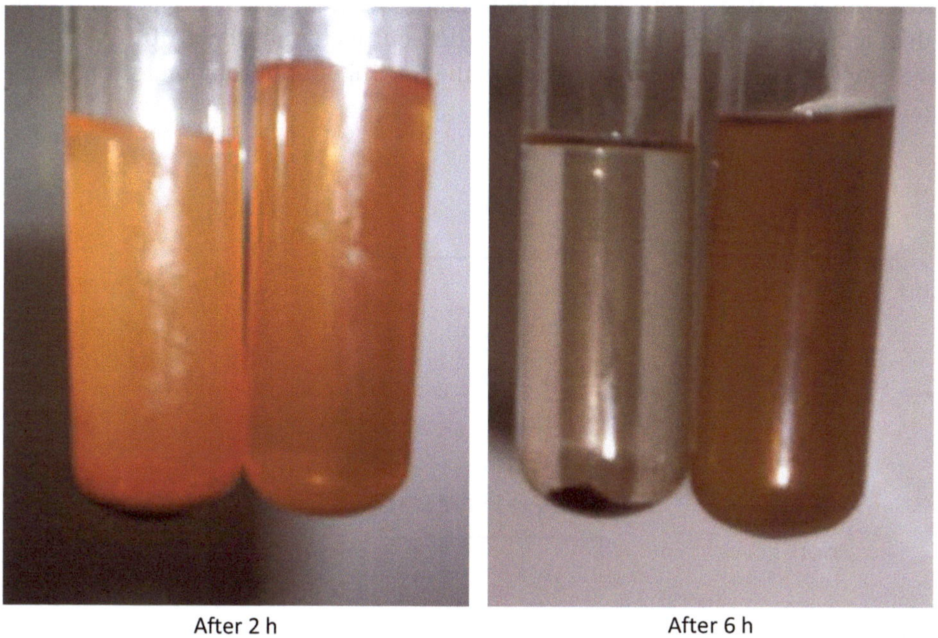

Figure 2.8: Photographs of magnetite nanoparticle dispersion in toluene. (Source: G.K. Raghuraman and R. Dhamodharan, Ph.D. thesis titled on "Attachment of polymer monolayers-A convenient approach to modify nanoparticulate and flat surfaces" submitted to Indian Institute of Technology, Madras, India, 2007) [6].

polymer brushes of PPEGMA-*b*-PNIPAM. The grafted PEGMA-*b*-PNIPAM was observed to decrease the nonspecific adsorption of protein on MNPs [144].

2.7.4 Titania nanoparticles

Titania (TiO_2) nanoparticles possess interesting optical, dielectric, and catalytic properties, which lead to industrial applications such as pigments, fillers, catalyst supports, and photocatalysts [28, 29]. In order to prepare of TiO_2 nanoparticles with precise size control, several techniques have been employed that include sol–gel process, solvothermal process, hydrothermal process, and so on. From the above-mentioned methods, the sol–gel method is normally used for the preparation of TiO_2 nanopowder in which the titanium isopropoxide was hydrolyzed in water at room temperature [145–147]. There has been increased interest in grafting/attaching polymeric materials on to the surface of titania nanoparticles, which can be used as fillers material in paints and ceramic industries and also used as fillers in polymer fibers, which can be used as a sensor [129]. Stable dispersion of TiO_2 NPs was prepared by grafting PMMA layer onto the surface both by SI conventional free radical polymerization as well as ATRP [148]. The titania nanoparticles were initially modified with a small molecule capable of initiating ATRP at ambient temperature and then ambient temperature ATRP was carried out using Cu(I)Br, PMDETA catalyst system with ethyl-2-bromoisobutyrate as the free initiator in solution to control the polymerization as shown in Scheme 2.15.

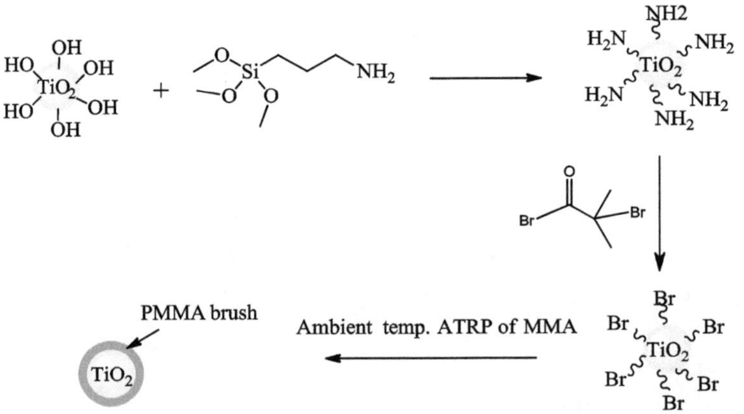

Scheme 2.15: SI-ATRP of MMA on titania nanoparticles. Reproduced with permission from G.K. Raghuraman, J. Rühe and R. Dhamodharan, Journal of Nanoparticle Research, 2008, 10, 3, 415. ©2008, Springer [148].

The SI free radical polymerization of MMA was carried out at 60 °C using hydridosilane-based azo initiator, which was immobilized on to the surface of titania. The core-shell structure of PMMA-grafted titania was confirmed by TEM analysis as shown in Figure 2.9. The image on the left shows that the titania particles are

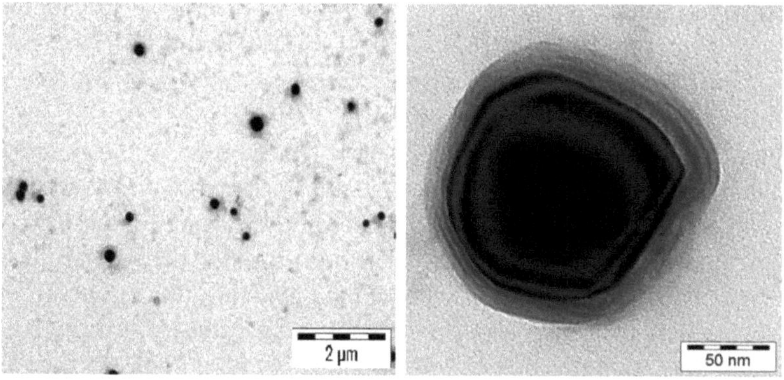

Figure 2.9: TEM images of PMMA-grafted titania nanoparticles. Reproduced with permission from G.K. Raghuraman, J. Rühe and R. Dhamodharan, Journal of Nanoparticle Research, 2008, 10, 3, 415. ©2008, Springer [148].

dispersed well, while that in the right establishes the formation of core-shell (TiO_2-PMMA) structures. Kim et al. covaltpdeently anchored hydrophilic zwitterionic polymer layer onto the surface of titania nanoparticles by functionalizing the surface with 3-methacryloxypropyl trimethoxysilane and then the polymerization was carried out from the surface of nanoparticles by emulsion polymerization process as shown in Scheme 2.16. Radical polymerization of 2-methacryroyloxyethyl phosphoryl choline (MPC) was performed to incorporate zwitterionic moieties onto the surface of particles. These phosphorylcholine groups are known to have an excellent biocompatibility [149].

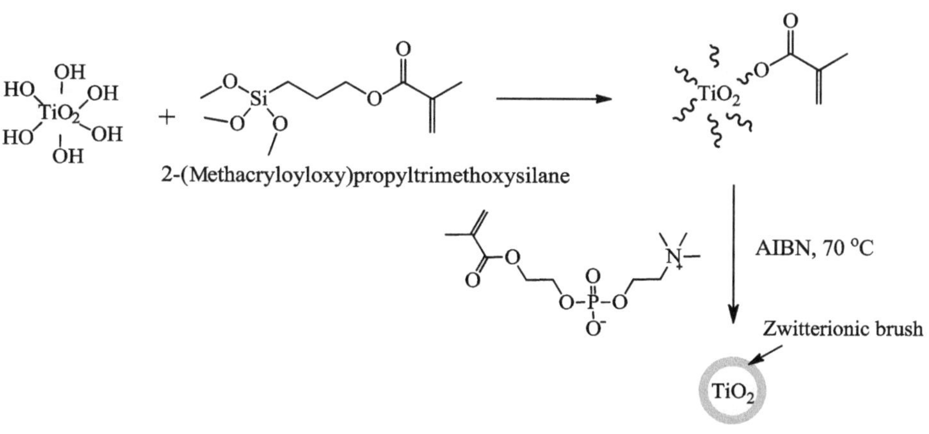

Scheme 2.16: Polymerization of 2-methacryroyloxyethyl phosphorylcholine onto the surface of titania nanoparticles. (Source: Y. Kim, S. An, S. Kim, J.H. Park, H.A. Son, H.T. Kim, K.-D. Suh and J.W. Kim, Polymer, 2013, 54, 21, 5609) [149].

2.7.5 Gold nanoparticles

Colloidal AUNPs have been utilized for centuries by artists due to the vibrant colors produced with visible light. When these nanoparticles are irradiated by an electromagnetic radiation, the oscillating electric field causes the electrons in the conduction band to oscillate. When the frequency of oscillation becomes equal to that of frequency of incident light, then resonance takes place and the absorption peak appears in the visible region known as the *surface plasmon absorption* or *surface plasmon resonance*. With the variation in the size of the nanoparticles, there is variation in the band gap and surface plasmon absorption that is responsible for vibrant colors in visible region. The scientific evaluation of colloidal gold did not begin until Faraday's work on the synthesis of colloidal gold where he reduced gold chloride using phosphorous and prepared pure sample of gold colloid, which he called "activated" gold [7]. The simplest method to prepare gold colloids was carried out by Turkevich in 1951 as described in Scheme 2.1 [10].

The AuNPs are extensively used in various applications, which include electronics, photodynamic therapy targeting tumor cells, therapeutic agent in drug delivery, sensors, biological imaging applications, and in catalysis [150–154]. In order to minimize aggregation, the surface modification of AuNPs with polymers, small molecules, and biological recognition molecules has been employed. AuNPs functionalized with thymine groups were assembled into spherical aggregates by diaminotriazine-functionalized random and block copolymers and the use of a diblock copolymer allowed size-controlled formation of AuNPs in solution as well as in thin film [155]. Dendrimers uniformly dispersed in a hydrophilic polymer network were used as a template for the formation of AuNPs. As the network was swollen in an aqueous solution of gold salt, the ions were attached to the dendrimer, where they were reduced, forming AuNPs in the dendrimer [156]. In these approaches, the polymeric materials have been used as stabilizing groups during the synthetic step. There were few attempts made to graft functionalized polymer such as poly(ethylene glycol)-based glycopolymer onto the surface of AuNPs to quantitatively investigate the reversible lectin-induced association of AuNPs in order to construct a colloidal sensor system applicable to bioassay and biorecognition [157]. Thiol-capped polystyrene (PSt) was synthesized and AuNPs were successfully incorporated into the polymer matrix, which resulted in AuNPs decorated with PS matrix [158]. Fukuda et al. have employed the SI LRP technique to the synthesis of an AuNP coated with a well-defined, high-density polymer brush. The ATRP initiator attached disulfide groups were synthesized and used in one-pot synthesis of AuNPs from chloroauric acid reduction as shown in Scheme 2.17. The initiator-modified AuNPs were subsequently subjected to SI-ATRP of MMA at 40 °C using Cu(I)Br and (−)Sparteine as catalyst system. The AuNPs exhibit surface plasmon absorption at 530 nm. The PMMA-grafted AuNPs with differing PMMA chain lengths exhibited surface plasmon absorption at about the same wavelength, but marginally blue

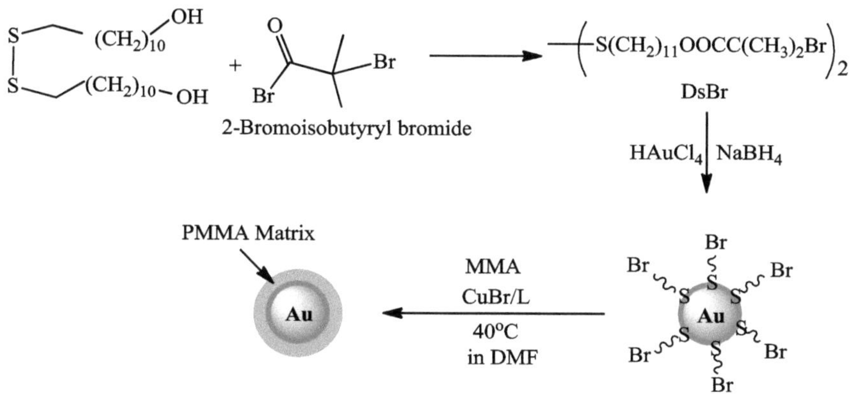

Scheme 2.17: SI-ATRP of MMA on AuNPs. (Source: K. Ohno, K. Koh, Y. Tsujii and T. Fukuda, Macromolecules, 2002, 35, 24, 8989. ©2002, ACS) [86].

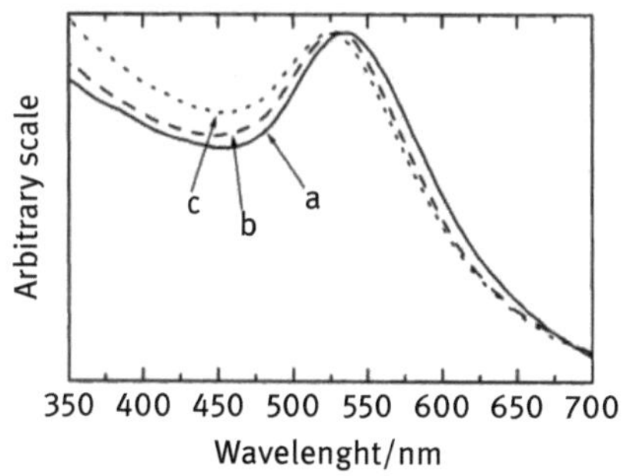

Figure 2.10: Absorption spectra of PMMA-AuNPs dissolved in THF at room temperature: M_n's of the graft polymers are (a) 6,000, (b) 12,000, and (c) 29,000. Reproduced with permission from K. Ohno, K. Koh, Y. Tsujii and T. Fukuda, Macromolecules, 2002, 35, 24, 8989. ©2002, ACS [86].

shifted, as shown in Figure 2.10. As the surface plasmon absorption of AuNP is known to be sensitive to the surrounding environment, it is evident that the PMMA chains were responsible for the change in the spectral shift [86].

Thermoresponsive polymer brushes on 20 nm colloidal gold were prepared by surface SI-ATRP of *N*-isopropylacrylamide (NIPAM) in aqueous media by Chakraborty et al. In their approach, stepwise functionalization of AuNPs were carried out with mercaptohexadecanoic acid, followed by reacting it with *N*-hydroxysuccinimide and then with 2-bromopropionylbromide. The surface ATRP initiator enabled effective growth of dense polymer chains from the particle surface using the "grafting-from" technique as shown in Scheme 2.18. These thermoresponsive polymer brush functionalization could be tailored for applications in drug delivery, medical diagnosis, and nanoactuators [159]. Polymer-modified AUNPs such as PEGylated (polyethylene glycol units) AUNPs find applications in tumor detection and gene therapy. When AuNPs were encapsulated with a thiol modified poly(ethylene glycol) units which allows for the compatibility and circulation in vivo and used in cancer

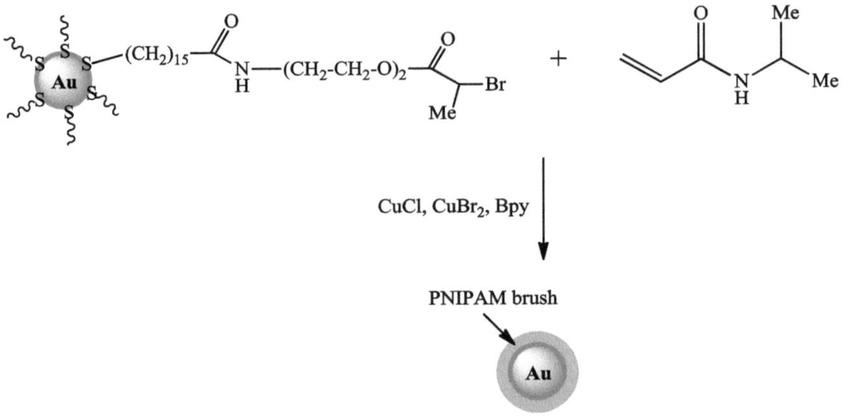

Scheme 2.18: Synthesis of thermoresponsive PNIPAM brushes on gold surfaces. (Source: S. Chakraborty, S.W. Bishnoi and V.H. Pérez-Luna, Journal of Physical Chemistry, C, 2010, 114, 13, 5947. ©2010, ACS) [159].

research to target tumors. In addition, to specifically target tumor cells, the PEGylated gold particles are conjugated with an antibody, and then using surface-enhanced Raman spectrum, these PEGylated AuNPs can detect the location of the tumor [160–162].

2.7.6 Silver nanoparticles

Silver is widely known as a catalyst for the oxidation of methanol to formaldehyde and ethylene to ethylene oxide [163]. Apart from AuNPs, other nanoparticles that exhibit surface plasmon resonance and have unique optical, electrical, and thermal properties are silver nanoparticles (AgNPs). AgNPs are used in biosensors and numerous assays where the AgNP materials can be used as biological tags for quantitative detection. They are incorporated in apparel, footwear, paints, wound dressings, appliances, cosmetics, and plastics for their antibacterial properties. AgNPs are normally composed of large percentage of silver oxide due to their large ratio of surface-to-bulk silver atoms. Although different shapes of nanoparticles can be constructed based on the application requirement, commonly used are spherical AgNPs. Chemical reduction is the most frequently applied method for the preparation of stable colloidal dispersions of AgNPs in water or organic solvents. Commonly used reductants are borohydride, citrate, ascorbate, and elemental hydrogen [164–166]. Initially, the reduction of silver salts or complexes leads to the formation of zero-valent Ag atoms, which undergo nucleation and growth steps and result in the formation of colloidal AgNPs, which were stabilized by stabilizing agents.

Grafting of polymer layers onto AgNPs attracts much attention for its wide range of applications. Li et al. synthesized anemone-shaped polymer brush on AgNP surface. Initially, an alternating copolymer of maleic anhydride (MA) and p-chloromethyl styrene (CMS) were prepared and then block copolymers of p-chloromethyl styrene–MA–CMS alternating copolymer were prepared, which were modified with 2-mercaptoethylamine by reacting with MA groups. Subsequently, AgNPs were synthesized by inverse micellization of Ag salt reduction in the presence of mercapto-modified block copolymer to obtain spherical AgNPs out-shelled with *p*-chloromethylstyrene groups. The surface *p*-chloromethyl groups of Ag nanoparticles were then used as an ATRP initiator to initiate polymerization of styrene at 110 °C to obtain anemone-shaped polymer brushes [167].

Polymer-encapsulated AgNPs were prepared using poly(St-*alt*-MA)-*graft*-PMMA copolymer that acts as a scaffold for the synthesis of size-confined AgNPs as shown in Scheme 2.19. The graft copolymer is synthesized via ambient temperature ATRP using the CuBr/PMDETA catalytic system at ambient temperature. The graft copolymer is hypothesized to function as a scaffold with the anhydride part interacting strongly with the silver ions, while the PMMA grafts function as a polymer brush that stabilizes the dispersion and prevents the particle aggregation due to the "polymer brush effect" [168].

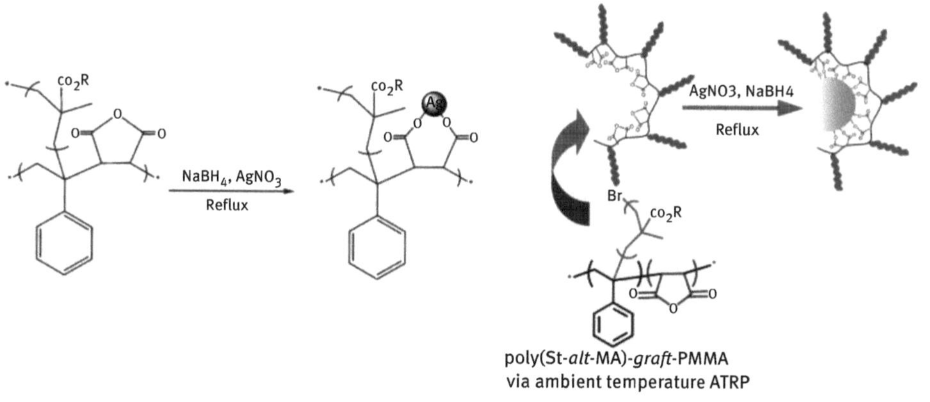

Scheme 2.19: Synthetic scheme and graphical representation of AgNP synthesis using polymeric scaffold. Reproduced with permission from A.V. Vivek, S.M. Pradipta and R. Dhamodharan, Macromolecular Rapid Communications, 2008, 29, 9, 737. ©2008, John Wiley & Sons [168].

Silver is known for its antimicrobial properties and has been used for years in the medical field for antimicrobial applications and even has been shown to prevent HIV binding to host cells. Enhanced antibacterial activities have been reported in AgNPs modified by polymers such as PVP. The antibacterial activities of PVP-modified AgNPs are significant because the polymer is most effective in stabilizing particles against aggregation. AgNPs may eventually be useful for the treatment of various diseases. The properties of AgNPs applicable to human treatments are under investigation in laboratory and animal studies, assessing potential efficacy, toxicity, and costs [169–172].

2.7.7 Cadmium sulfide nanoparticles

Semiconductor nanocrystals or quantum dots have attracted much attention both in fundamental and industrial research due to their unique size-dependent optical and electronic properties and their study provides an opportunity to observe the evolution of collective behavior of the bulk from the discrete nature of molecular properties. In particular, CdS nanoparticles have been extensively investigated due to the their strong quantum confinement effects that result in significant variation in their electrical and optical properties and also due to their exciting utilization in the fields of light-emitting diodes, electrochemical cells, lasers, catalytic applications, and biological labeling [173–176]. Nanocrystalline CdS is a II–VI semiconductor, which shows size-dependent properties. Bulk CdS has a band gap of 2.42 eV, with a melting point around 1,600 °C, whereas the band gap of a 2 nm sized CdS nanoparticles is 3.57 eV, with a melting point around 400 °C [177, 178]. Due to its large value of band gap, which allows light emission between blue and red wavelengths, CdS has been extensively studied and it is used as window material in hetero-junction solar cells. In p–n junction solar cells, CdS is used as n-type material along with p-type materials like gallium arsenide, indium phosphide, and cadmium telluride [179]. CdS has three types of crystal structures, namely, hexagonal wurtzite, cubic zinc blend, and high-pressure rock-salt phase. Among these, the hexagonal wurtzite has been intensively investigated because it is the most stable of the three phases and can be easily synthesized [177]. CdS nanoparticles have been prepared using chemical precipitation methods, solvothermal methods, laser ablation methods, hydrothermal methods, photochemical methods, one-pot synthesis methods, mesoporous copolymer template methods, and so on [180–182]. In the chemical precipitation method, dilute ammonium disulfide solution was added to Cd $(NO_3)_2$ solution in the presence of nonionic surfactant polyoxyethylene nonaphenyl ether, which results in the formation of CdS quantum dots as evident from the development of a bright yellow color [183]. The surface functionalization of these nanoparticles with polymer layers results in a hybrid materials. The surface functionalization also serves to passivate the surface and improves their properties, as shown by the increase in photoluminescence quantum yield for organic or inorganic capped CdS nanoparticles [184]. Patten et al. synthesized novel core-shell semiconductor nanoparticle (or quantum dot)/polymer hybrids, that is, CdS/SiO_2/PMMA composite material and their use in forming photoluminescent films in which the nanoparticles are evenly dispersed on the microscopic level [88]. In their approach, at first the surface of the CdS nanoparticles were functionalized with tetraethoxysilane to introduce silica layer, which is subsequently functionalized with 3-(2-bromopropionyloxy) propyl dimethylethoxysilane to introduce ATRP initiator onto the surface as shown in Scheme 2.20. SI-ATRP of MMA was then carried with the immobilized initiator and $NiBr_2(PPh_3)_2$ catalytic system.

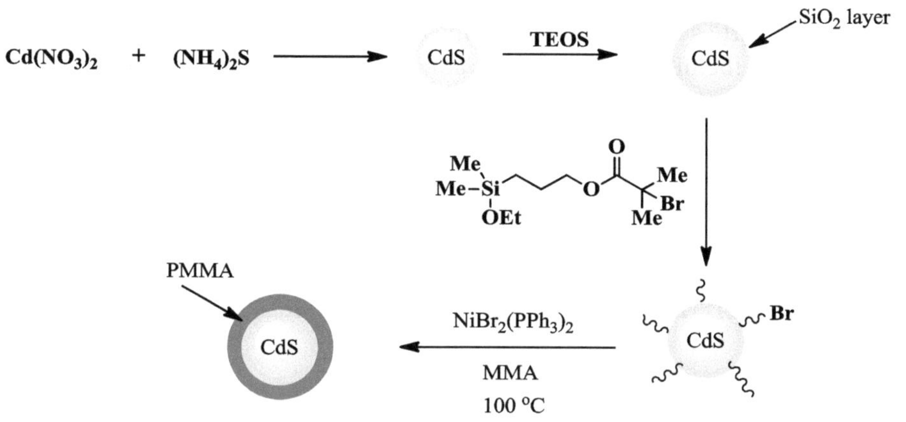

Scheme 2.20: Schematic representation of grafting PMMA layer onto CdS/SiO_2 surface. (Source: S.C. Farmer and T.E. Patten, Chemistry of Materials, 2001, 13, 11, 3920. ©2001, ACS) [88].

When these hybrid nanoparticles (CdS/SiO_2/PMMA) were subjected to film casting on glass slide, it forms continuous films, which were transparent and had a faint yellow tint due to the presence of the CdS quantum dots. The cast film was then subjected to illumination with 365 nm UV light, and under this light, all regions of the film were observed to emit orange-red light. These hybrid nanoparticles are versatile structures in that they have components whose structures and compositions can be altered to introduce new properties to the material (i.e., the luminescence of the inorganic core or the conductivity of the polymer arms) and to allow for self-assembly of the material.

2.8 Polymer clay nanocomposites

A nanocomposite is defined as a composite material in which one of the components is in the nanometer size scale (<100 nm) at least in one dimension. A well-known example for nanocomposites comes from nature, that is, bones, shells, and wood are made of combination of materials such as carbohydrates, lipids, and proteins [185, 186]. A composite material can be custom tailored to have specific properties that will meet special requirements. In a composite material, one phase is matrix, while the other is filler material. Based on the nature of matrix phase, they were divided into polymeric, ceramic, and metallic composites. Polymers have advantages over other materials because of their unique advantages such as ease of production, light weight, and ductility. In polymer nanocomposites, the filler materials are at nanoscale and its nanoscale dispersion within the polymer matrix leads to the superior properties than those of bulk polymer phase. The properties of the polymer composites depend strongly on the mechanical properties of the filler material, their adhesion with polymer, and most importantly, the aspect ratio of the filler material [187]. Clays are a group of nanofillers that have been widely used for the preparation of polymer nanocomposites. Clay minerals belong to a main group of silicates with layered structure known as *layered silicates*. Polymer–clay composites based on layered silicates can be classified into three types, namely, conventional composites, intercalated composites, and exfoliated composites based on the extent of separation of silicate layers as shown in Figure 2.11 [188].

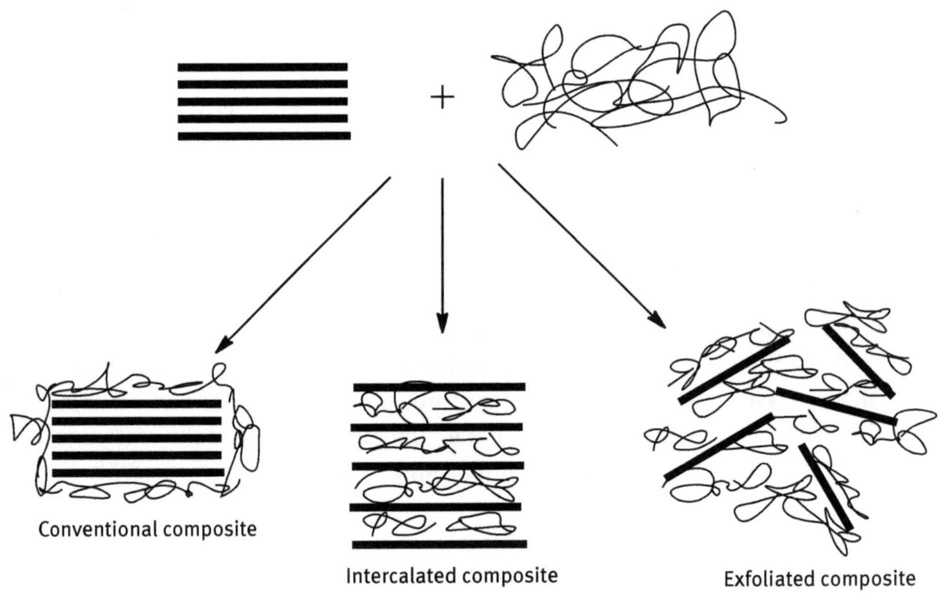

Figure 2.11: Schematic representation of different types of composites. Reproduced with permission from B. Chen, J.R.G. Evans, H.C. Greenwell, P. Boulet, P.V. Coveney, A.A. Bowden and A. Whiting, Chemical Society Reviews, 2008, 37, 3, 568. ©2008, RSC [188].

The distance between a plane in the unit layer and the corresponding plane in the next unit layer is defined as the basal plane spacing, d_{001}. If the polymer layer does not enter between these two unit layers and the basal plane spacing remains unchanged, and it is called "conventional composite." If the polymer layers enter the galleries causing an increase in d_{001}, but the clay layers remain stacked, the composite is "intercalated." If the clay layers are completely pushed apart to create a disordered array, then the composite is called "exfoliated composites," and based on the arrangements, it is called *ordered* or *disordered exfoliation*. Based on the structure, clays are generally classified as allophane, kaolinite, halloysite, smectite, illite, chlorite, vermiculite, attapulgite–palygorskite–sepiolite, and mixed layered minerals [189, 190]. Polymer–clay nanocomposites are mainly based on smectite clays because of their swelling properties, which result from their capacity to host water and organic molecules between silicate layers, high cation exchange capacities, high aspect ratio, and large surface area [191, 192]. Smectite clays consist of units in the form of "sheets" or "platelets," made up of two silica tetrahedral layers with a central alumina or magnesia octahedral layer. Hydrated exchangeable cations are found in the spaces between lattices. The layers are held together by van der Waals and electrostatic forces and the absence of primary chemical bonds allows the intercalation of water or polar organic molecules, causing the lattice to expand [190].

Polymer–clay nanocomposites were prepared mainly by in situ polymerization, solution methods, and melt processing. A wide range of polymers have been employed to produce clay nanocomposites that include polyglycols, polyethylene oxide, polyvinyl chloride, polyolefins, PS, polyamides, PMMA, polyesters, epoxies, and polylactic acid [193–205]. In in situ polymerization process, clay (in suspension) is mixed to a monomer solution in the presence of initiator and then subjected to heating to obtain polymer–clay nanocomposites. In the case of solution process, the clay (in suspension) is dispersed in polymer solution and then subjected to heating if appropriate. In melt processing method, polymer melt is directly mixed with clay to allow the migration of macromolecules into clay and then subjected to annealing process. The melt-processing method is normally employed for bulk processing. Clays can also be modified by initiators bearing a cationic group to interact with clays by cation exchange, followed by the surface-initiated polymerization (SIP). The SIP induces exfoliation of ordered silicate layers by the growth of polymeric chains initiated from the clay surface [206–208].

Although the syntheses of polymer clay nanocomposites are well studied, a great challenge is to synthesize polymer/clay materials with controlled architectures and narrow dispersities of the polymeric phase. The control over the polymer architectures can be obtained through RDRP techniques. The first successful attempt was in 1999, when Weimer et al. used a silicate-anchored initiator to elaborate exfoliated PS/montmorillonite (PS/MMT) materials via TEMPO-mediated polymerization [206]. Cherifi et al. synthesized polymer clay nanocomposites by in situ synthesis in which the poly(BA-*grad*-MMA)/clay materials were prepared by performing polymerization in the presence of either the raw MMT or the initiator-modified MMT [210]. Ma et al. synthesized PMMA/MMT organic–inorganic hybrid materials by conventional free radical polymerization and RAFT polymerization [211]. Shen and coworkers prepared polymer-grafted MMT composite material that possesses a hard MMT backbone and PS brush, which was prepared by NMP using TEMPO radical as shown in Scheme 2.21. The anchoring of the methacrylatoethytrimethyl ammonium chloride (DMC) group onto the surface of MMT by ion-exchange reaction and introduction of alkoxyamine initiator (NMP initiator) were carried out in one-step process by reacting the TEMPO, benzoyl peroxide, and DMC onto the surface of MMT. The NMP of styrene was then carried out from MMT surface at 130 ° C to obtain PS brush with controlled molecular weights and narrow dispersities [209]. A wide variety of polymer clay nanocomposites were prepared for different applications and the current focus on this topic is to understand the nanocomposite formation that allows tailoring the structure and to assess quantitatively the degree of intercalation and exfoliation in nanocomposites that provides the comprehensive information on the structure.

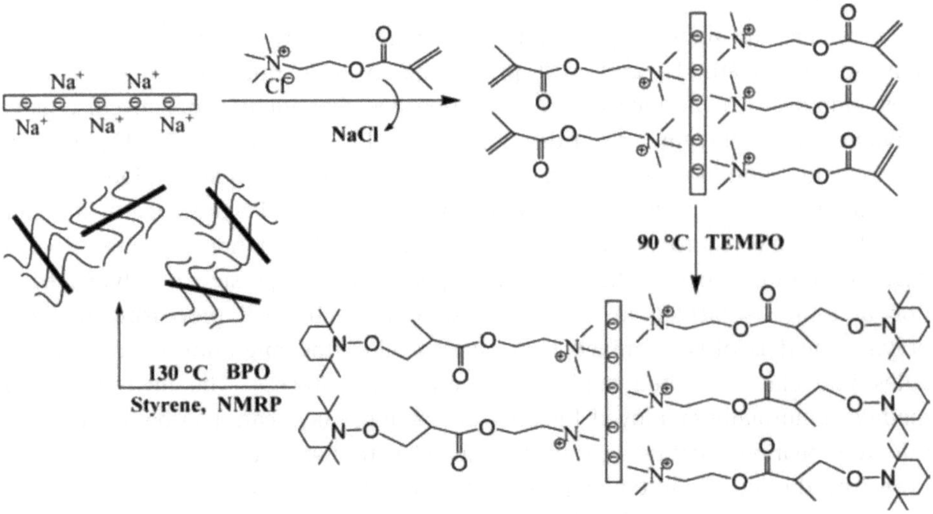

Scheme 2.21: The nitroxide-mediated polymerization of styrene from the alkoxyamine modified MMT surface. Reproduce with permission from Y. Shen, Y. Wang, J. Chen, H. Li, Z. Li and C. Li, Journal of Applied Polymer Science, 2010, 118, 2, 1198. ©2010, John Wiley & Sons [209].

To conclude, RDRP methods enable the controlled synthesis of polymer–nanohybrid materials with precision and control that is unique. In view of the wide range of applications that the nanohybrid materials offer, these methods acquire large significance.

References

[1] P. Innocenzi and B. Lebeau, Journal of Material Chemistry, 2005, 15(35–36), 3821.
[2] C.-L, Lin, M.-Y. Yeh, C.-H. Chen, S. Sudhakar, S.-J. Luo, Y.-C. Hsu, C.-Y. Huang, K.-C. Ho and T.-Y. Luh, Chemistry of Materials, 2006, 18(17), 4157.
[3] E.S. Kang, M. Takahashi, Y. Tokuda and T. Yoko, Journal of Materials Research, 2006, 21(5), 1286.
[4] Y.-J. Kwark, J.P. Bravo-Vasquez, H.B. Cao, H. Deng and C.K. Ober, Journal of Photopolymer Science and Technology, 2005, 18(4), 481.
[5] K. Matyjaszewski, Nonlinear Optics, Quantum Optics, 2003, 30(3–4), 167.
[6] G.K. Raghuraman and R. Dhamodharan, Ph.D. thesis titled on "Attachment of polymer monolayers-A convenient approach to modify nanoparticulate and flat surfaces" submitted to Indian Institute of Technology, Madras, India, 2007.
[7] M. Faraday, Philosophical Transactions Royal Society, London, 1857, 147, 145.
[8] J. Turkevich and G. Kim, Science, 1970, 169(3948), 873.
[9] J. Turkevich, Gold Bulletin, 1985, 18(3), 86.
[10] J. Turkevich, P.C. Stevenson and J. Hillier, Discussions of the Faraday Society, 1951, 11, 55.
[11] H. Bönnemann and R.M. Richards, European Journal of Inorganic Chemistry, 2001, 2001(10), 2455.
[12] K.M. Brunner, G.E. Harper, K. Keyvanloo, B.F. Woodfield, C.H. Bartholomew and W.C. Hecker, Energy & Fuels, 2015, 29(3), 1972.
[13] V. Stengl, J. Subrt, P. Bezdicka, P.M. Marikova and S. Bakardjieva, Diffusion and Defect Data-Solid State Data, Part. B: Solid State Phenomena, 2003, 90–91, 121.
[14] L. Zhou, S. Wang, D. Xu and Y. Guo, Industrial & Engineering Chemistry Research, 2014, 53(1), 481.
[15] J.R. Kim, K.Y. Lee, M.J. Suh and S.K. Ihm, Catalysis Today, 2012, 185(1), 25.
[16] A. Zielnska-Jurek, J. Reszczynska, E. Grabowska and A. Zaleska, Microemulsions: An Introduction to Properties and Applications, Ed., R. Najjar, InTech, Rijeka, 2012, 229.
[17] D. Chen and L. Gao, Journal of Colloid and Interface Science, 2004, 279(1), 137.
[18] Y. Zhu, J. Chen, N. Zhao, W. Lin, C. Lai and Q. Wang, Materials Letters, 2015, 160, 171.
[19] W. Zhang, F. Liu, Q. Li, Q. Shou, J. Cheng, L. Zhang, B.J. Nelson and X. Zhang, Physical Chemistry Chemical Physics, 2012, 14(47), 16331.
[20] D. Holzinger and G. Kickelbick, Chemistry of Materials, 2003, 15(26), 4944.
[21] M. Niederberger and G. Garnweitner, Chemistry-A European Journal, 2006, 12(28), 7282.
[22] L. Naszályi, F. Bosc, A. El Mansouri, A. van der Lee, D. Cot, Z. Hórvölgyi and A. Ayral, Separation and Purification Technology, 2008, 59(3), 304.
[23] W. Huang, W. Qian and M.A. El-Sayed, Journal of the American Chemical Society, 2006, 128 (41), 13330.
[24] S. Schelm, G.B. Smith, P.D. Garrett and W.K. Fisher, Journal of Applied Physics, 2005, 97(12), 124314.
[25] S. Odenbach, Journal of Physics: Condensed Matter, 2003, 15(15), S1497.
[26] M.T. Lopez-Lopez, J.D.G. Duran, A.V. Delgado and F. Gonzalez-Caballero, Journal of Colloids and Interface Science, 2005, 291(1), 144.
[27] G.D. Moeser, K.A. Roach, W.H, Green, P.E. Laibinis and T.A. Hatton, Industrial Engineering and Chemical Research, 2002, 41(19), 4739.
[28] A.M. El-Toni, S. Yin and T. Sato, Journal of Colloidal Science, 2006, 300(1), 123.
[29] N.S. Allen, M. Edge, G. Sandoval, J. Verran, J. Stratton and J. Maltby, Photochemical and Photobiological Sciences, 2005, 81(2), 279.
[30] A. Trovarelli, C. Leitenburg, M. Boaro and G. Dolcetti, Catalysis Today, 1999, 50(2), 353.

[31] E.L. Brosha, R. Mukundan, D.R. Brown, F.H. Garzon and J.H. Visser, Solid State Ionics, 2002, 148(1–2), 61.
[32] F. Mueller and P. Winter, inventors; EP 1541652, A1 20050615, 2005.
[33] A.P. Alivisatos, Science, 1996, 271, 933.
[34] W.E. Buhro and V.L. Colvin, Nature Materials, 2003, 2, 138.
[35] Z.A. Peng and X. Peng, Journal of American Chemical Society, 2001, 123, 183.
[36] M.A. Reed, J.N. Randall, R.J. Aggarwal, R.J. Matyi, T.M. Moore and A.E. Wetsel, Physical Review Letters, 1988, 60, 535.
[37] C. Wang, M. Shim and P. Guyot-Sionnest, Science, 2001, 291, 2390.
[38] C. Larpent, F. Brisse-Le Menn and H. Patin, Journal of Molecular Catalysis, 1991, 65(3), L35.
[39] M.N. Vargaftik, V.P. Zargorodnikov, I.P. Stolarov, I.I. Moiseev, D.I. Kochubey, V.A. Likholobov, A.L. Chuvilin and K.I. Zarnaraev, Journal of Molecular Catalysis, 1989, 53, 315.
[40] M.N. Vargaftik, V.P. Zargorodnikov, I.P. Stolarov, I.I. Moiseev, D.I. Kochubey, V.A. Likholobov, A.L. Chuvilin, V.I. Zaikosvsky, K.I. Zamaraev and G.I. Timofeeva, Journal of Chemical Society, Chemical Communications, 1985, 14, 937.
[41] M.T. Reetz, R. Breinbauer and K. Wanninger, Tetrahedran Letters, 1996, 37(26), 4499.
[42] M. Beller, H. Fischer, K. Kuhlein, C.-P. Reisinger and W.A. Herrmann, Journal of Organometallic Chemistry, 1996, 520(1–2), 257.
[43] M.T. Reetz and E. Westermann, Angewandte Chemie International Edition, 2000, 39(1), 165.
[44] E.J.W. Verwey and J.T.G. Overbeek, Transactions of the Faraday Society, 1946, 42, B117.
[45] K.D. Belfield and L. Zhang, Chemistry of Materials, 2006, 18(25), 5929.
[46] D.N. Muraviev, M.I. Pividory, J.L.M. Soto and S. Alegret, Solvent Extraction and Ion Exchange, 2006, 24(5), 731.
[47] A.B. Lowe, B.S. Sumerlin, M.S. Donovan and C.L. McCormick, Journal of the American Chemical Society, 2002, 124(39), 11562.
[48] P.J. Flory, Principles of Polymer Chemistry, Cornell University Press, 1953.
[49] P.J. Flory, Journal of the American Chemical Society, 1940, 62(6), 1561–1565.
[50] M. Szwarc, Nature, 1956, 178, 1168.
[51] M. Szwarc, M. Levy and R. Milkovich, Journal of the American Chemical Society, 1956, 78(11), 2656.
[52] K. Matyjaszewski and T.P. Davies, Handbook of Radical Polymerization, Wiley Publishers, 2002.
[53] K. Matyjaszewski and J. Xia, Chemical Reviews, 2001, 101(9), 2921.
[54] T. Otsu and M. Yoshida, Macromolecular Rapid Communications, 1982, 3(2), 127.
[55] M.K. Georges, R.P.N. Veregin, P.M. Kazmaier and G.K. Hamer, Macromolecules, 1993, 26(11), 2987.
[56] M. Kato, M. Kamigaito, M. Sawamoto and T. Higashimura, Macromolecules, 1995, 28(5), 1721.
[57] J.S. Wang and K. Matyjaszewski, Journal of the American Chemical Society, 1995, 117(20), 5614.
[58] J. Chiefari, Y.K. Chong, F. Ercole, J. Krstina, J. Jeffery, T.P.T. Le, R.T.A. Mayadunne, G.F. Meijs, C.L. Moad, G. Moad, E. Rizzardo and S.H. Thang, Macromolecules, 1998, 31(16), 5559.
[59] V. Percec, T. Guliashvili, J.S. Ladislaw, A. Wistrand, A. Stjerndahl, M.J. Sienkowska, M.J. Monteiro and S. Sahoo, Journal of the American Chemical Society, 2006, 128(43), 14156.
[60] S.H. Subramanian, R.P. Babu and R. Dhamodharan, Macromolecules, 2008, 41(1), 262.
[61] T. Otsu, Journal of Polymer Science, 1956, 21(99), 559.
[62] T. Otsu, K. Nayatani, I. Muto and M. Imai, Macromolecular Chemistry and Physics, 1958, 27 (1), 142.
[63] T. Otsu, Journal of Polymer Science Part A: Polymer Chemistry, 2000, 38(12), 2121.

[64] H. Fischer, Journal of Polymer Science Part A: Polymer Chemistry, 1999, 37(13), 1885.
[65] H. Fischer, Chemical Reviews, 2001, 101(12), 3581.
[66] G. Moad, E. Rizzardo and D.H. Solomon, Macromolecules, 1982, 15(3), 909.
[67] K.A. Moffat, G.K. Hamer and M.K. Georges, Macromolecules, 1999, 32(4), 1004.
[68] M.S. Kharasch and E.V. Jensen, Science, 1945, 102(2640), 128.
[69] D.P. Curran, Radical addition reactions, Comprehensive organic synthesis, Pergamon, Oxford, 1991, Vol. 4, 716.
[70] B.M. Rosen and V. Percec, Journal of Polymer Science Part A: Polymer Chemistry, 2007, 45 (21), 4950.
[71] G. Lligadas and V. Percec, Journal of Polymer Science Part A: Polymer Chemistry, 2007, 45 (20), 4684.
[72] G. Lligadas and V. Percec, Journal of Polymer Science Part A: Polymer Chemistry, 2008, 46 (10), 3174.
[73] G. Lligadas, B.M. Rosen, C.A. Bell, M.J. Monteiro and V. Percec, Macromolecules, 2008, 41 (22), 8365.
[74] B.M. Rosen and V. Percec, Chemical Reviews, 2009, 109(11), 5069.
[75] N. Haridharan, V. Ramkumar and R. Dhamodharan, Acta Crystallographica, Section E-Structure Reports Online, 2012, E68, O811.
[76] N. Haridharan and R. Dhamodharan, Journal of Polymer Science Part A: Polymer Chemistry, 2011, 49(4), 1021.
[77] W. Zhang, W. Zhang, Z. Zhang, Z. Cheng, Y. Tu, Y. Qiu and X. Zhu, Journal of Polymer Science Part A: Polymer Chemistry, 2010, 48(19), 4268.
[78] Q. Zhang, Z. Zhang, W. Wang, J. Zhu, Z. Cheng, N. Zhou, W. Zhang and X. Zhu, Journal of Polymer Science Part A: Polymer Chemistry, 2012, 50(7), 1424.
[79] S.T. Milner, Science, 1991, 21, 905.
[80] E.V. Mackor and J.H. Van der Waals, Journal of Colloids and Interface Science, 1952, 7(5), 535.
[81] O. Prucker and J. Rühe, Macromolecules, 1998, 31(3), 591.
[82] G. Boven, M.L.C.M. Oosterling, G. Challa and A.J. Schouten, Polymer, 1990, 31(12), 2377.
[83] M. Ryosuke, Y. Kazuya, O. Hideyuki and A. Takahara, Chemistry of Materials, 2003, 15(1), 3.
[84] S. Blomberg, S. Ostberg, E. Harth, A.W. Bosman, B.V. Horn and C.J. Hawker, Journal of Polymer Science Part A: Polymer Chemistry, 2002, 40(9), 1309.
[85] T.K. Mandal, M.S. Fleming and D.R. Walt, Chemistry of Materials, 2000, 12(11), 3481.
[86] K. Ohno, K. Koh, Y. Tsujii and T. Fukuda, Macromolecules, 2002, 35(24), 8989.
[87] M. Eizo, Y. Shinpei, N. Tsedev, T. Yoshinobu, T. Fukuda and M. Takano, Polymer, 2004, 45(7), 2231.
[88] S.C. Farmer and T.E. Patten, Chemistry of Materials, 2001, 13(11), 3920.
[89] R. Matsuno, H. Otsuka and A. Takahara, Soft Matter, 2006, 2(5), 415.
[90] C. Li and B.C. Benicewicz, Macromolecules, 2005, 38(14), 5929.
[91] R. Jordan, N. West, A. Ulman., Y.-M. Chou and O. Nuyken, Macromolecules, 2001, 34(6), 1606.
[92] A.Y. Fadeev and T.J. McCarthy, Journal of the American Chemical Society, 1999, 121(51), 12184.
[93] A. Y. Fadeev and T.J. McCarthy, inventors; US 6,673,459 B2, 2004.
[94] V. A. Ogarev and S. L. Selector, Progress in Organic Coatings, 1992, 31, 135.
[95] M. Rohwerder and M. Stratmann, MRS Bulletin, 1999, 24, 43.
[96] Z. Mekhalif, J. Riga, J.J. Pireaux and Delhalle, Langmuir, 1997, 13, 2285.
[97] G. Herbert, Journal of European Ceramic Society, 1994, 14, 205.
[98] R.K. Iler, The Chemistry of Silica, Wiley, New York, 1979.

[99] C. Payne and H. Bergna, Ed., The Colloid Chemistry of Silica, American Chemical Society, Washington, DC, 1994.
[100] S. Edmondson, V.L. Osborne and W.T.S. Huck, Chemical Society Reviews, 2004, 33(1), 14.
[101] A. Rungta, B. Natarajan, T. Neely, D. Duked, L.S. Schadler and B.C. Benicewicz, Macromolecules, 2012, 45(23), 9303.
[102] W. Huang, G.L. Baker and M.L. Bruening, Angewandte Chemie International Edition, 2001, 40 (8), 1510.
[103] S. Munirasu, G.K. Raghuraman, J. Rühe and R. Dhamodharan, Langmuir, 2011, 27(21), 13284.
[104] W. Stöber, A. Fink and E. Bohn, Journal of Colloids and Interface Science, 1968, 26(1), 62.
[105] T. von Werne and T.E. Patten, Journal of the American Chemical Society, 2001, 123(31), 7497.
[106] A. Ramakrishnan, R. Dhamodharan and J. Rühe, Journal of Polymer Science Part A: Polymer Chemistry, 2006, 44(5), 1758.
[107] J.B. Mamani, A.J. Costa-Filho, D.R. Cornejo, E.D. Vieira and L.F. Gamarra, Materials Characterization, 2013, 81, 28.
[108] K. Petcharoen and A. Sirivat, Materials Science and Engineering: B, 2012, 177(5), 421.
[109] C.S.S.R. Kumar, C. Leuschner, E.E. Doomes, L. Henry, M. Juban and J. Hormes, Journal of Nanoscience and Nanotechnology, 2004, 4(3), 245.
[110] J.H. Jang and H.B. Lim, Microchemical Journal, 2010, 94, 148.
[111] H. Iida, K. Takayanagi, T. Nakanishi and T. Osaka, Journal of Colloid and Interface Science, 2007, 314(1), 274.
[112] S.M. Montemayor, L.A. Garcia-Cerda and J.R. Torres-Lubián, Material Letters, 2005, 59(8–9), 1056.
[113] R. Betancourt-Galindo, O. Ayala-Valenzuela, L.A. Gracia-Cerda, O.R. Fernandez, J. Matutes-Aquino, G. Ramos and H. Yee-Madeira, Journal of Magnetism and Magnetic Materials, 2005, 294(2), e33.
[114] V. Sreeja and P.A. Joy, Materials Research Bulletin, 2007, 42(8), 1570.
[115] K. Simenoidis, S. Mourdikoudis, M. Moulla, I. Tsiaoussis, C. Martinez-Boubeta and M. Angelakeris, Journal of Magnetism and Magnetic Materials, 2007, 316(2), e1.
[116] W. Zhou, K. Tang, S. Zeng and Y. Qi, Nantechnology, 2008, 19(6), 602.
[117] P.D. Cozzoli, E. Snoeck, M.A. Garcia, C. Giannini, A. Guagliardi and A. Cervellino, Nano Letters, 2006, 6(9), 1966.
[118] M.A. Correa-Duarte, M. Giersig, N.A. Kotov and L.M. Liz-Marzan, Langmuir, 1998, 14(22), 6430.
[119] Z.-M. Huang, Y.-Z. Zhang, M. Kotaki and S. Ramakrishna, Composites Science and Technology, 2003, 63, 2223.
[120] R. Matsuno, K. Yamamoto., H. Otsuka and A. Takahara, Chemistry of Materials, 2003, 15(1), 3.
[121] R. Matsuno, K. Yamamoto., H. Otsuka and A. Takahara, Macromolecules, 2004, 37(6), 2203.
[122] C. Zhijun, Y. Qingxiang, P. Kai and G. Yabing, Journal of Applied Polymer Science, 2011, 119 (6), 3582.
[123] I. Garcia, A. Tercjak, N.E. Zafeiropoulos, M. Stamm and I. Mondragon, Journal of Polymer Science, Part A: Polymer Chemistry, 2007, 45(20), 4744.
[124] K. Babu and R. Dhamodharan, Nanoscale Research Letters, 2008, 3, 109.
[125] K. Babu and R. Dhamodharan, Nanoscale Research Letters, 2008, 4, 1090.
[126] H. Dong, J. Huang, R.R. Koepsel, Y. Penglin, A.J. Russell and K. Matyjaszewski, Biomacromolecules, 2011, 12(4), 1305.
[127] G.K. Raghuraman and R. Dhamodharan, Journal of Nanoscience and Nanotechnology, 2006, 6, 2018.
[128] Y. Sun, X. Ding, Z. Zheng, X. Cheng, X. Hu and Y. Peng, European Polymer Journal, 2007, 43 (3), 762.

[129] F. Galeotti, F. Bertini, G. Scavia and A. Bolognesi, Journal of Colloids and Interface Science, 2011, 360(2), 540.
[130] Y. Zhou, S. Wang, B. Ding and Z. Yang, Chemical Engineering Journal, 2008, 138(1–3), 578.
[131] Y. Li and B.C. Benicewicz, Macromolecules, 2008, 41(21), 7986.
[132] L. Chang, Y. Li, J. Chu, J. Qi and X. Li, Analytical Chimica Acta, 2010, 680(1–2), 65.
[133] Y. Zhou, S. Wang, K. Xie, Y. Dai and W. Ma, Applied Surface Science, 2011, 257(24), 10384.
[134] S. Wang, Y. Zhou, J. Peng, H. Niu, X. Zhang and F. Yang, Chemical Engineering Journal, 2011, 173(3), 873.
[135] H. Hayashi and K. Torii, Journal of Material Chemistry, 2002, 12(12), 3671.
[136] C.-S. Kim, B.K. Moon, J.-H. Park, B.-C. Choi and H.-J. Seo, Journal of Crystal Growth, 2003, 257 (3–4), 309.
[137] J.N. Nian and H. Teng, Journal of Physical Chemistry B, 2006, 110(9), 4193.
[138] G.K. Raghuraman, J. Rühe and R. Dhamodharan, Journal of Nanoparticle Research, 2008, 10 (3), 415.
[139] Y. Kim, S. An, S. Kim, J.H. Park, H.A. Son, H.T. Kim, K.-D. Suh and J.W. Kim, Polymer, 2013, 54 (21), 5609.
[140] D. Huang, F. Liao, S. Molesa, D. Redinger and V. Subramanian, Journal of Electrochemical Society, 2003, 150(7), 412.
[141] T. Stuchinskaya, M. Moreno, M.J. Cook, D.R. Edwards and D.A. Russell, Photochemical and Photobiological Sciences, 2011, 10(5), 822.
[142] S.D. Brown, P. Nativo, J.A. Smith, D. Stirling, P.R. Edwards, B. Venugopal, D.J. Flint, J.A. Plumb, D. Graham and J.N. Wheate, Journal of the American Chemical Society, 2010, 132 (13), 4678.
[143] S.D. Perrault and W.C.W. Chan, Proceedings of the Natural Academy of Sciences, USA, 2010, 107, 11194.
[144] D.T. Thompson, Nano Today, 2007, 2, 40.
[145] B.L. Frankamp, O. Uzun, F. Ilhan, A.K. Boal and V.M. Rotello, Journal of the American Chemical Society, 2002, 124(6), 892.
[146] F. Groehn, G. Kim, B.J. Bauer and E.J. Amis, Macromolecules, 2001, 34(7), 2179.
[147] H. Otsuka, Y. Akiyama, Y. Nagasaki and K. Kataoka, Journal of the American Chemical Society, 2001, 123(34), 8226.
[148] M.K. Corbierre, N.S. Cameron, M. Sutton, S.G.J. Mochrie, L.B. Lurio, A. Ruehm and R.B. Lennox, Journal of the American Chemical Society, 2001, 123(42), 10411.
[149] S. Chakraborty, S.W. Bishnoi and V.H. Pérez-Luna, Journal of Physical Chemistry C, 2010, 114 (13), 5947.
[150] X. Qian, P. Xiang-Hong, O.A. Dominic, Y.-G. Qiquin, Z.C. Georgia, M.S. Dong, Y. Lily, N.Y. Andrew, D.W. May and N. Shuming, Nature Biotechnology, 2008, 26(1), 83.
[151] A.Y. Sajjadi, A. Suratkar, K. Mitra and M.S. Grace, Journal of Nanotechnology and Engineering Medicine, 2012, 3(2), 021002.
[152] J. Conde, F. Tian, Y. Hernandez, C. Bao, D. Cui, K.-P. Janssen, M.R. Ibarra, P.V. Baptista, T. Stoeger and J.M. de la Fuente, Biomaterials, 2013, 34(31), 7744.
[153] A. Nagy and G. Mestl, Applied Catalysis A: General, 1999, 188(1–2), 337.
[154] A. Tao, P. Sinsermsuksaku and P. Yang, Angewandte Chemie International Edition, 2006, 45 (28), 4597.
[155] P.C. Lee and D. Meisel, Journal of Physical Chemistry, 1982, 86(17), 3391.
[156] U. Nickel, A.Z. Castell, K. Poppl and S. Schneider, Langmuir, 2000, 16(23), 9087.
[157] Z. Ming-Qiang, W. Liu-He, Z. Wei-Cheng, F.-S. Du., Z.-C. Li and F.-M. Li, Polymer Preprints, 2002, 43(2), 68.

[158] A.V. Vivek, S.M. Pradipta and R. Dhamodharan, Macromolecular Rapid Communications, 2008, 29(9), 737.
[159] N. Nino-Martinez, G.A. Martinez-Castanon, A. Aragon-Pina, F. Martinez-Gutierrez, J.R. Martinez-Mendoza and F. Ruiz, Nanotechnology, 2008, 19(6), 065711/1.
[160] V. Alt, T. Bechert, P. Steinruecke, M. Wagener, P. Seidel, D. Elvira, D. Eugen and R. Schnettler, Biomaterials, 2004, 25(18), 4383.
[161] A.D. Russell and W.B. Hugo, Progress in Medicinal Chemistry, 1994, 31, 351.
[162] L. Kvitek, A. Panacek, J. Soukupova, M. Kolar, R. Vecerova, R. Prucek, H. Mirka and Z. Radek, Journal of Physical Chemistry C, 2008, 112(15), 5825.
[163] C.B. Murray, D.J. Norris and M.G. Bawendi, Journal of the American Chemical Society, 1993, 115(19), 8706.
[164] J.L. Zhao, J.A. Bardecker, A.M. Munro, M.S. Liu, Y.H. Niu, I.K. Ding, J.D. Luo, B.Q. Chen, A.K.Y. Jen and D.S. Ginger, Nano Letters, 2006, 6(3), 463.
[165] J.S. Jang, U.A. Joshi and J.S. Lee, Journal of Physical Chemistry C, 2007, 111(35), 13280.
[166] M. Bruchez, M. Moronne, P. Gin, S. Weiss and A.P. Alivisatos, Science, 1998, 281(5385), 2013.
[167] R. Banerjee, R. ayakrishnan and P. Ayyub, Journal of Physics: Condensed Matter, 2000, 12 (50), 10647.
[168] A.N. Goldstein, C.M. Echerand and A.P. Alivisatos, Science, 1992, 256(5062), 1425.
[169] N. Romeo, A. Bosio, V. Canevari and A. Podesta, Solar Energy, 2004, 77(6), 795.
[170] V. Singh and P. Chauhan, Journal of Physics and Chemistry of Solids, 2009, 70(7), 1074.
[171] A. Phuruangrat, T. Thongtem and S. Thongtem, Journal of Experimental Nanoscience, 2009, 4(1), 47.
[172] H. Tong and Y.J. Zhu, Nanotechnology, 2006, 17, 845.
[173] S.-Y. Chang, L. Liu and S.A. Ascher, Journal of the American Chemical Society, 1994, 116(15), 6739.
[174] Y. Tian, T. Newton, N.A. Kotov, D.M. Guldi and J.H. Fendler, Journal of Physical Chemistry, 1996, 100(21), 8927.
[175] E.P. Giannelis, Advanced Materials, 1996, 8(1), 29.
[176] F. Hussain, M. Hojjati, M. Okamoto and R.E. Gorga, Journal of Composite Materials, 2006, 40 (17), 1511.
[177] N. Sheng, M.C. Boyce, D.M. Parks, G.C. Rutledge, J.I. Abes and R.E. Cohen, Polymer, 2004, 45 (2), 487.
[178] B. Chen, J.R.G. Evans, H.C. Greenwell, P. Boulet, P.V. Coveney, A.A. Bowden and A. Whiting, Chemical Society Reviews, 2008, 37(3), 568.
[179] R.A. Schoonheydt, T.J. Pinnavaia, G. Lagaly and N. Gangas, Pure and Applied Chemistry, 1999, 71(12), 2367.
[180] R.E. Grim, Clay Mineralogy, 2nd Edition, McGraw-Hill Book Company, New York, 1968.
[181] T.J. Pinnavaia and G.W. Beall, Polymer-clay nanocomposites, John Wiley & Sons Ltd., Chichester, 2000.
[182] T.J. Pinnavaia, Science, 1983, 220(4595), 365.
[183] B. Chen, A.A. Bowden, H.C. Greenwell, P. Boulet, P.V. Coveney, A. Whiting and J.R.G. Evans, Journal of Polymer Science Part B: Polymer Physics, 2005, 43(14), 1785.
[184] B. Chen and J.R.G. Evans, Polymer International, 2005, 54(5), 807.
[185] B. Lepoittevin, M. Devalckenaere, N. Pantoustier, M. Alexandre, D. Kubies, C. Calberg, R. Jerome and P. Dubois, Polymer, 2002, 43(14), 4017.
[186] E. Ruiz-Hitzky and P. Aranda, Advanced Materials, 1990, 2(11), 545.
[187] C.Y. Wan, Y. Zhang, Y.X. Zhang, X.Y. Qiao and G.M. Teng, Journal of Polymer Science Part B: Polymer Letters, 2004, 42(2), 286.

[188] S.K. Swain and A.I. Isayev, Polymer, 2007, 48(1), 281.
[189] R.A. Vaia, H. Ishii and E.P. Giannelis, Chemistry of Materials, 1993, 5(12), 1694.
[190] A. Okada and A. Usuki, Macromolecular Materials and Engineering, 2006, 291(12), 1449.
[191] C. Zeng and L.J. Lee, Macromolecules, 2001, 34(12), 4098.
[192] X. Huang, S. Lewis, W.J. Brittain and R.A. Vaia, Macromolecules, 2000, 33(6), 2000.
[193] T. Lan, P.D. Kaviratna and T.J. Pinnavaia, Journal of Physics and Chemistry of Solids, 1996, 57 (6–8), 1005.
[194] S.S. Ray, K. Yamada, M. Okamoto, Y. Fujimoto, A. Ogami and K. Ueda, Chemistry of Materials, 2003, 15(7), 1456.
[195] M.W. Weimer, H. Chen, E.P. Giannelis and D.Y. Sogah, Journal of the American Chemical Society, 1999, 121(7), 1615.
[196] X. Fan, Q. Zhou, C. Xia, W. Cristofoli, J. Mays and R.C. Advincula, Langmuir, 2002, 18(11), 4511.
[197] X. Fan, C. Xia and R.C. Advincula, Langmuir, 2003, 19(10), 4381.
[198] Y. Shen, Y. Wang, J. Chen, H. Li, Z. Li and C. Li, Journal of Applied Polymer Science, 2010, 118 (2), 1198.
[199] N. Cherifi, A. Benaboura, M. Save and L. Billion, Journal of Polymer Science Part A: Polymer Chemistry, 2012, 50(19), 3976.
[200] J. Ma, M. Lu, C. Cao and H. Zhang, Journal of Macrmolecular Science Part A: Pure and Applied Chemistry, 2013, 50(6), 653.
[201] M. Baum and W.J. Brittain, Macromolecules, 2002, 35(3), 610.
[202] C. Perruchot, M.A Khan, A Kamitsi, S.P. Armes, T. von Werne and T.E. Patten, Langmuir, 2001, 17(15), 4479.
[203] N. Ayres, S.G. Boyes and W.J. Brittain, Langmuir, 2007, 23(1), 182.
[204] B. Mu, R. Shen and P. Liu, Colloids and Surfaces B: Biointerfaces, 2009, 74(2), 511.
[205] E.P. Giannelis, The Journal of the Minerals, Metals and Materials Society, 1992, 44(3), 28.
[206] Q. Zhou, S. Wang, X. Fan, R. Advincula and J. Mays, Langmuir, 2002, 18(8), 3324.
[207] A. Juang, O.A. Scherman, R.H. Grubbs and N.S. Lewis, Langmuir, 2001, 17(5), 1321.
[208] J. Pyun and K. Matyjaszewski, Chemistry of Materials, 2001, 13(10), 3436.
[209] T. Meyer, S. Spange, S. Hesse, C. Jager and C. Bellmann, Macromolecular Chemistry and Physics, 2003, 204(4), 725.
[210] M. Husseman, E.E. Malmstrom, M. McNamara, M. Mate, D. Mecerreyes, D. G. Benoit, J.L. Hedrick, P. Mansky, E. Huang, T.P. Russell and C.J. Hawker, Macromolecules, 1999, 32(5), 1424.
[211] J. Lahann and R. Langer, Macromolecular Rapid Communications, 2001, 22(12), 968.

Sachin S. Patil, Prakash S. Sane, Dnyaneshwar V. Palaskar and Prakash P. Wadgaonkar

3 Synthesis of functionally terminated polymers by atom transfer radical polymerization (ATRP) and their applications

3.1 Introduction

Telechelic polymers are defined as polymers bearing reactive functional group(s) at the chain end(s) that can react selectively with other molecules/reagents to form a new chemical bond. The term "telechelic" was first introduced by Uraneck et al. in 1960 from the Greek words "tele" meaning "end" and "chelos" meaning "claw" to describe polybutadiene-bearing carboxyl or hydroxyl end groups designed to study the curing reactions for improved elastic properties [1]. The first report on the synthesis of telechelic polymer and the properties of obtained architecture was published in 1947 [2]. However, the concept was not fully understood till 1960 [1]. In the last few decades, there have been considerable advancements in synthesis of telechelic polymers and their applications [3]. Telechelic polymers have attracted a considerable attention due to their utility as chain extenders, cross linkers [4, 5], and as precursors to design different macromolecular architectures for various applications in the form of engineering plastics [6], viscosity modifiers [7], thermosetting resins and rubbers [8], compatibilizers in blends [9], emulsifiers [10], thermoplastic elastomers [11], self-healing materials [4], surfactants [12], surface modifiers [13], lubricants in mechanical and biological systems [14], drug [15] and gene [16] delivery vehicles, and so on.

Depending on the nature of terminal group(s), telechelic polymers are classified as under:

1) Homo-telechelic polymers: Polymers bearing same functional groups at both the terminals are termed as *homo-telechelics* [17–21].
2) Hetero-telechelic polymers: Polymers bearing different functional end groups are referred to as *hetero-telechelic polymers* [18, 19, 22–27].
3) Semi-telechelic polymers: Polymers bearing functional group only at the one chain end are categorized as *semi-telechelic polymers* [28, 29].

Schematic representation of different types of telechelic polymers is shown in Figure 3.1.

Sachin S. Patil, Reliance corporate park, Reliance Industries Limited, Ghansoli, Navi Mumbai, Maharashtra, India
Prakash S. Sane, Asian Paints Limited, Research & Technology Center, Navi Mumbai, India
Prakash P. Wadgaonkar, Polymer Science and Engineering Division, CSIR-National Chemical Laboratory, Pune, Maharashtra, India
Dnyaneshwar V. Palaskar, ELANTAS Beck India Limited, Pune, Maharashtra, India

https://doi.org/10.1515/9783110643695-003

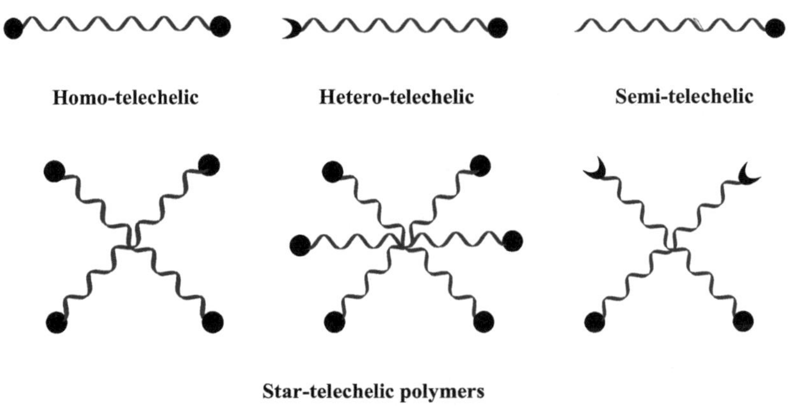

Figure 3.1: Schematic representation of different types of telechelic polymers.

The earlier popular methods for the synthesis of telechelic polymers involved the conventional free radical polymerization using initiators containing appropriate functional groups. Free radical polymerizations are relatively insensitive to impurities and can be performed at moderate temperatures. However, the disadvantages associated with conventional free radical polymerization methods are 1) poor control over the molecular weight, 2) broad molecular weight distribution, and 3) difficulties in the preparation of well-defined polymers with predetermined functionality. The uncontrolled behavior of free radical polymerization makes the technique of limited utility with respect to synthesis of well-defined functionally terminated polymers.

The essential requirements for utility of polymers in terms of certain applications are the precise control of polymer structure, molecular weight, and molecular weight distribution. Telechelic polymers with precise control of end functionalities, controlled molecular weight, and narrow molecular weight distribution can be obtained by various controlled/living polymerization methods such as anionic [30], cationic [31], group transfer [32], and reversible deactivation radical polymerization (RDRP) methods, namely, reversible addition-fragmentation chain transfer (RAFT) [33], nitroxide-mediated radical polymerization (NMP) [34], atom transfer radical polymerization (ATRP) [34] and its variants [35, 36], and so on. Telechelic polymers prepared by these methods are free from the complications of nonuniformity of molecular weight distribution and hence these methods are mostly preferred to achieve the controlled molecular designs. However, the ionic polymerization methods such as anionic and cationic are highly sensitive to the environmental conditions and other reactive functional groups present in the monomers or initiator involved in the polymerization process and often require the complex reaction set up.

The combination of fast initiation and absence of termination reactions are the ideal conditions for a "living" polymerization. These controlled/living systems usually proceed via formation of an intermediate active propagating species, which has a dynamic equilibrium between growing and dormant species like in anionic and cationic systems. The establishment of such a dynamic equilibrium between propagating radicals and dormant species is a critical step to all RDRP systems. RDRPs provide the ability to control the molecular weight, molecular weight distribution, composition, site specific functionality, and topology of polymers [37, 38].

3.2 Atom transfer radical polymerization (ATRP)

Among all the controlled/living polymerization methods, ATRP is a robust and widely used method due to its tolerance to the environmental conditions and reactive functional groups present on the initiators and monomers involved in the polymerization process. ATRP overcomes the drawbacks of both, that is, the sensitivity and reactivity of ionic polymerizations and uncontrolled behavior of conventional radical polymerizations [39]. ATRP was discovered independently by Matyjaszewski [40] and Sawamoto [41] in 1995. A large number of book chapters, monographs, and review articles are available pertaining to the synthesis of macromolecular architectures by ATRP [38, 42].

Scheme 3.1 depicts the mechanism of ATRP process. Initiation step involves the abstraction of halide radical from the organic halide (R–X) by transition metal complex, creating a radical that subsequently adds to double bond of the monomer to generate another radical that reversibly deactivates again to form dormant species [43]. Chain transfer and termination reactions have mostly been eliminated in this polymerization method [44].

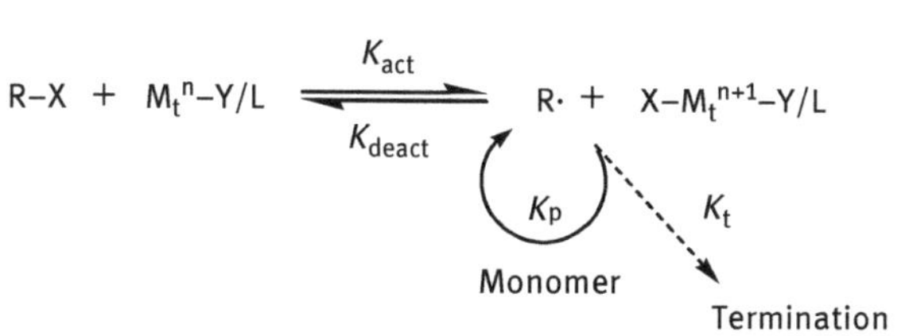

Scheme 3.1: Schematic representation of ATRP mechanism. Reproduced with permission from K. Matyjaszewski and J. Xia, Chemical Reviews, 2001, 101, 9, 2921. ©2001, ACS [39].

To obtain better control on ATRP reaction, initiation step should be faster than the propagation, that is, $K_i > K_p$ meaning the initiation rate constant (K_i) should be greater than propagation rate constant (K_p). Polymerization reactions above 95% conversion may have the chances of radical coupling in ATRP processes that can be avoided by maintaining the conversion of reactions below 95%. It is well known that less than 5% terminations are observed in the ATRP reactions [45]. The selection of a suitable catalyst is an important factor in ATRP that enables the system to establish a reversible equilibrium between active and dormant species and in enhancing the rate of initiation by minimizing the possibility of radical–radical terminations [46]. Initiators used in ATRP generally contain active carbon–halide bond, that is, alkyl halides R–X (X = Cl, Br) [45, 47]. Structural adjustment of the alkyl group (R) and the leaving group (X) of an initiator in order to make the R–X bond more labile than the propagating polymer–halide bond provides a handle to vary the rate of initiation in the ATRP system. The number-average molecular weight (M_n) of polymers prepared by ATRP depends on the ratio of concentration of monomer (M) to the concentration of initiator and conversion of the reaction (eq. (3.1)):

$$M_n = ([M]_0/[RX]_0) \times \text{Conversion} \times \text{Molecular weight of monomer (M)}. \qquad (3.1)$$

The initiator used in ATRP may contain one or more α-halo-ester moieties. The architecture of the prepared polymers can be varied from linear (using alkyl halides with a single halogen atom) to star (multiple halogen atoms in the initiator) depending on the initiator structure and the exact number of halogen atoms present on them.

3.2.1 General observations on the initiator structure in ATRP

Two parameters are important for a successful ATRP initiating system. 1) Initiation should be faster as compared to propagation and 2) equilibrium has to shift at the dormant side resulting into the minimization of probable side reactions. Analogous to the "living" carbocationic systems, the main factors that determine the overall rate constants are the equilibrium constants rather than the absolute rate constants of addition [48]. There are several general considerations for the choice of ATRP initiators: 1) The presence of radical stabilizing group in the order of CN > C(O)R > C(O)OR > Ph > Cl > Me [49]. Multiple radical stabilizing groups increase the activity of the alkyl halide, for example, carbon tetrachloride, benzhydryl derivatives, and malonates. The malonate with two geminal ester groups generate radicals faster than 2-bromoisobutyrate, which leads to the lower dispersities. 2) Degree of substitution on generated initiator radical species in the order of primary < secondary < tertiary. Tertiary alkyl halides are better initiators than secondary ones, which are better than primary alkyl halides. These results have been partially confirmed by the measurements of activation rate constants [50–52]. 3) The presence of weak carbon–halide bond in the ATRP initiator. The order of bond strength in the alkyl halides is R–Cl > R–Br > R–I. Thus, alkyl chlorides should be the least efficient and alkyl iodides the most efficient initiators due to the faster leaving ability of radical species (Cl < Br < I). However, the use of alkyl iodides requires special precautions. They are light sensitive and can form thermodynamically unstable nonisolable metal iodide complexes (e.g., CuI_2) with an unusual reactivity. The R–I bond may possibly be cleaved heterolytically and there are potential complications in the ATRP process by degenerative transfer [53, 54]. Mostly, bromide and chloride containing initiators are frequently used. In general, halide in the initiator and in the metal salt (e.g., RBr/CuBr) is same; however, the halogen exchange can sometimes be used to obtain better polymerization control [55]. In a mixed halide initiating system (R–X/Mt–Y where X–Br and Y–Cl) the bulk of the polymer chains are terminated with chloride due to the stronger carbon–chloride bond.

3.2.2 Initiator efficiency

The lowering of initiator efficiency for a particular initiating system may be due to the factors such as 1) side reactions of functionality in the initiator, 2) initiator concentration in the reaction medium, that is, the ratio of monomer to the initiator, and 3) the reaction temperature. Xu et al. [56] studied the ATRP of styrene in the presence of CuCl/PMDETA as catalyst using 5-chloromethyl-2-hydroxybenzaldehyde as an initiator. The initiator efficiency obtained was in the range 0.08–0.66 due to the hydrogen-bonding interactions of phenolic hydroxyl group. Similarly, Marsh et al. [57] studied the initiator efficiency of unprotected uridine derived initiator in ATRP of methyl methacrylate (MMA) using CuBr/*N*-(*n*-pentyl)-2-pyridylmethanimine as a catalyst. They found the decrease in initiator efficiency with increasing molecular weight of the polymer. Matyjaszewski et al. [58] observed the lower initiator efficiency in ATRP of styrene using α-halocarboxylic acids in the presence of CuBr/PMDETA as the catalyst, which was attributed to the intramolecular cyclization resulting in the formation of γ-butyrolactone. Mei et al. [59] studied ATRP of styrene in the presence of CuBr/bipyridine as the catalyst using coumarin-functionalized initiator. The initiator efficiency was found to decrease at higher temperature.

Furthermore, compatibility of the initiator, monomer, and ligand plays an important role to obtain better efficiency. The initiator and ligand with long chains are efficient for polymerization of long chain containing monomers such as lauryl methacrylate and stearyl methacrylate [60]. The structural resemblance of initiator, monomer, and catalytic system makes the synthesis more efficient. Polymerization by long chain containing initiators requires the long chain containing ligand that helps in increasing initiator efficiency by well solubilizing the copper catalyst in the reaction medium [60].

3.3 Synthetic approaches for telechelic polymers

Telechelic polymers can be obtained by ATRP using four different approaches: 1) functional initiator approach by using an appropriate initiator [61], 2) the use of functional initiator followed by the postpolymerization transformation of terminal halide into other useful functional group [62], 3) chemical modification of a functional group incorporated via functional initiator into other useful functionalities employing the strategies such as deprotection/selective hydrolysis, and 4) radical–radical coupling process. These approaches lead to the formation of either homo- or hetero-telechelic polymers depending on the choice of initiator.

Figure 3.2 represents the different approaches for the synthesis of telechelic polymers by ATRP.

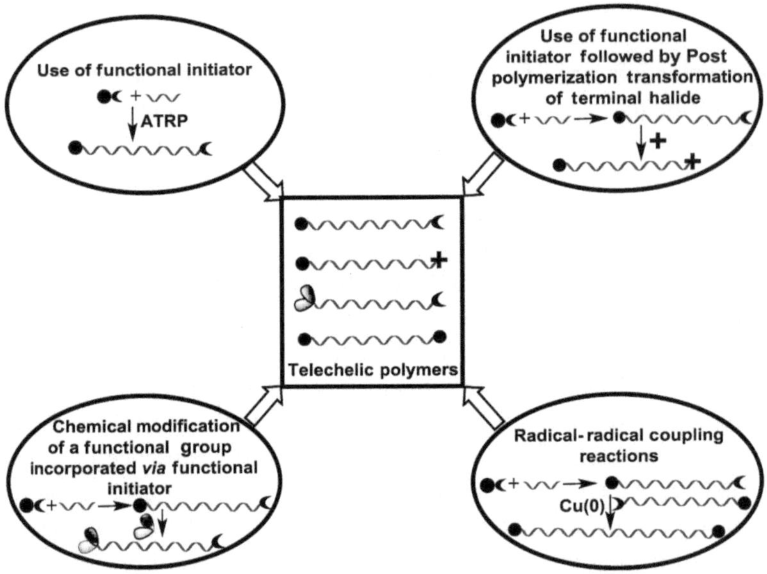

Figure 3.2: Different approaches for the synthesis of telechelic polymers by ATRP.

The synthesis of telechelic polymers by the above-mentioned approaches are summarized later.

3.3.1 Functional initiator approach

Among the above-mentioned approaches, the functional initiator approach has been considered as highly efficient due to high-end group fidelity and the possibility to introduce a wide variety of functional groups by selecting an appropriate functional initiator. The initiator is an important component in ATRP due to its key role in controlling the equilibrium and the rate of initiation. The basic criteria for selection of initiator are the quantitative and fast initiation as compared to the propagation with less probability of side reactions. The amount of initiator determines molecular weight of the polymer and its efficiency determines the number of chains initiated.

The incorporation of functional group at one chain end and halogen at the other chain end provides a route to synthesize end-functional polymers using functional initiator approach. A range of ATRP initiators are commercially available for the synthesis of linear and star architectures [63]. In the recent past, clickable functional initiators are frequently used due to the high efficiency, high yields of reactions, and requirement of simpler purification methods due to the virtue of their chemistry [64]. The obtained clickable polymers are further modified either to other reactive groups to make them suitable for polycondensations in the form of macromonomers or to perform other organic transformations to synthesize different macromolecular architectures and to change their properties [42].

The essential requirement of incorporation of a functional group by initiator approach is that the functional group must not interfere with rapid formation of the transition metal catalyst/ligand complex, which is required to provide a fast reversible activation and deactivation of the growing polymer chains in the ATRP reaction. Most of the successful initiators employed in ATRP are organic halides with a potentially active carbon–halogen bond, which can easily generate a radical species through electronic and steric effects of their substituent [39]. An organic halide with structural similarity to that of the dormant chain end of polymer is preferentially used so that activity of the carbon–halogen bond in the initiator is similar to that of the dormant polymer terminal. A large number of initiators with different functionalities have been compiled in an excellent review article published by Matyjaszewski et al. in 2001 [39] and Tasdelen et al. in 2011 [34]. Summarized herein is a plethora of functional ATRP initiators reported in the literature for the synthesis of functional polymers wherein one of the functional group is either bromide or chloride.

3.3.1.1 Functional ATRP initiators for the synthesis of hetero-telechelic polymers

Polymers obtained using functional ATRP initiators retain halogen (mostly bromide) at one of the terminals; hence, these polymers are termed as *hetero-telechelic* (α–ω functionalized) polymers. Polymers with a functionality at the initiation side are called as *α-functionalized polymers*; if two functional groups are on the same side, those polymers are referred to as *α, α′-functionalized polymers*. If both the functional groups are same, then such polymers are called as *α,α′-homo-*, and if they are different, the polymers are termed as *α,α′-hetero-bifunctionalized polymers*. Synthesis of α,α′-homo [61, 65, 66] and α,α′-hetero-bifunctionalized [67] polymers with controlled molecular weight have been reported using the functional initiator approach. Furthermore, α,α′-homo-bifunctionalized polymers with functional groups such as dihydroxyl [68, 69], dicarboxylic acid [70], aromatic difluoro [71], aromatic dibromo [72], and tert-amino [73] have also been prepared successfully using the initiator approach.

Halogenated alkanes, such as chloroform, trichloroethanol, or carbon tetrachloride (CCl_4), were among the first studied ATRP initiators [40, 41]. Benzyl halides are useful initiators for the polymerization of styrene and its derivatives due to the stabilization of generated radical by phenyl ring and their structural resemblance. However, they failed in the polymerization of more reactive monomers such as MMA [74]. Improvement of initiation efficiency for MMA using benzylic halides is possible by employing the concept of halogen exchange [55].

α-Haloesters, such as ethyl 2-bromoisobutyrate and ethyl 2-bromopropionate, are the widely employed commercial initiators for controlling the ATRP and are considered as the standard initiators. In general, α-haloisobutyrates produce initiating radicals faster than the corresponding α-halopropionates due to better stabilization of the generated radicals after the halogen abstraction step. Thus, slow initiation generally occurs if α-halopropionates are used to initiate the polymerization of methacrylates. In contrast, α-bromopropionates are good initiators for the ATRP of acrylates due to their structural resemblance. Sawamoto and coworkers [75] examined three different α-bromoesters to obtain better control using ruthenium-based catalyst. A variety of functionalities, such as hydroxyl, epoxy, allyl, vinyl, γ-lactone, and so on, have been introduced onto the α-end of the polymer using α-haloester initiators [76, 77]. Apart from the first studied nonfunctionalized ATRP initiators, the range of initiators that have been used successfully in ATRP are categorized and discussed below based on the functional groups, while a separate section is devoted to initiators containing "clickable" groups.

3.3.1.1.1 Hydroxyl-containing initiators

Hydroxyl group containing ATRP initiators are well studied and can be synthesized by esterification or amidation of corresponding dihydroxy or amino-hydroxy precursors. Generally, the hydroxyl groups are well-known initiators for ring-opening polymerization (ROP) of ε-caprolactone and ethylene oxide. Aliphatic hydroxyl-terminated polymers prepared by ATRP have been widely used for the synthesis of diblock copolymers such as polystyrene-*b*-poly(L-lactide) [78] and poly(methyl methacrylate)-*b*-poly(ε-caprolactone) [79] and star copolymers such as poly(tBMA)$_2$-poly(ε-caprolactone)$_2$ [80] by controlled ROP of corresponding lactides/lactones. The number of arms in the star copolymer depends on the number of hydroxyl groups present at the initiator end. A range of star copolymers with hydroxyl groups at the periphery were also reported by Matyjaszewski [81] and other research groups [82].

3.3.1.1.2 Acid-containing initiators

It is well known that acid-containing ATRP initiators quench the ligands employed for the stabilization of metal complex that results into the loss of control of molecular weight and molecular weight distribution. In such cases, the well-defined polymers with terminal acid groups can be prepared by using protected α-halocarboxylic acid or the initiators with acid and halide groups away from each other [76]. Protection of carboxylic acid or the presence of remote halogen suppresses an intramolecular cyclization reaction between α-halocarboxylic acid and styrene to form γ-butyrolactones. Such reaction is known to occur under ATRA condition [83, 84].

3.3.1.1.3 Amine-containing initiators

Generally the ligands used for complex formation with the metal catalysts in ATRP reactions are mostly bidentate and multidentate amine derivatives. The preparation of primary amine- functionalized polymers by ATRP requires indirect methods such as protection–deprotection chemistry. To reduce the possible interaction between the primary amine group of the initiator and the catalyst system, protected primary amine initiators have been used, which after the deprotection afford primary amine-functionalized polymers [85–94]. Percec et al. [95] reported the synthesis of a series of the well-defined primary amine end functionalized polymers using different *N*-chloro-substituted amides, lactams, carbamates, and imides as protected primary amine-functionalized initiators. The synthesis of primary amine end-functionalized polymers can also be achieved using phthalimido-functionalized initiators, followed by hydrazinolysis of the phthalimido end group [96–99]. Primary amine-functionalized polymers can also be prepared using functionalized initiators, whereby the functional group is in the form of a precursor of the amine group, namely, the nitro [100, 101]. After the polymerization process, the nitro group can easily be converted to the corresponding aromatic amine group by the reduction using zinc/acetic acid [102–111]. Haddleton et al. [112] and Blazquez et al. [113] reported the use of nonprotected aromatic primary amine-functionalized initiators in ATRP for the preparation of amine-functionalized polymers.

3.3.1.1.4 Ketone-containing initiators

Ketone-functionalized initiators, namely, α-bromoketones, have been used to initiate the controlled polymerization of MMA catalyzed by Ni{*o*,*o*′-$(CH_2\text{-}NMe_2)_2C_6H_3$}Br [114] and $Ni(PPh_3)_4$ [115]. Multihalogenated alkyl ketones, namely, CCl_3COCH_3 and $CHCl_2COPh$, are among the best initiators. The stronger electron-withdrawing nature of the carbonyl group of ketone induces further polarization of the carbon–halide bond, which is attributed to the faster initiation observed with the ketones than with the ester counterparts.

3.3.1.1.5 Nitrile-containing initiators

Nitrile-bearing initiators, namely, α-halonitriles, are the fast radical generators in ATRP of acrylonitrile monomer due to the structural similarity and the presence of strong electron-withdrawing cyano group. Moreover, the radical generated after halogen abstraction is sufficiently reactive, which leads to the fast initiation through rapid radical addition to the monomer. Among the initiators studied for the polymerization of acrylonitrile monomers catalyzed by copper complexes, 2-bromopropionitrile resulted in the polymers with lowest dispersities [116] and is also the initiator of choice in the iron-mediated ATRP of MMA [117]. However, α-halonitriles have not been used in ruthenium-catalyzed ATRP as the cyano-group deactivates the catalyst by forming a strong complex with ruthenium [117].

3.3.1.1.6 Sulfonyl halides as initiators

The halides attached to sulfonyl group, that is, sulfonyl chlorides yield a much faster rate of initiation than the propagation [118]. The apparent rate constants of initiation are about four (for styrene and methacrylates) and three (for acrylates) orders of magnitude higher than those for propagation. As a result, well-controlled polymerizations of a large number of monomers have been obtained in copper-catalyzed ATRP [88]. Various arenesulfonyl chloride ATRP initiators-bearing functional groups, such as carboxylic acid, hydroxyl, diazo, amine, on the benzene ring have been used for the synthesis of well-defined PS, PMMA, and PBA [119]. The polymerization reactions were performed using copper (I) chloride and bipyridine as a catalytic system. Aliphatic and aromatic sulfonyl chlorides tolerate other functional groups present in their structures and initiate the polymerization in quantitative manner.

3.3.1.1.7 Quaternary ammonium group-containing initiators

Quaternary ammonium group (ionic liquid)-containing initiators have been used for ATRP of different monomers. In addition, ionic liquids have also been used as ligands in ATRP and as catalysts in ROP. The well-defined end-functional PMMA and PBA were synthesized from imidazolium-based ATRP initiators [120, 121]. The use of different cationic and anionic counter ions affects the physical and chemical properties of resulting polymers. For example, the incorporation of the imidazolium cations in PMMA chain enhances the solubility of polymer in the solvent such as methanol [121].

3.3.1.1.8 Initiators-containing clickable functionalities

Click reactions are considered as efficient reactions for polymer modifications due to quantitative yields and simpler purification of the final polymers. The different click reactions such as thiol-ene, copper-catalyzed alkyne-azide click (CuAAC), Diels–Alder, and its variants are widely used for the synthesis of functional polymers and different polymer architectures. Clickable polymers are desirable for efficient synthesis of different macromolecular architectures. The combination of RDRP and CuAAC allows the efficient synthesis of a wide range of polymers such as functional, graft, star-shaped polymers, polymeric networks, and so on. In many cases, the catalyst used in ATRP, that is, CuBr/PMDETA (*N, N′, N′, N″, N″*-pentamethyldiethylene triamine) is same as that for the alkyne–azide click reaction that allows the simultaneous (one-pot) process of polymerization and click reaction.

3.3.1.1.8.1 Alkyne-functionalized initiators

The first report on the combination of RDRP with alkyne-azide click chemistry was published by VanHest and co-workers [62] in 2005, showing the facile approach toward block copolymers. The alkyne functionality in propargyl 2-bromoisobutyrate initiator was protected with a trimethylsilyl group (TMS) to prevent possible side reactions under the polymerization conditions, such as 1) complexation with the copper catalyst [122, 123], 2) subsequent homo-coupling of alkynes [124], 3) chain transfer by hydrogen abstraction from the alkyne [125] and 4) interference with propagating radicals leading to crosslinking [126]. Nevertheless, the TMS group was found to be unstable under the polymerization conditions using CuBr/PMDETA as catalyst that leads to a loss of protecting group up to 70%. The loss was ascribed to a nucleophilic attack to the TMS group by one of the nitrogen atom of PMDETA. As a consequence, the weak nucleophilic ligand 2,2′-bipyridine (bpy) was chosen to reduce the deprotection, though it is not the optimum catalyst for ATRP reactions and does not avoid the decomposition completely [127]. Another strategy uses the more stable tri-isopropylsilyl group (TIPS) instead of TMS, revealing no loss during the polymerization [127]. This might be due to the bulky character of the protecting group that hinders the nucleophilic attack of the metal/ligand complex. The alkyne-functionalized initiators bearing either chloride or bromide as an initiating species were frequently employed for the polymerization of (meth)acrylates, styrene, and *N*-isopropylacrylamide, wherein the terminal alkyne was protected with TMS [128–130] or in unprotected form [27, 131–135]. Recently, Miller et al. [136] showed the requirement of protection of propargyl group with detailed description of side reactions of alkyne group occurring during the ATRP. In these cases, the undesired chain transfer and termination reactions were suppressed by performing the reaction for short time and for lower conversions. Furthermore, side reactions involving the alkyne functionality were suppressed employing lower alkyne concentrations using high monomer to initiator concentration ratio [102] or by performing the reactions at relatively lower temperatures [137].

3.3.1.1.8.2 Azido-functionalized initiators

The well-known method for the synthesis of azido-functional ATRP initiator is esterification/amidation of an azide-containing alcohol or amine with halo-isobutyryl halide. Since the azide functionality can easily and efficiently be introduced by the substitution of the mediating halide with sodium azide, one can avoid the requirement of azide-functionalized initiators. However, the degree of azide functionalization obtained using the initiator approach is higher compared to the transformation route. The reason is the nature of RDRPs, that is, termination reactions occur to some extent and hence the polymer chains cannot retain 100% of the halide at the ω-terminal. As a consequence, the quantitative azide functionalization at the polymer chain end is difficult. The azide moiety in the initiator can be used without protection during polymerization although some side reactions were described: 1) Cyclization reactions between the azide and the propagating radical that causes low initiator efficiency [105] and 2) 1,3-cycloaddition of azides with the double bond of the monomer occurs in the absence of a catalyst at high temperatures and long reaction times at which the monomer concentration decreases. The rate of cycloaddition of azide with monomer is in the order of acrylates > acrylamides, methacrylates > styrenes [138]. To reduce the side reactions to a negligible extent, short reaction time [104] and low temperatures [139] are preferably used. It was demonstrated that the side reactions involving the azido group [140] were completely suppressed by performing the ATRP at room temperature.

3.3.1.1.8.3 Maleimido-functionalized initiators

Maleimide is one of the components of Diels–Alder click reaction frequently used in combination with controlled radical polymerizations in the field of materials science [141–145]. Maleimido-functionalized ATRP initiators have been used in their protected form because the activated double bond present in maleimide undergoes the ATRP reaction. For example, maleimide has been protected in the form of its adduct with furan [86]. The preclick approach has been used to protect the maleimido group and the initiator was used for the homo-polymerization of MMA [146, 147]. The protection of maleimido group with anthracene requires comparatively higher temperature, which may affect the other thermally unstable groups present in the initiator structure.

3.3.1.1.8.4 Anthracene-functionalized initiators

Like furan, anthracene takes part in the Diels–Alder click reactions as a diene. The anthracene-functionalized ATRP initiator was synthesized by esterification of 9-anthracenemethanol with 2-bromoisobutyryl bromide and was used for the polymerization of MMA [148]. Click reaction between anthracene and maleimide requires the higher temperature (110 °C) as compared to furan–maleimide reaction (60 °C). Energy barrier to occur anthracene–maleimide reaction is higher and as a result retro Diels–Alder reaction between anthracene and maleimide requires higher temperature (above 220 °C).

3.3.1.1.8.5 Pyridyl-disulfide-functionalized initiators

The disulfide-containing ATRP initiators are mainly used for the introduction of redox-sensitive disulfide linkage in the polymers, which show the dynamic covalent bond nature. Pyridyl-disulfide-functionalized polymers have been reported and are used in pyridyl-disulfide exchange click couplings. The reaction is widely employed in bioconjugations of drug, proteins, and so on [149, 150]. It is a metal-free reaction and is considered as a click coupling reaction in a broader sense. Thayumanavan and coworkers [151] introduced a disulfide linkage in the PNIPAM-based block copolymer by using pyridyl-disulfide-containing ATRP initiator. However, side reactions in terms of chain coupling and transfer involving the disulfide have been observed at high conversions.

3.3.1.1.8.6 Aminoxy-functionalized initiators

It is known that aminoxy-functional group interferes in the ATRP reactions. To avoid the direct interaction of aminoxy group in the polymerization reaction, it should be protected either by *t*-Boc-protection or by phthalimido-protection. Aminoxy-functionalized poly(hydroxyethyl methacrylate) (PHEMA), poly(ethylene glycol methacrylate) (PEGMA), PNIPAM, and polystyrene were reported by Maynard and co-workers using *t*-butoxycarbonyl (*t*-Boc) [85, 152] or phthalimido-protected [97] aminoxy-functionalized ATRP initiators, respectively. The absence of termination reactions was found during the polymerization performed using these protected aminoxy-functionalized initiators. The aminoxy group can easily be obtained from *t*-Boc-protected polymers by reaction with trifluoroacetic acid and from phthalimido-protected polymers by reaction with hydrazine hydrate after the polymerization. Generally, aminoxy-functionalized polymers undergo the rapid aldehyde-aminoxy click reaction without the additional external reagents. The formed oxime linker between the polymers has been considered as dynamic covalent bond [153] and the approach is applicable for the synthesis of pH-responsive polymers as the oxime functional group shows pH responsive behavior [154] and is useful for protein conjugations [155].

3.3.1.1.8.7 Allyl/vinyl ether/norbornene-containing initiators

This class of ATRP initiators contains double bond that can react with thiol or tetrazine in thiol-ene and tetrazine-norbornene click reactions, respectively. Vinyl ethers are electron-rich functional groups than simple alkenes. Allyl and vinyl ether containing initiators have been examined for ATRP of acrylates, methacrylates, and styrene. It has been found that the vinyl ether undergoes side reactions at higher conversions (above 80% conversions) for polymerization of methacrylates and styrene, while for acrylates side reactions occur even at lower conversions [156, 157]. Thiol-ene reaction is well known and is a simple coupling tool for the conjugation of small organic/drug molecules to the polymer backbone and can be performed thermally or photochemically by irradiating the material with ultraviolet radiations. The reaction is considered as efficient only for the conjugation of small molecules to the polymers, but for polymer–polymer conjugations lower efficiency has been shown by Junkers and coworkers due to the side reactions occurring from the generated thiyl radicals [158]. It is known that norbornene shows higher reactivity for thiol-ene reaction as compared to allyl group [159]. Norbornene-containing ATRP initiators have been used for the synthesis of polymers for thiol-ene click reactions with enhanced conjugation efficiencies or for the synthesis of graft copolymers by ring-opening metathesis polymerization [160]. Thiol-norbornene click reaction has been used to develop the materials with high glass transition temperatures (T_g) [161]. Furthermore, norbornene functional polymers undergo the additive-free, inverse electron demand tetrazine-norbornene click reaction [162].

3.3.1.1.8.8 Epoxide-containing initiators

Epoxy-terminated polymers can be prepared either by direct esterification of glycidol or by epoxidation of olefin-containing alcohols, followed by esterification with α-bromo-acid halides. PMMA with mono-, di-, and tri-epoxide end groups were prepared by these methods. Thiol-epoxy click reaction of these PMMA with hydroxyl functional thiols was performed to obtain hydroxyl-terminated PMMA [163]. Protection and deprotection strategies were avoided due to the tolerance of epoxide group in ATRP conditions.

The representative examples of functional ATRP initiators with alcohol, acid, aldehyde, amine, azide, alkyne, epoxide, nitro, allyl, vinyl ether, norbornene, and so on are listed in Table 3.1.

Although a large number of reactive functional groups have been introduced at polymer chain ends by ATRP technique, there are still unresolved issues in incorporating certain functional groups such as aziridine, furan, isocyanate, isothiocyanate, and so on using ATRP initiator approach due to synthetic complications. However, some of these functional groups have been introduced in the polymer backbone as pendant groups using respective functional monomers [189, 190].

3.3.1.2 Difunctional ATRP initiators for synthesis of homo-telechelic polymers

Polymers prepared by ATRP using difunctional initiator retain bromide at their terminals. As the functional groups at α- and ω-ends are same, these polymers are termed as homo-telechelic polymers. These bromide end groups can either be replaced by nucleophilic reagents or transformed into other desired functionalities by some efficient organic reactions.

Cyclopropenone-masked dibenzocyclooctyne difunctional ATRP initiator was synthesized and employed for the polymerization of styrene, *t*-butyl acrylate, and MMA to obtain homo-telechelic PS (M_n-2,400, Đ-1.06), P*t*BA (M_n-2,860, Đ-1.06), and PMMA (M_n-4,800, Đ-1.11) with terminal bromide groups (Scheme 3.2) [191]. Polymerization of acrylates using free dibenzocyclooctyne functionalized ATRP

Table 3.1: Representative examples of functional ATRP initiators for the synthesis of α, ω-hetero-telechelic polymers.

Sr. no.	Functionality	ATRP initiator	Polymer (polymerization conditions)	Ref.
1.	Alcohol	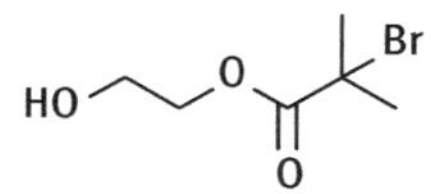	PBA, M_n-7,100, $Đ$-1.10 (CuBr, $CuBr_2$, PMDETA/50 °C)	[81 82 164]
		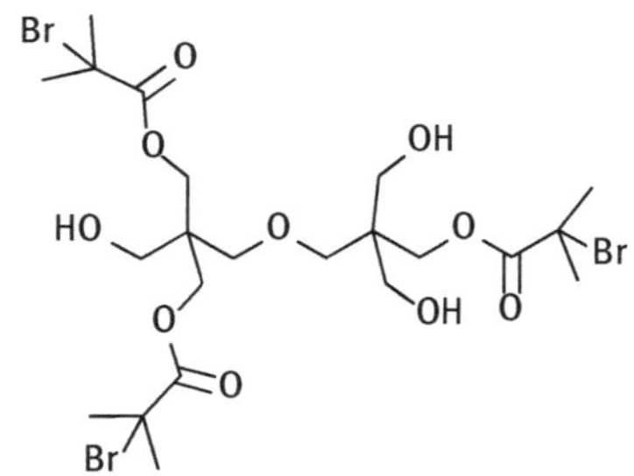	PDEAEMA and PPEGMA ($CuBr_2$, HMTETA, $Sn(Oct)_2$)	
		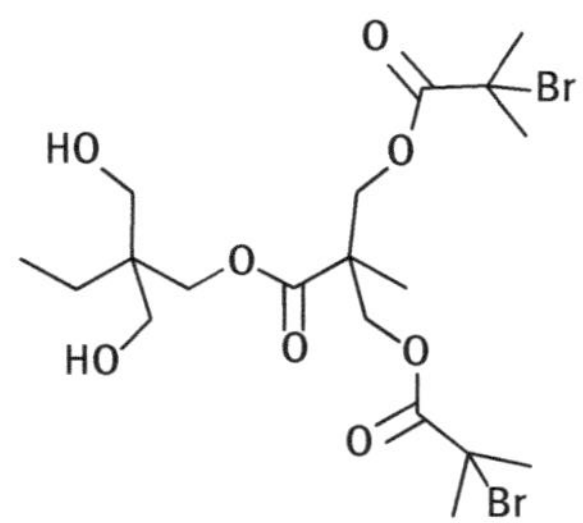	Polystyrene, M_n-8,740, $Đ$-1.15 (CuBr, bpy/110 °C)	

(continued)

Table 3.1 (continued)

Sr. no.	Functionality	ATRP initiator	Polymer (polymerization conditions)	Ref.
2.	Acid	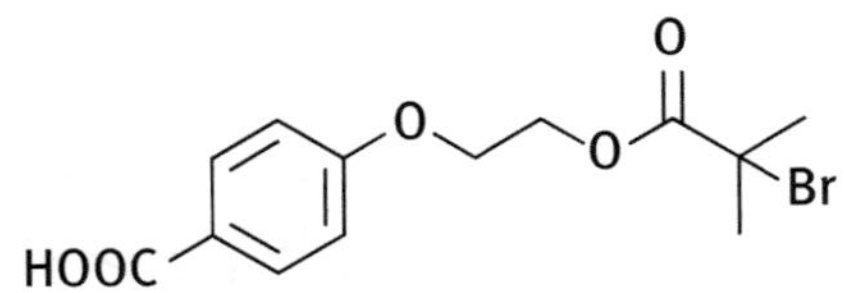	Polystyrene, M_n-12,000, $Đ$-1.10 (CuBr, PMDETA/ 110 °C)	[76 165]
		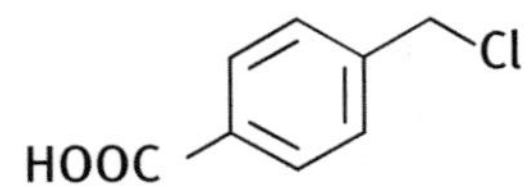	Polystyrene, M_n-3,800–4,800, $Đ$-1.18–1.54 (CuCl, bpy/110–140 °C)	
3.	Aldehyde	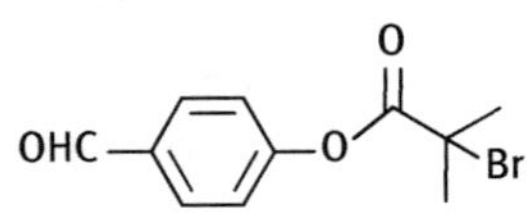	PMMA, M_n-6,440, $Đ$-1.10 (CuBr, *N*-(*n*-octyl)-2-pyridylmethanimine/90 °C)	[112 166 167]
		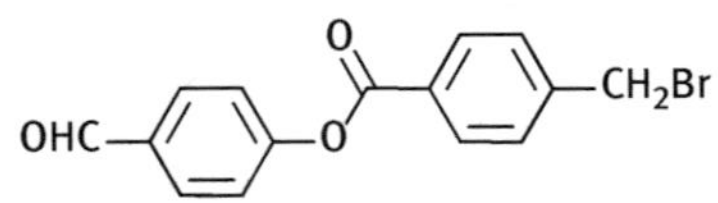	Polystyrene, M_n-2,100, $Đ$-1.38 (CuBr, bpy/110 °C)	
		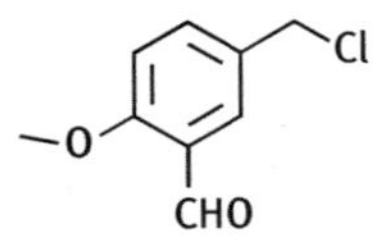	Polystyrene, (CuCl, PMDETA/ 80–120 °C)	

4.	Amine	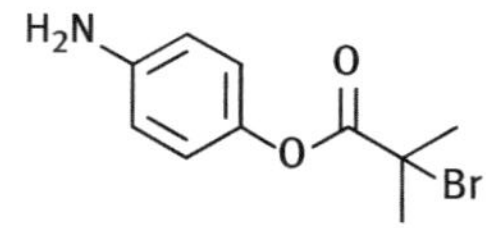	PMMA, M_n-9,340, $Đ$-1.19 (CuBr, *N*-(*n*-octyl)-2-pyridylmethanimine/ 90 °C)	[112 168]
		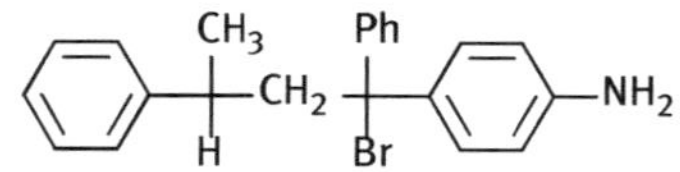	Polystyrene, M_n-3,300, $Đ$-1.08 (CuBr, PMDETA/ 130 °C)	
5.	Alkyne	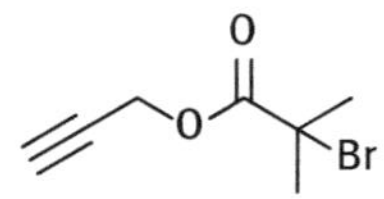	PDMAEMA, M_n-10,400, $Đ$-1.17 (CuBr, HMTETA/ 60 °C)	[104 169, 170 127]
		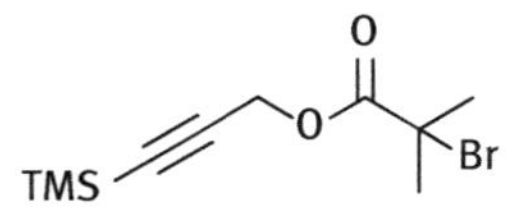	Polystyrene, M_n-4,650, $Đ$-1.08 (CuBr,$CuBr_2$, PMDETA/ 80 °C) and PtBA, M_n-4,200, $Đ$-1.11 (CuBr,$CuBr_2$, PMDETA/ 60 °C)	
		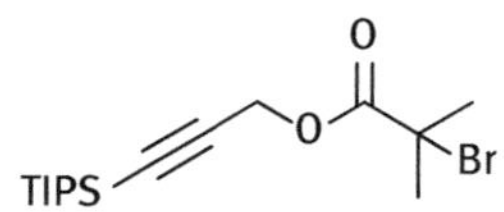	PMA, M_n-7,170, $Đ$-1.18 (CuBr, PMDETA/ 90 °C)	

(continued)

Table 3.1 (continued)

Sr. no.	Functionality	ATRP initiator	Polymer (polymerization conditions)	Ref.
6.	Epoxy	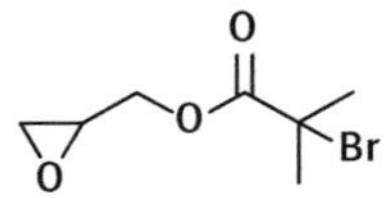 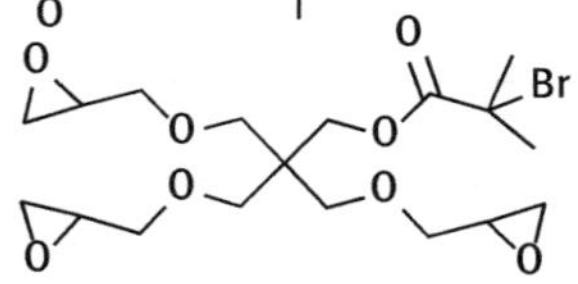	PMMA, M_n-12,100, $Đ$-1.5 (CuBr, PMDETA/ 35 °C)	[163 171 172]
		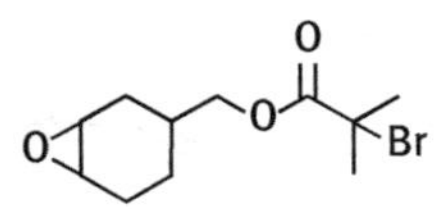	Polystyrene (CuBr, bpy, 110 °C)	
		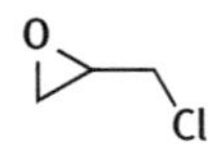	PHFMA, M_n-6,550, $Đ$-1.29 ($CuBr_2$, Cu, PMDETA/ 90 °C)	
7.	Nitro	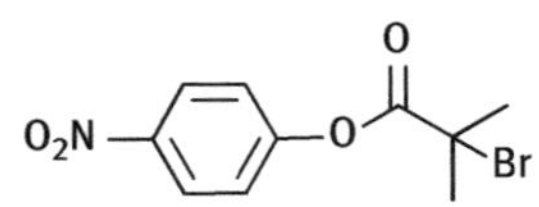	PMMA, M_n-6,350, $Đ$-1.09 (CuBr, *N*-(*n*-octyl)-2-pyridylmethanimine/ 90 °C)	[112 173]
		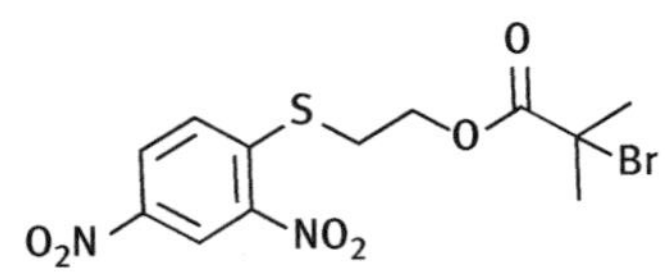	PMMA, M_n-3,800, $Đ$-1.28, $NiBr_2$ $(PPh_3)_2$, 85 °C)	

8.	Azido	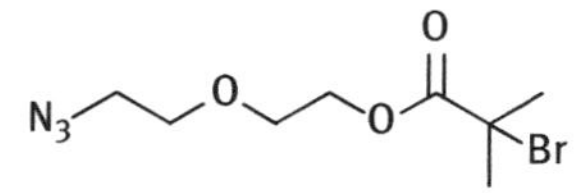	PDMAEMA, M_n-8,800, $Đ$-1.12 (CuBr, HMTETA/ 60 °C)	[104 174 175]
		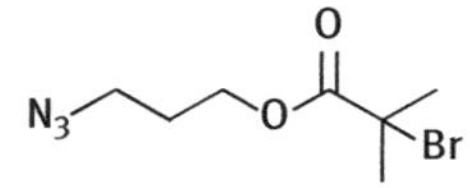	Polystyrene, M_n-1,200–4,550, $Đ$-1.2–1.3 (CuBr, bpy/130 °C)	
		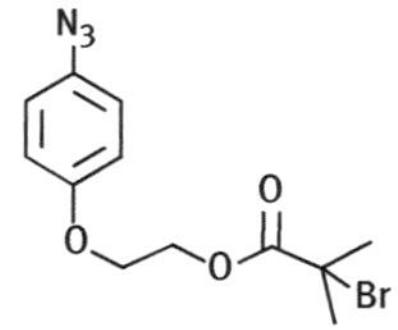	PNIPAM, M_n-12,400, $Đ$-1.19 (CuBr, Me_6TREN/ 25 °C)	
9.	Allyl	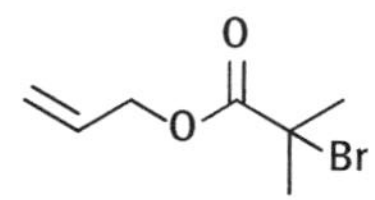	PDMAEMA, M_n-8,600, $Đ$-1.34 (CuBr, BA_6TREN/ 60 °C)	[176 177]
		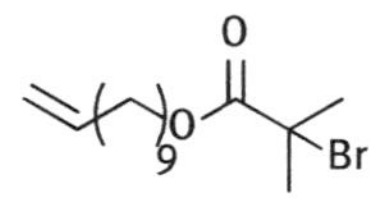	Polystyrene, M_n-6,000, $Đ$-1.08 (CuBr, PMDETA/ 90 °C) and PMMA, M_n-7,500, $Đ$-1.2 (CuBr, PMDETA)	

(continued)

Table 3.1 (continued)

Sr. no.	Functionality	ATRP initiator	Polymer (polymerization conditions)	Ref.
10.	Anhydride	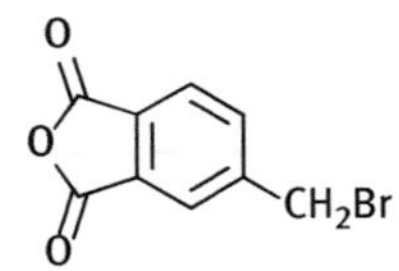	Polystyrene, M_n-7,100–12,000, $Đ$-1.31–1.42 (CuBr, bpy/130 °C)	[178]
11	Lactone	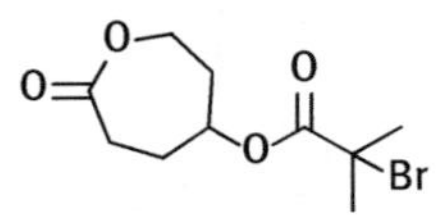	PMMA, M_n-2,200–3,800, $Đ$-1.15–1.22 ($NiBr_2(PPh_3)_2$)	[179 45]
		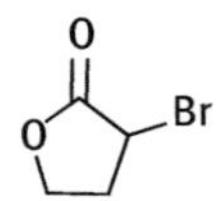	PMA, M_n-3,330, $Đ$-1.13 (CuBr, dNbpy/110 °C) and polystyrene, M_n-4,050, $Đ$-1.17 (CuBr, dNbpy/110 °C)	
12.	Vinyl ether	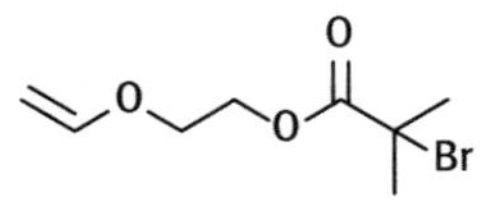	PMMA, (silica gel-CuBr, HMTETA)	[180 156]
		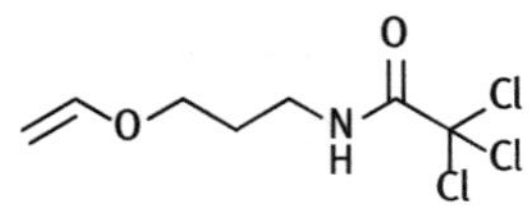	Polystyrene, PDMAEMA, PMMA, PMA, PBA (CuBr, HMTETA)	

13	Oxazoline	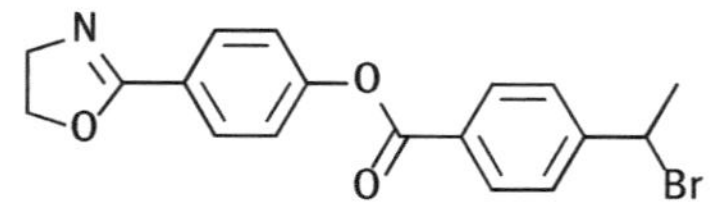	Polystyrene (CuBr, bpy/110 °C)	[181 181, 182 181]
		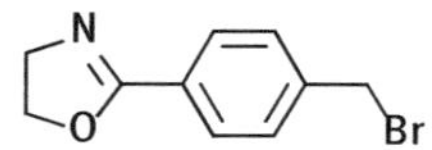	Polystyrene, M_n-3,700, $Đ$-1.22 (CuBr, bpy/diphenyl ether, 110 °C)	
		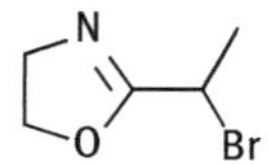	Polystyrene, M_n-2,100, $Đ$-1.09 (CuBr, PMDETA/ 70 °C)	
14	Carbonate	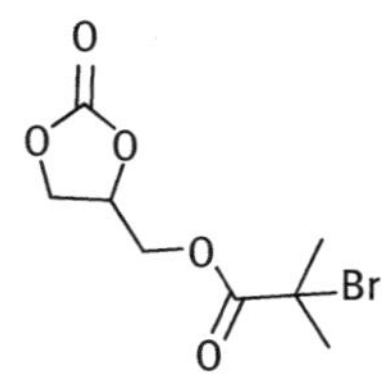	PMMA, M_n-1,740–3,500, $Đ$-1.17–1.30 (CuCl, PMDETA/60 °C)	[183]
15	Dibenzo-cycloctyne	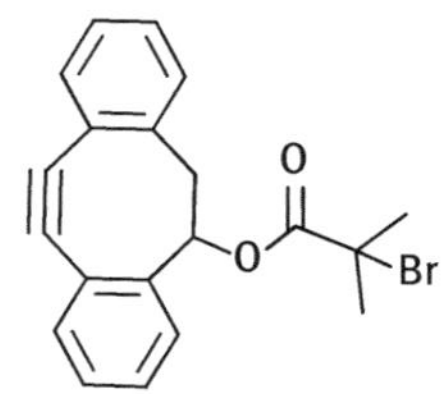	PMMA, M_n-3,860, $Đ$-1.17 (CuBr, PMDETA/ 60 °C) and polystyrene, M_n-2,330, $Đ$-1.07 (CuBr, PMDETA/ 90 °C)	[184]

(continued)

Table 3.1 (continued)

Sr. no.	Functionality	ATRP initiator	Polymer (polymerization conditions)	Ref.
16	Protected maleimide	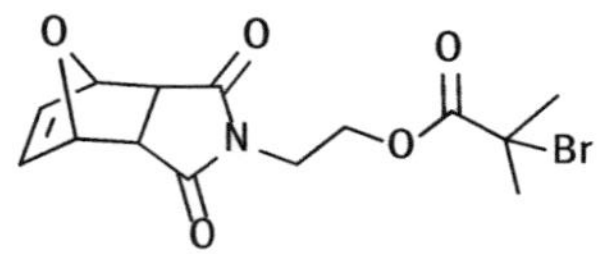	PMMA, M_n-5,500–6,300, Đ-1.12–1.13 (CuCl, PMDETA/40 °C)	[185]
17	Anthracene	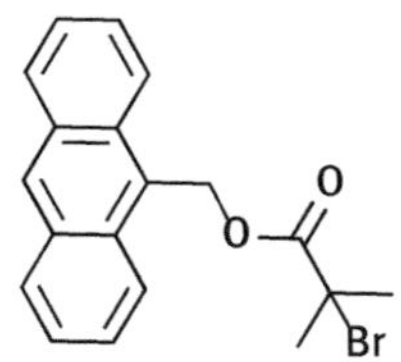	Polystyrene, M_n-3,240, Đ-1.13 (CuBr, PMDETA/110 °C)	[185]
18	Benzophenone	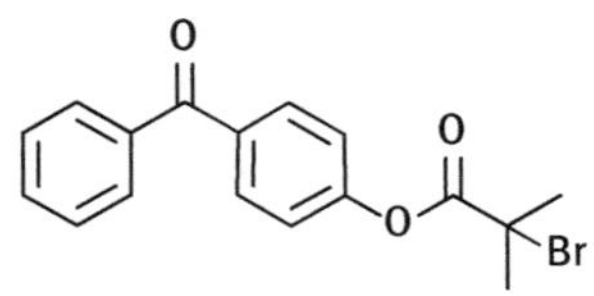	PMMA, M_n-2,900, Đ-1.31 (CuCl, PMDETA/70 °C) and polystyrene, M_n-1,800, Đ-1.13 (CuBr, PMDETA/110 °C)	[186]

19	Succinimide	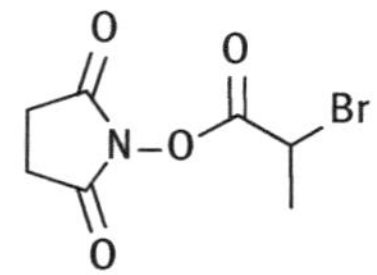	Glycopolymers, M_n-4,500–10,200, $Đ$-1.10–1.31 (CuBr, *N*-(*n*-octyl)-2-pyridylmethanimine/ 70 °C)	[187 188]
		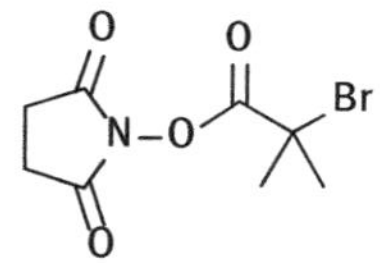	PEGMA, M_n-6,400, $Đ$-1.14 (CuBr, *N*-(ethyl) 2- pyridyl-methanimine, 30 °C)	
20	Imidazolium (Ionic-liquid)	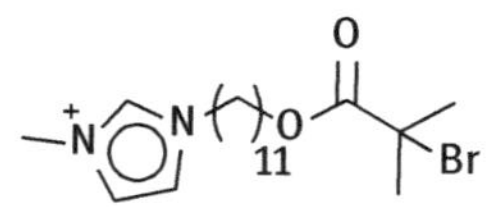	PBA, M_n-5,000–11,900, $Đ$- 1.16–1.21 (CuBr, PMDETA/75 °C) and polystyrene M_n-16,400, $Đ$-1.25 (CuBr, PMDETA/ 105 °C)	[120]

(continued)

Table 3.1 (continued)

Sr. no.	Functionality	ATRP initiator	Polymer (polymerization conditions)	Ref.
21	α,α′-Bisallyloxy	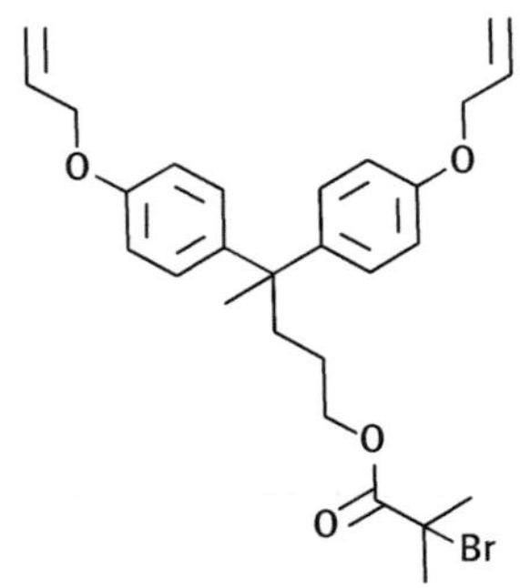	PMMA, M_n-9,000–21,500, Đ- 1.23–1.34 (CuBr, PMDETA/60 °C) and polystyrene M_n-14,500–29,600, Đ-1.06–1.09 (CuBr, PMDETA/110 °C)	[65 61]
		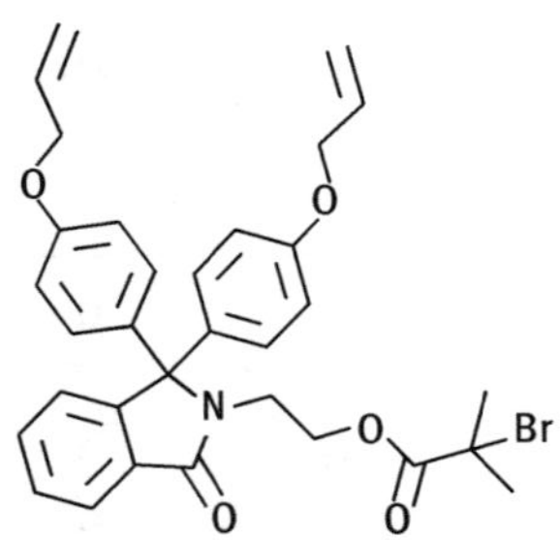	Polystyrene, M_n-4,800–11,700, Đ-1.05–1.09 (CuBr, PMDETA/80 °C)	

22	α,α′- Bisaldehyde	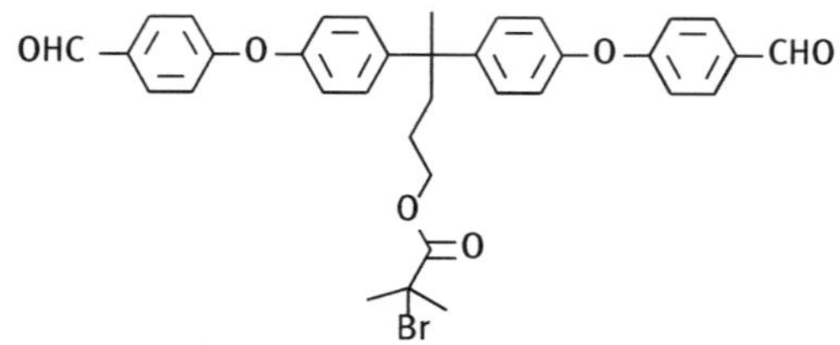	Polystyrene, M_n-2,200–31,500, Đ- 1.11–1.16 (CuBr, PMDETA/110 °C)	[66]
23	α,α′- Aldehyde, allyloxy	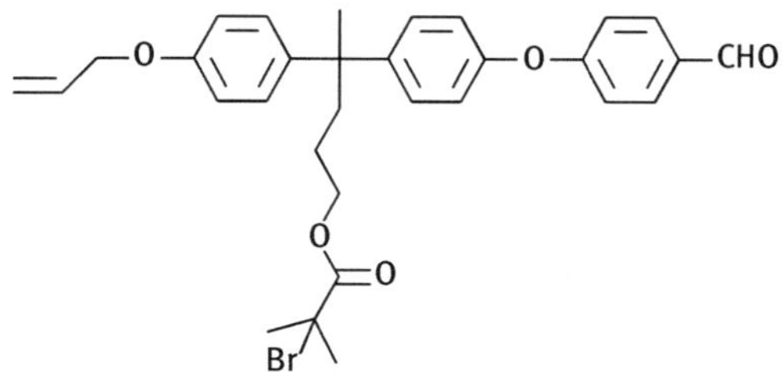	PMMA, M_n-3,700–19,700, Đ- 1.19–1.25 (CuBr, PMDETA/80 °C)	[67]

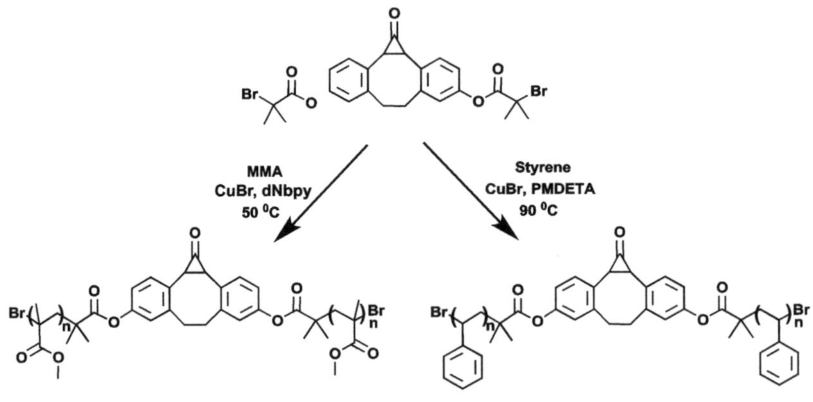

Scheme 3.2: Synthesis of homo-telechelic polystyrene and PMMA using cyclopropenone-masked dibenzocyclooctyne difunctional ATRP initiator.

initiator showed nonliving behavior. The use of cyclopropenone-masked initiator helped in controlling the ATRP of acrylates and methacrylates. Dibenzocyclooctyne functionality can easily, efficiently, and quantitatively be regenerated from the masked cyclopropenone after the polymerization by irradiating the polymer with UV-light and can be used further for the metal-free strain promoted alkyne-azide click reaction.

Chen et al. [192] attempted the synthesis homo-telechelic PMMA using difunctional ATRP initiator obtained by the reaction of ethylene glycol with 2-bromoisobutyryl bromide (Scheme 3.3). ABA triblock copolymer with PMMA as glassy middle block and amide containing dynamic hydrogen bond forming polyacrylate as outer blocks has been synthesized from these homo-telechelic PMMA. The triblock copolymer self-assembled into a nanostructure, which exhibited the improved mechanical properties and toughness as compared to the previously reported systems showing self-healing behavior based on hydrogen bonding (supramolecular). The mechanical properties were tuned by varying the chain length of outer amide containing hydrogen bonding soft block.

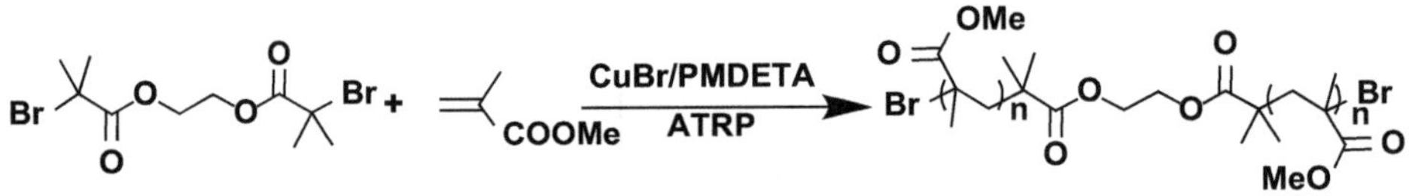

Scheme 3.3: Synthesis of homo-telechelic PMMA from ethylene glycol-based ATRP initiator. Reproduced with permission from Y. Chen and Z. Guan, Chemical Communications, 2014, 50, 74, 10868. ©2014, RSC [192].

Synthesis of molecular architectures with complex topologies such as rotaxanes and catenanes are challenging due to the synthetic difficulties. The design of rotaxanes and catenanes requires the use of chemical templates based on hydrogen bonding and donor–acceptor systems. The synthesis of catenane – the interlocked molecular architectures – consisting of two interlocked macrocycles was demonstrated by Advincula and coworkers [193, 194] (Figure 3.3). The design was based on the supramolecular "metal–ligand complex" template approach via ATRP and ATRC using a difunctional ATRP initiator bearing 1,10-phenanthroline as a complexing ligand by taking the advantage of its complex formation with metal and the postpolymerization cyclization via ATRC resulted in the formation of catenane.

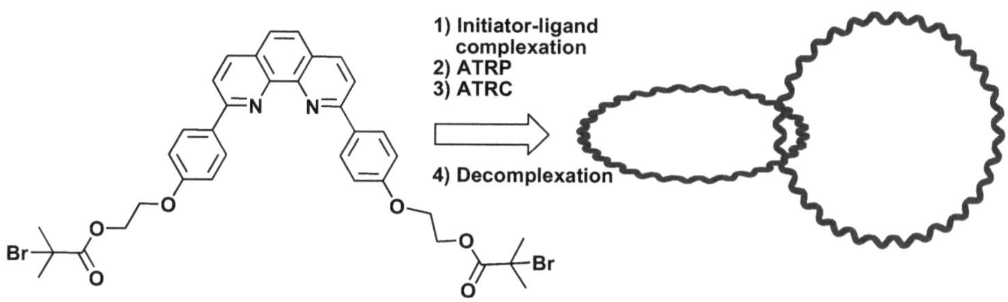

Figure 3.3: 1,10-Phenanthroline containing ATRP initiator for catenane formation via homotelechelic polymer. Reproduced with permission from A. Bunha, M.C. Tria and R. Advincula, *Chemical Communications*, 2011, 47, 32, 9173. ©2011, RSC [193].

3.3.1.3 Multifunctional ATRP initiators for synthesis of star-telechelic polymers

It is well known that the star copolymers show unique properties in solution as well as in bulk in terms of hydrodynamic volume, viscosity, rheological and mechanical properties, and lower critical solution temperature (LCST) as compared to their linear counterparts with same molecular parameters. Various approaches have been explored such as "core-first," "arm-first," "coupling onto," and so on to obtain the star architectures involving the use of multifunctional initiators. Multifunctional initiators based on the compounds with multiple reactive sites have been demonstrated to be useful. For example, a range of star-telechelic PNIPAM was demonstrated by Lyngso et al. [195] using different multifunctional initiator core molecules (Scheme 3.4). The effect of star architecture with different number of arms, arm lengths, and arm configurations on the LCST behavior of PNIPAM has been investigated. The structural modification of such star-telechelic polymer to 3-arm star-block copolymer may be of interest to carry hydrophobic drugs to the target.

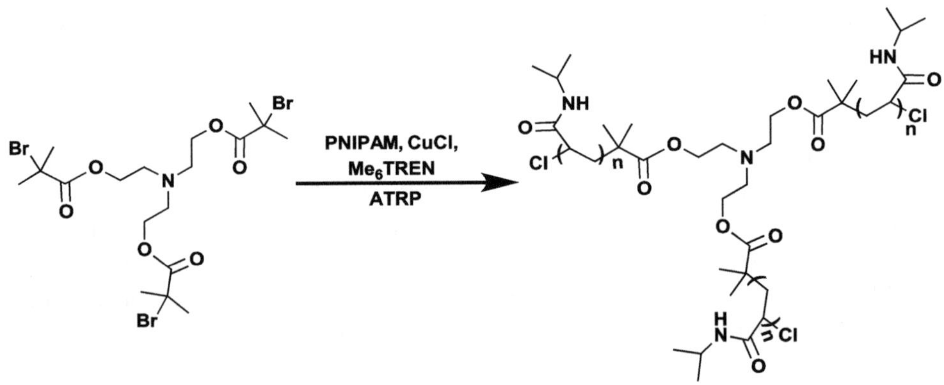

Scheme 3.4: Trifunctional ATRP initiator for 3-arm star telechelic PNIPAM. Reproduced with permission from J. Lyngsø, N.A. Manasir, M.A. Behrens, K. Zhu, A.L. Kjøniksen, B. Nyström and J.S. Pedersen, Macromolecules, 2015, 48, 7, 2235. ©2015, ACS [195].

Pentaerythritol is one of the widely used precursors for designing the initiators for star architectures. Four-armed multifunctional ATRP initiator, namely, pentaerythritol tetrakis(2-bromoisobutyate), was designed from pentaerythritol to synthesize star telechelic P*n*BA (M_n-70,000, Đ-1.08) (Scheme 3.5) [196]. Star telechelic P*n*BA with varying arm numbers, arm length, and compositions has been studied using ATRP and proved the technique as amenable to design the star-block copolymers for utilization in viscosity modification and adhesive technologies.

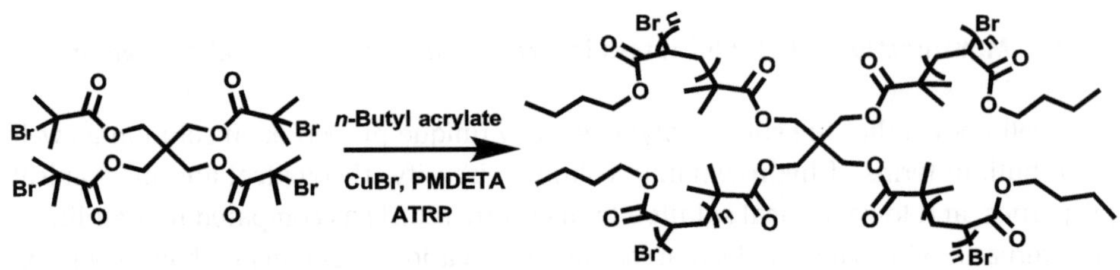

Scheme 3.5: Multi-functional ATRP initiator from pentaerythritol for 4-arm star telechelic polymer. Reproduced with permission from K. Matyjaszewski, P.J. Miller, D.C. Pyun, G. Kickelbick and S. Diamanti, Macromolecules, 1999, 32, 20, 6526. ©1999, ACS [196].

Phosphazene-based organic–inorganic hybrid multifunctional ATRP initiator was designed with six initiating sites to synthesize 6-arm star telechelic PNIPAM (Figure 3.4) [195]. Effect of the 6-arm architecture on polymer behavior in water has been investigated. Star-block polystyrene-*b*-poly(3-hexylthiophene) (PS-*b*-P3HT) [197] was synthesized employing a 6-arm initiator based on triphenylene using combination of ATRP and click reaction (Figure 3.4). Effect of star architecture on microphase

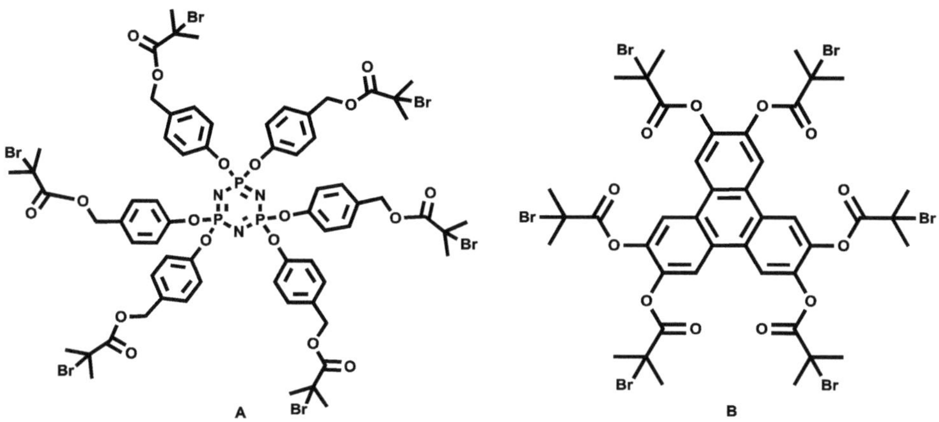

Figure 3.4: Multifunctional ATRP initiators for 6-arm star telechelic polymers.

separation morphology has been studied by making thin films of polymer. Microphase separation of star-block PS-*b*-P3HT showed larger crystalline domains due to aggregation of P3HT as compared to the linear block copolymer counterparts.

The commercially available hexakis(bromomethyl)benzene has been directly used to obtain a six-armed star polystyrene using the CuBr/bpy system [198]. Multifunctional polymer additive was reported using functional initiator approach by Narrainen et al. [199]. A series of multifunctional initiators carrying multiple fluroalkyl groups were synthesized and were used for controlled radical polymerization of vinyl monomers such as styrene and methyl methacrylate to prepare polymer additives carrying fluroalkyl groups at the chain end. Furthermore, temperature and pH dual-responsive 18-arm star-shaped poly(*N*-isopropylacrylamide)-*co*-poly(itaconamic acid) copolymers was prepared using an 18-arm multifunctional ATRP initiator designed from cyclodextrin (Figure 3.5) [200]. LCST of the star-shaped copolymer was

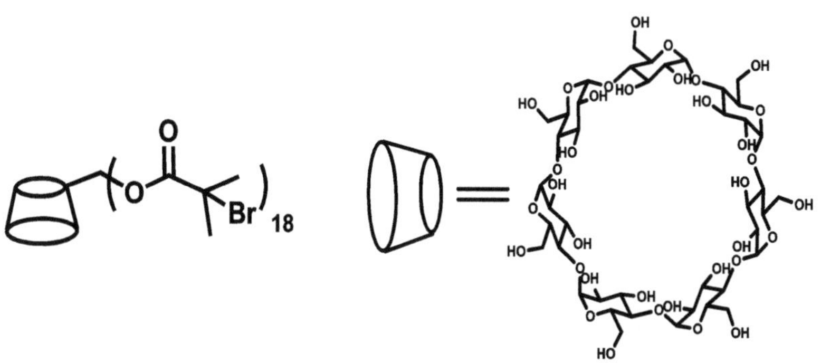

Figure 3.5: Multifunctional ATRP initiator based on cyclodextrin for 18-arm star telechelic polymer.

increased with increasing molar fraction of poly(itaconamic acid). Self-assembled aggregates of this star copolymer encapsulate drug in the cavity of cyclodextrin core. Hydrophobic cavity of cyclodextrin helps in drug encapsulation and showed drug release in the intestine due to the pH responsive nature of star polymers. These star copolymers have been utilized for intestinal drug delivery applications.

3.3.2 Use of functional initiator followed by postpolymerization transformation of terminal halide

Polymers prepared by ATRP retain halide (bromide) functional group at one of the chain ends that can be transformed into other useful functional groups using organic transformations. The approach involves the use of functional initiator, followed by postpolymerization functionalization of the halogen end group of the polymer into the desired functional groups to obtain α, ω-bifunctionalized polymers [166, 201, 202]. Different routes of postpolymerization transformation of terminal halide obtained by ATRP are represented in Scheme 3.6 [194].

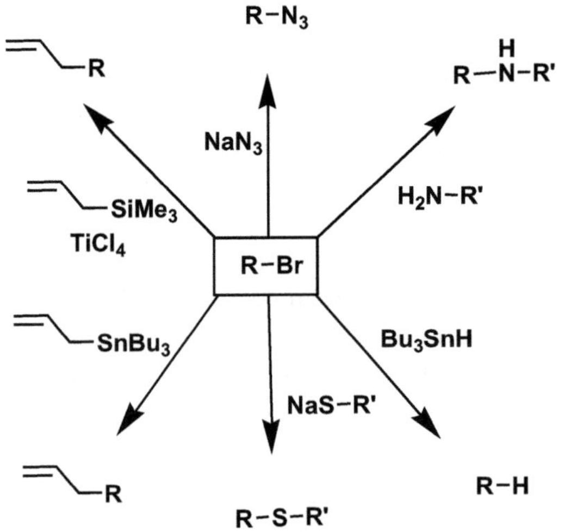

Scheme 3.6: Different routes of postpolymerization transformation of terminal halide. Reproduced with permission from K. Pangilinan and R. Advincula, Polymer International, 2014, 63, 5, 803. ©2014, John Wiley & Sons [194].

A prerequisite to reach a high degree of functionality by end group transformation is the stability of halogenated chain end throughout the polymerization, that is, termination must be avoided. However, it is well known that termination reactions in ATRP cannot be completely ruled out under certain reaction conditions. Less than

5% of terminations always occur, resulting into the lack of bromide end group on some polymer chains. As a result, quantitative transformation of terminal bromide cannot be achieved. The transformation of terminal halide to other functional groups can be achieved by two ways.

3.3.2.1 Nucleophilic substitution

Nucleophilic substitution approach was utilized to transform the halogen end group of the polymer into different useful functional groups such as azido [203], (Scheme 3.7), thiol, hydroxyl, carboxylic acid, amine, and so on [204, 205].

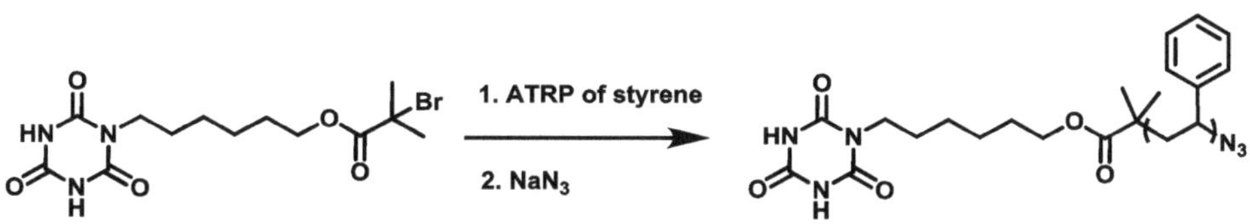

Scheme 3.7: Hetero-telechelic polystyrene with cyanuric acid and azido group by postpolymerization transformation. Reproduced with permission from O. Altintas, P.K. Sidenstein, H. Gliemann and C. Barner-Kowollik, Macromolecules, 2014, 47, 17, 5877. ©2014, ACS [203].

The use of alkyne-containing ATRP initiator followed by transformation of bromide to azido group (Scheme 3.8) was mainly useful for the synthesis of two architectures: 1) cyclic polymers and 2) linear step growth polymers via click polymerization. The cyclic polymers such as cyclic PNIPAM [137] were prepared under high-dilution conditions to avoid competing linear polymer formation. These cyclic PNIPAM showed broad thermal phase transition range, lower LCST and enthalpy change as compared to their linear counterparts due to the absence of chain ends and conformational restrictions of polymer backbone.

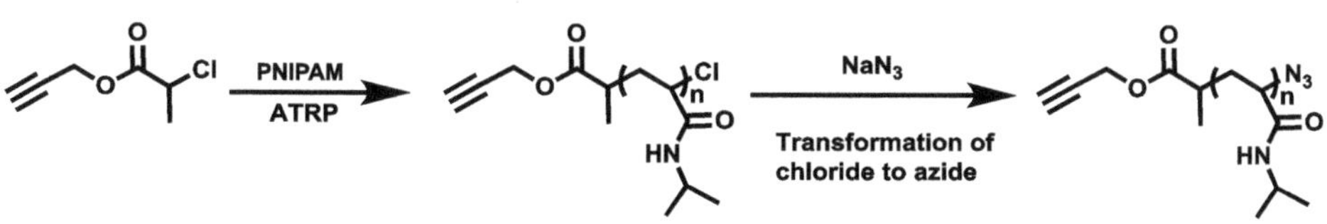

Scheme 3.8: Transformation of terminal halide to azido by nucleophilic substitution. Reproduced with permission from J. Xu, J. Ye and S. Liu, Macromolecules, 2007, 40, 25, 9103. ©2007, ACS [137].

Matyjaszewski and coworkers [27] attempted the synthesis of α-propargyl ω-azido and α,ω- diazido telechelic polystyrene by transformation approach. The linear step growth polymers were obtained from these telechelic polystyrene by alkyne-azide click homo-coupling and coupling with bispropargyl ether, respectively, via click polymerization. The obtained polystyrene showed broad dispersity due to the competing intramolecular cyclization of polymer chains in *N,N*-dimethylformamide to form cyclic polystyrene fractions.

A range of functionalities such as dihydroxyl and diamino have been introduced at the ω-chain end via nucleophilic substitution of bromide end group prepared by ATRP [206–208]. Babin et al. [208] introduced dihydroxyl and diamino groups onto the ω-chain end via nucleophilic substitution of bromide end group with appropriate amino-diols or triamine, respectively (Scheme 3.9). The dihydroxyl chain ends were further modified to ATRP initiators for *t*-butyl acrylate polymerization to obtain water-soluble, pH-responsive polystyrene-poly(acrylic acid)$_2$ star copolymer, while the amino groups were utilized directly for *N*-carboxy anhydride polymerization to obtain water-soluble, pH-responsive polystytrene-poly(L-glutamic acid)$_2$ star copolymer to study the effect of stimuli on the self-assembly.

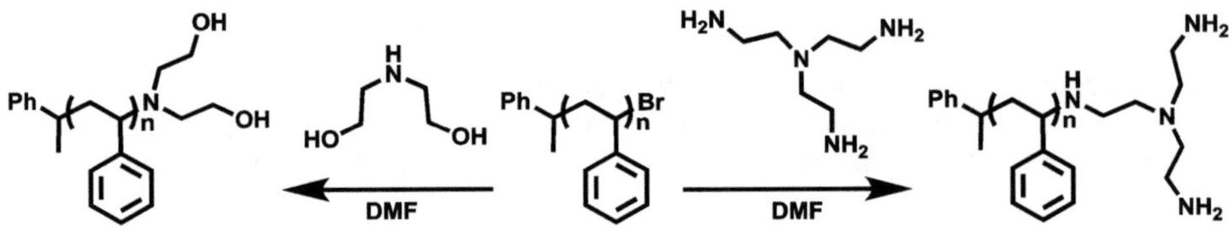

Scheme 3.9: Transformation of terminal halide by nucleophilic substitution. Reproduced with permission from J. Babin, C. Leroy, S. Lecommandoux, R. Borsali, Y. Gnanou and D. Taton, Chemical Communications, 2005, 15, 15, 1993. ©2005, RSC [208].

Dialkynyl functionalities have been introduced at ω-chain end via the transformation of the terminal bromide group of polystyrene prepared by ATRP into azido group followed by the CuAAC click reaction with trialkyne [209] (Scheme 3.10).

Similarly, dialkynyl functionalities were introduced at α and ω-chain ends (Scheme 3.11) to synthesize homo-telechelic polystyrene using the same strategy [169], which on alkyne-azide click reactions afforded amphiphilic second- and third-generation dendrimers with acetal linkages at the periphery, which can selectively be cleaved in acidic pH for the application of drug delivery in biomedical science.

The same strategy was used for the synthesis of ω-hydroxyl- ω′-alkynyl polystyrene via the click reaction of ω-azido polystyrene with 3,5-bis(propargyloxy) benzyl alcohol [210]. The reaction of bromine end group with amine nucleophile is slower than that with sodium azide. However, the basicity of the amine is greater than that of azide, which increases the possibility of side reactions. ω-Amino functionalization

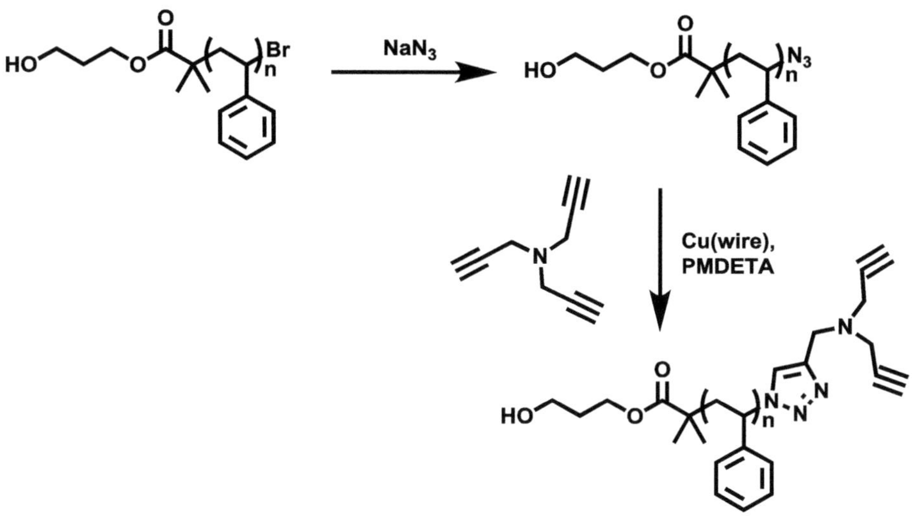

Scheme 3.10: Synthesis of hetero-telechelic polystyrene by postpolymerization transformation of bromides to dialkyne. Reproduced with permission from C.N. Urbani, C.A. Bell, M.R. Whittaker and M.J. Monteiro, Macromolecules, 2008, 41, 4, 1057. ©2008, ACS [209].

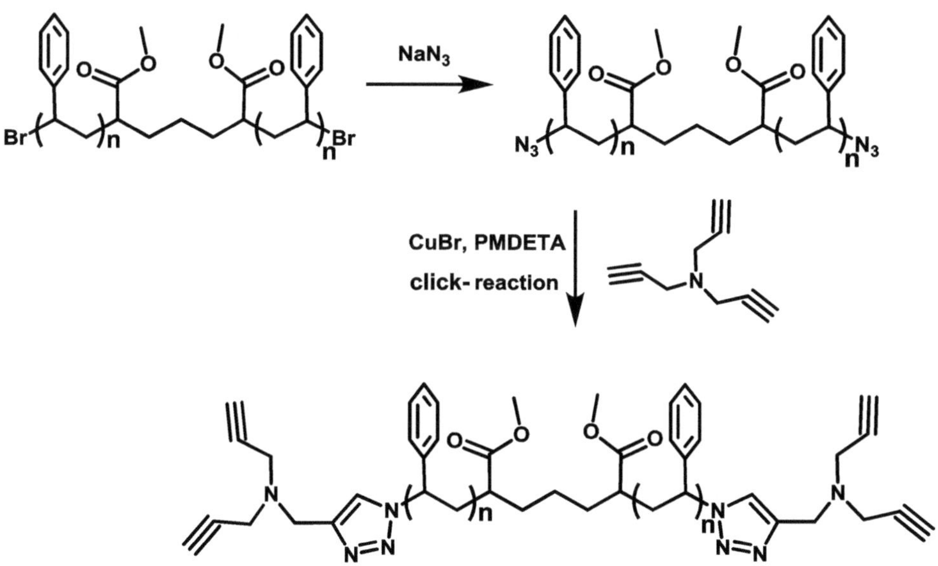

Scheme 3.11: Synthesis of homo-telechelic polystyrene by postpolymerization transformation of bromides to dialkyne. Reproduced with permission from C.N. Urbani, C.A. Bell, D. Lonsdale, M.R. Whittaker and M.J. Monteiro, Macromolecules, 2008, 41, 1, 76. ©2008, ACS [169].

on polystyrene has been attempted by Postma et al. [99] using potassium phthalimide as a nucleophile, which leads to the formation of double bond at the ω-terminal due to the elimination of bromide. Furthermore, general procedure for ω-amino functionalization is the conversion of the azido group into the phosphoranimine, followed by the hydrolysis to amino end group [211]. The direct displacement of bromide from polystyrene, poly(acrylate)s, and poly(methacrylate)s with hydroxide ion was not successful and was accompanied by significant elimination. The bromide end group has also been replaced by acetate group using sodium acetate. The cationic groups at the ω-terminal were obtained using phosphines as nucleophiles. For example, tri-*n*-butylphosphine and triphenylphosphine were employed and their feasibility and selectivity with model alkyl halides was investigated [212]. Boyer et al. reported thiol terminated poly(*t*-butyl acrylate) using nucleophilic substitution of bromide with sodium methanethiolsulfonate, followed by the generation of thiol by hydrolysis using triethylamine (Scheme 3.12) [204]. These thiol-terminated P*t*BAs were amenable for protein conjugation by thiol-ene or thiol-maleimide Michael addition reactions.

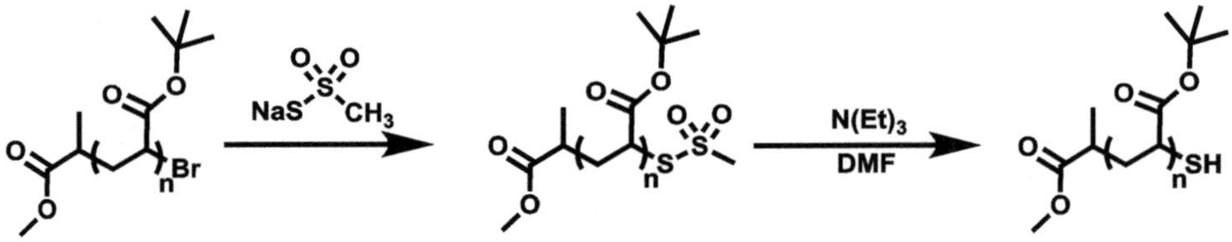

Scheme 3.12: Transformation of terminal halide to thiol using sodium methanethiolsulfonate followed by hydrolysis. Reproduced with permission from C. Boyer, A.H. Soeriyadi, P.J. Roth, M.R. Whittaker and T.P. Davis, Chemical Communications, 2011, 47, 4, 1318. ©2011, RSC [204].

3.3.2.2 Electrophilic addition

Halide end groups on the polymers can also be transformed by the addition of electrophiles. For example, Nakagawa et al. [213] successfully transformed the bromide end group of polystyrene by titanium tetrachloride catalyzed electrophilic addition of allyltrimethylsilane.

3.3.3 Chemical modification of a functional group incorporated via functional initiator

If the functional group in the initiator is unstable or capable of participating in side reactions under ATRP conditions, such group should be protected. Functional groups such as maleimide, amine, alkyne, acid, and so on should be protected using preclick strategy or simple organic reaction. After completion of polymerization, the protected group (*t*-butoxycarbonyl (*t*-Boc) for amine and *t*-butyl (*t*-Bu) for acid) can be deprotected to obtain desired functionality. Functional groups such as acid and amine interfere in ATRP; thus, the telechelic polymers with carboxylic acid and amine groups [214] can be obtained using the masked initiators that after polymerization can selectively be hydrolyzed to obtain the desired functionality (Scheme 3.13).

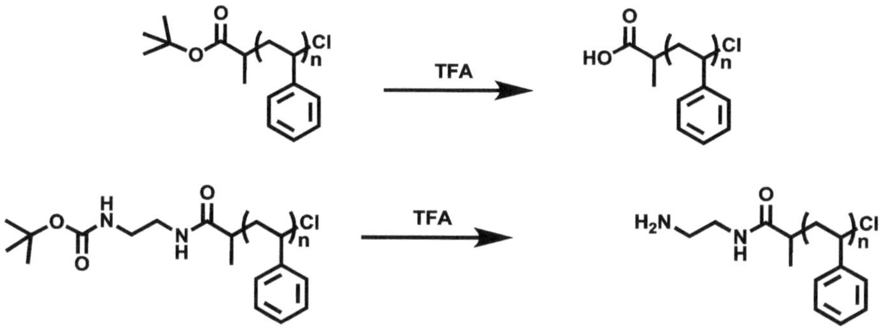

Scheme 3.13: *t*-Boc-protected carboxylic acid and amine-functionalized polystyrene and their chemical modification to corresponding acid and amine. Reproduced with permission from J. Hegewald, J. Pionteck, L. Haubler, H. Komber and B. Voit, Journal of Polymer Science Part A: Polymer Chemistry, 2009, 47, 15, 3845. ©2009, John Wiley & Sons [214].

Similarly, Haddleton and coworkers [215] obtained the amino telechelic poly(ethyleneglycol methacrylate) (PEGMA) using *t*-Boc-protected amine containing ATRP initiator followed by acid hydrolysis (Scheme 3.14).

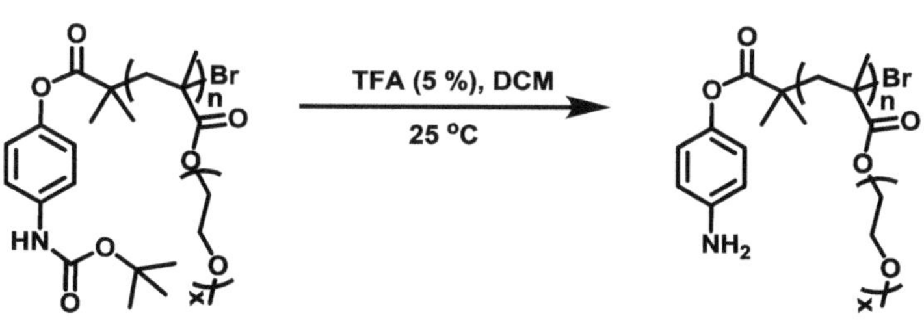

Scheme 3.14: *t*-Boc-protected initiator for ATRP and its chemical modification to amine. Reproduced with permission from M.W. Jones, R.A. Strickland, F.F. Schumacher, S. Caddick, J.R. Baker, M.I. Gibson and D.M. Haddleton, Journal of the American Chemical Society, 2012, 134, 3, 1847. ©2012, ACS [215].

α-Primary amine-functionalized polystyrene has been synthesized by Postma et al. [99] by employing the phthalimido-masked amino-ATRP initiator and post-polymerization hydrazinolysis of phthalimido end groups. The bromide end group was removed by radical-induced reduction using tri-*n*-butylstannane before hydrazinolysis to avoid the reaction of hydrazine hydrate on bromide end. This approach has been investigated as applicable for α-functionalization because the ω-functionalization led to the elimination of bromide to generate alkene at the terminal.

Caprolactone-containing ATRP initiators, namely, 2-chloro- or 2-bromo-caprolactone after ATRP can be hydrolyzed or their nucleophilic ring opening led to the formation of telechelic polymers. Bai and coworkers reported a facile strategy to synthesize hetero-bifunctionalized polystyrene combining ATRP and ring opening modification using α-bromocaprolactone as the initiator (Scheme 3.15) [216]. Polystyrenes with α-hydroxyl, α′-carboxylic acid and α-hydroxyl, α′- amine hetero-telechelic groups have been prepared using amine and water as nucleophiles by ring-opening modification of caprolactone incorporated via functional initiator and the approach is considered to be useful for the synthesis of miktoarm star copolymers.

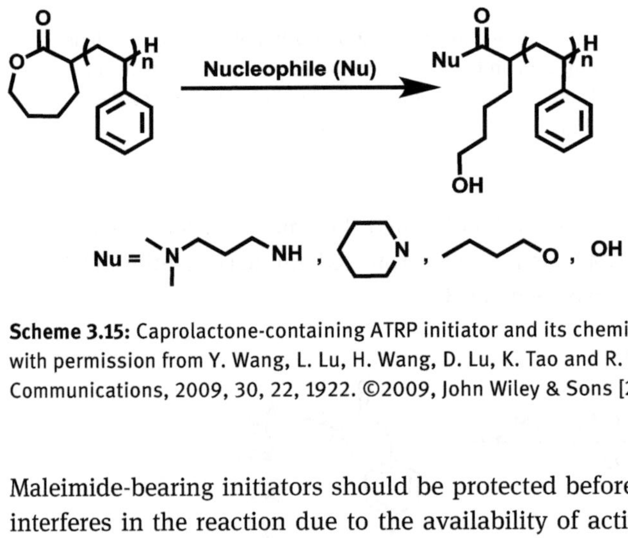

Scheme 3.15: Caprolactone-containing ATRP initiator and its chemical modification. Reproduced with permission from Y. Wang, L. Lu, H. Wang, D. Lu, K. Tao and R. Bai, Macromolecular Rapid Communications, 2009, 30, 22, 1922. ©2009, John Wiley & Sons [216].

Maleimide-bearing initiators should be protected before ATRP as maleimide group interferes in the reaction due to the availability of activated alkene in their structure. Yang et al. [217] synthesized maleimide-functionalized PNIPAM by protecting the initiator in the form of its adduct with furan before the polymerization using pre-click approach (Scheme 3.16). Thermo-sensitive electrospun fibers were fabricated by Michael addition reaction of thiol functionalized polysiloxane fibers on these maleimide-functionalized PNIPAM. These graft copolymer fibers showed the thermo-sensitive behavior to the environment due to temperature-sensitive PNIPAM brushes on the surfaces.

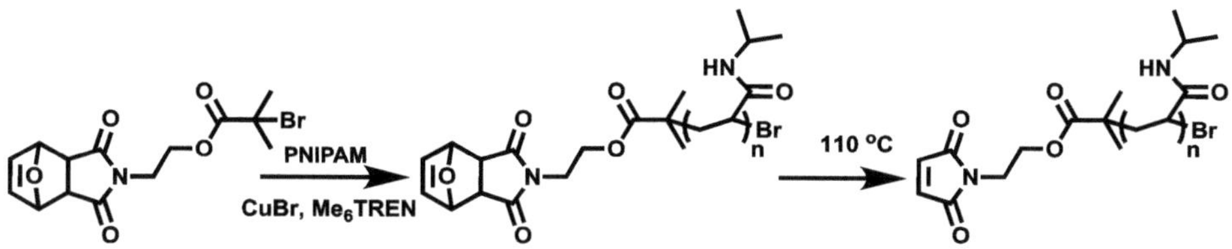

Scheme 3.16: Furan-protected maleimide-functionalized ATRP initiator and its thermal deprotection. Reproduced with permission from H. Yang, Q. Zhang, B. Lin, G. Fu, X. Zhang and L. Guo, Journal of Polymer Science Part A: Polymer Chemistry, 2012, 50, 20, 4182. ©2012, John Wiley & Sons [217].

3.3.4 Radical–radical coupling reactions

It is one of the widely used approaches to obtain homo-telechelic polymers. Sometimes, halide end groups in the polymer are not desired during the processing of polymer. The terminal halides can be removed by 1) adding double bond containing reactants and 2) direct replacement of bromide terminals to form carbon–carbon bond using activated halides. Both the methods are performed in the presence of metal catalysts such as copper or manganese.

The one-pot procedure for the replacement of halide by allyl group after the polymerization has been attempted using allyl tri-*n*-butylstannane, which tolerates the other functional groups such as acetals, epoxides, hydroxyl, and so on present in the polymer [218]. Monomers such as allyl alcohol and 1, 2-epoxy-5-hexene are nonpolymerizable with the available ATRP systems because of the slow activation step due to the formation of unstable radicals. However, using radical coupling approach, these less reactive groups have been incorporated at the polymer end [219]. Other nonpolymerizable monomers incorporated at the chain end include the divinyl benzene and ethylene on MMA [220] and maleic anhydride on polystyrene [221].

Furthermore, the direct replacement of terminal halides using activated halides has been investigated. Visible light irradiation of bromo-terminated polymers prepared by ATRP in the presence of catalysts such as dimanganese decacarbonyl ($Mn_2(CO)_{10}$) and activated halide leads to the formation of telechelic polymer via radical–radical coupling (Scheme 3.17) [222]. The obtained polystyrene bearing ester functionalities at the terminal showed double molecular weight compared to that of precursor.

Synthesis of aldehyde–telechelic polystyrene was attempted by Durmaz et al. [223] combining ATRP and atom transfer radical cross-coupling (ATRCC). The bromo-terminated polystyrene prepared by ATRP was cross-coupled with aldehyde-functionalized initiator (Scheme 3.18). Detailed investigation of the reaction revealed that the efficiency of aldehyde functionalization was found to be 0.85

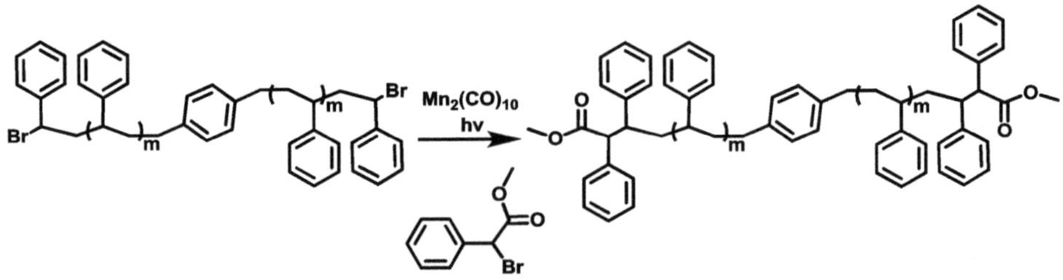

Scheme 3.17: Synthesis of homo-telechelic polymers by radical–radical coupling approach using dimanganese decacarbonyl catalyst. Reproduced with permission from B. Iskin, G. Yilmaz and Y. Yagci, Macromolecular Chemistry and Physics, 2013, 214, 1, 94. ©2013, John Wiley & Sons [222].

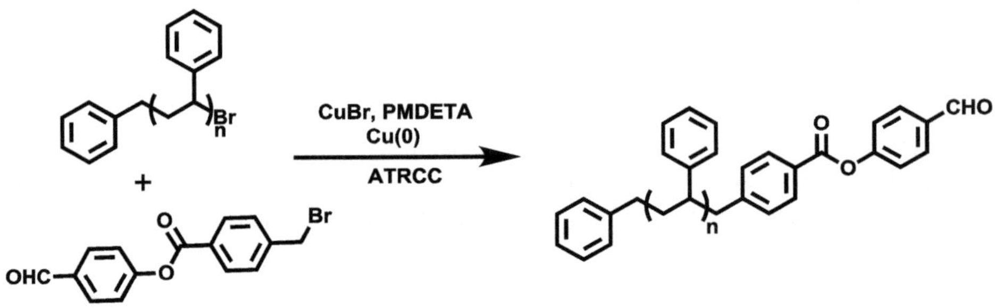

Scheme 3.18: Synthesis of aldehyde telechelic polystyrene combining ATRP and ATRCC.

due to the undesired self-coupling of species generated from polymer and from functional initiator. Influence of different factors such as concentrations of polystyrene, initiator, ligand, and catalyst on the coupling process has also been studied.

Similarly, Yagci and coworkers [166] synthesized homo-telechelic polystyrene with acid, aldehyde, aromatic hydroxyl, and dimethyl amino groups at α- and ω-terminals under copper (0)-mediated reductive conditions by ATRC process with double molecular weight compared to that of precursors (Scheme 3.19). Polymers prepared by this approach were demonstrated to be attractive for affording telechelic polymers due to the absence of potential side reactions.

The other way of radical coupling by UV-irradiation was developed to synthesize block-copolymer involving the use of benzophenone–benzhydrol systems. Taskin et al. [186] developed a simple strategy for synthesis of different telechelic block copolymers by radical–radical coupling process using benzophenone/benzhydrol initiating system by irradiating the mixture of benzophenone and benzhydrol terminated polymer with UV-light. On irradiation the ketene radicals were generated by hydrogen abstraction of benzophenone moieties from benzhydrol and

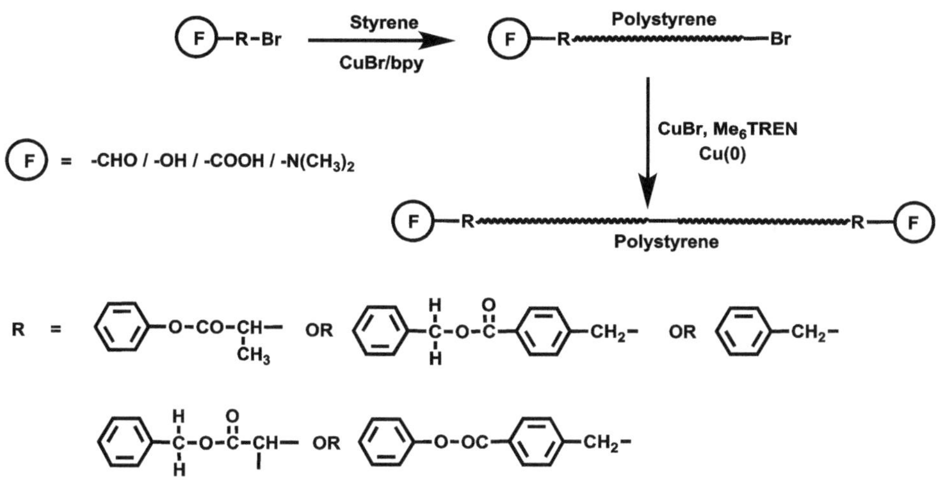

Scheme 3.19: Synthesis of homo-telechelic polymers by copper (0)-mediated ATRC. Reproduced with permission from S. Yurteri, I. Cianga and Y. Yagci, Macromolecular Chemistry and Physics, 2003, 204, 14, 1771. ©2003, John Wiley & Sons [166].

coupled with each other to form block copolymers. The key features of this coupling process are 1) useful at low temperatures and 2) useful for introducing the reactive sites at definite positions. Influence of other parameters such as solvent and the type of polymers has also been examined.

3.4 Advantages of ATRP over other CRP methods

Although RDRP methods, namely, RAFT, NMP, ATRP, and so on are widely used for controlled synthesis of telechelic polymers and different macromolecular architectures, each method has some advantages and some limitations. Compared to all these methods, ATRP has been considered as the most advantageous on account of the following [49].

1) The main advantage of ATRP over other methods is the commercial availability of all necessary reagents used in ATRP including alkyl halide, ligand, and transition metal.
2) A variety of monomers including (meth)acrylates, styrene, and NIPAM can be polymerized using different transition metal complexes such as copper, nickel, iron, palladium, and rhodium at wide range of temperatures.
3) The wide availability of ATRP initiators by simple, efficient, and convenient syntheses compared to initiators required for other RDRP methods makes ATRP an attractive technique.

4) Ligands can be easily modified to adjust the activity and solubility of metal catalyst in organic solvent, which could minimize the difficulty of solubility of metal complex in the reaction medium.
5) The reagents used in ATRP are less toxic and have no potential odor like in RAFT polymerization.
6) A range of functionalities can be introduced using different functional initiators and polymers with wide range of architectures such as block, star, graft, and so on including the topologically complex architectures such as rotaxanes, catenanes, inclusion complex containing polymers can be prepared using simple synthetic strategies.
7) Conducting polymerization in aqueous medium allows the controlled/living polymerization of hydrophilic and ionic monomers by ATRP, which cannot be otherwise polymerized in organic media.
8) Replacing organic solvents by aqueous reaction media is additionally helpful from economic and environmental point of view.
9) ATRP shows smallest sensitivity to the steric effects that could be helpful in polymerizing the sterically hindered monomers.
10) Polymer prepared by ATRP retains halide at the terminal that allows the easier and efficient synthesis of functional polymers by transformation route, which otherwise cannot be possible with other RDRP methods.

The main limitation of ATRP over other methods is the use of toxic copper metal catalyst. Although copper is most commonly used metal catalyst, less toxic metals such as iron can be used to perform the reaction. The methods have been developed to remove the residual metal catalyst including the use of supported catalyst and hybrid catalytic systems that reduces the residual metal to less than 5 ppm making the technique industrially favorable. Furthermore, the new versions of ATRP such as activators-generated electron transfer, activators-regenerated electron transfer (ARGET) and initiators for continuous activators generation ATRP, and so on [49]. overcome the limitations of ATRP by enabling the ATRP reactions with use of minimum amount (ppm level) of metal catalyst.

3.5 Applications of functionally terminated polymers

Since last few decades, polymers terminated with hydroxyl, carboxyl, amino, aldehyde, nitro, and anhydride functional groups have been much explored using ATRP. Apart from these functional polymers, clickable-telechelic polymers with different functionalities are of particular interest as they can be utilized for subsequent quantitative reactions leading to new macromolecular architectures. Click reactions such as CuAAC, furan-maleimide, anthracene-maleimide, and so on have been explored for the polymer conjugations. Although CuAAC reactions are associated with some limitations in biomedical applications due to the toxicity of copper catalysts, these are currently widely used in polymer chemistry. Polymers prepared by strain-promoted alkyne-azide click reaction have been used for medical applications as it is a metal-free click reaction. Architectures such as diblock, triblock, graft, star, cross-linked, and cyclic copolymers can be obtained from the corresponding homo/hetero-telechelics (Figure 3.6).

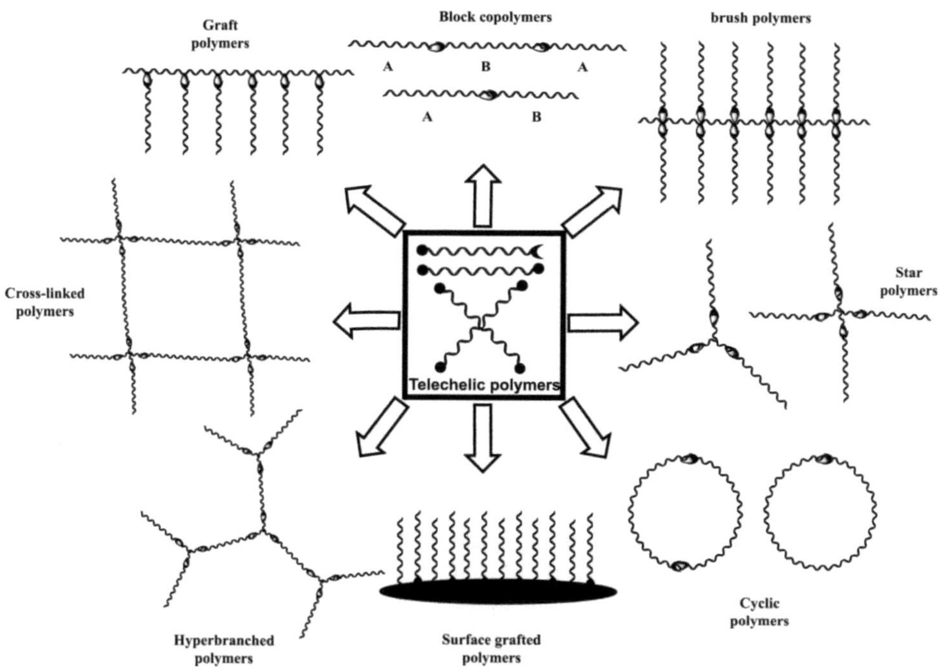

Figure 3.6: Different macromolecular architectures derived from telechelic polymers.

3.5.1 Synthesis of block/star block copolymer rubbers and thermoplastic elastomers

A well-known commercial thermoplastic elastomer is a styrene–butadiene–styrene with glassy polystyrene domains surrounded by a soft polybutadiene matrix, which provides the properties of elongation due to polybutadiene and recyclability due to polystyrene domains. Utilization of bio-derived materials for the synthesis of block copolymer to mimic the properties of styrene–butadiene–styrene thermoplastic elastomer is a focused area of research. Tang and coworkers [224] synthesized a plant oil-derived thermoplastic elastomer and enhanced their mechanical properties using triazolinedione as a cross-linker. Star-block copolymer thermoplastic elastomer involving poly(*n*-butyl acrylate) and polyacrylonitrile (P*n*BA-*b*-PAN) blocks was designed by Matyjaszewski and coworkers [11] (Figure 3.7).

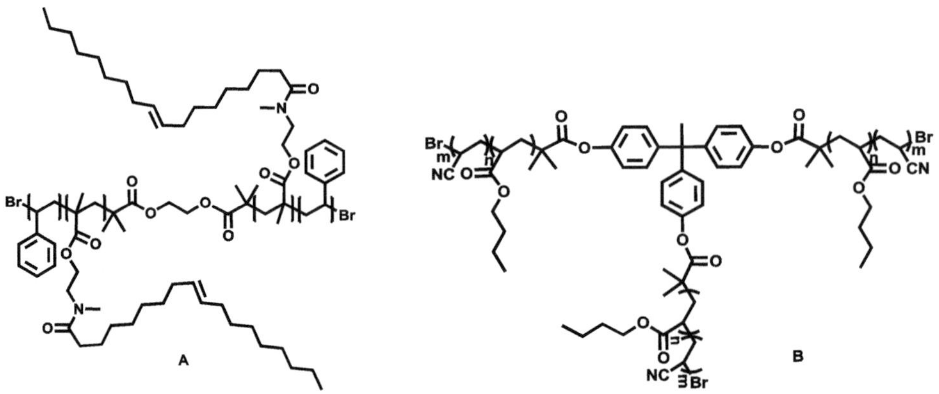

Figure 3.7: Structures of A) PS-*b*-PSBMA-*b*-PS triblock copolymer and B) poly(*n*-butyl acrylate)-*b*-polyacrylonitrile (P*n*BA-*b*-PAN) star block copolymer for the synthesis of thermoplastic elastomers.

3.5.2 Synthesis of cyclic polymers – catenanes and rotaxanes

Cyclic polymers show unique thermal, mechanical, and hydrodynamic properties as compared to other architectures. Some of the telechelic polymers prepared by ATRP have been used for synthesis of cyclic polymers. High dilution of reaction is a mandatory requirement for cyclic polymer formation. The cyclic polymers can be prepared by two approaches: 1) ring expansion and 2) ring closure approach, which includes the single- or double-click reaction strategies and atom transfer radical coupling processes [23, 203, 225, 226]. Durmaz et al. [23] examined the cyclic polymer formation using double-click reaction strategies involving sequential CuAAC and anthracene–maleimide click reactions. The cyclizations were carried out at higher dilutions (in the range of 4×10^{-5}–7×10^{-5} M) and the efficiency of cyclization was found to be in the range of 77–85%. Figure 3.8 represents the cyclic polymers prepared by anthracene–maleimide click reaction and ATRC. Cyclic polystyrene was reported by Tillman and coworkers [227] by combining ATRP and ATRC. Linear polystyrene was prepared from difunctional ATRP initiator and the terminal bromide ends were coupled using ATRC under high-dilution conditions.

Catenanes and rotaxanes are the well-known examples involving the cyclic macromolecular topologies. The extensive entanglements present in these types of macromolecular architectures give strength and toughness to the materials. Catenanes – the

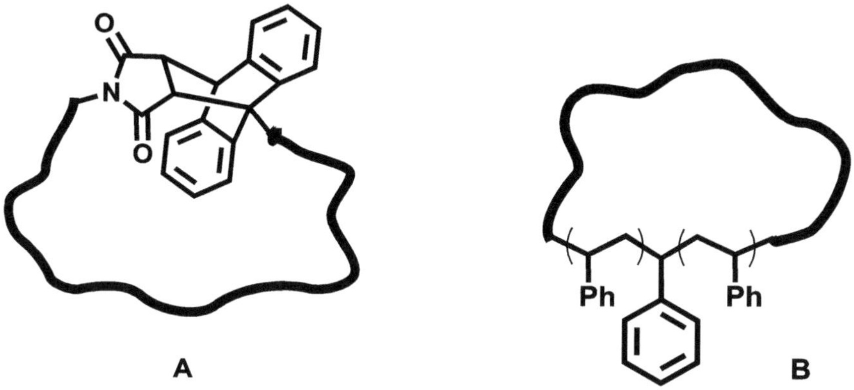

Figure 3.8: Cyclic polymers by A) anthracene–maleimide click reaction and B) ATR coupling process.

interlocked molecular architectures – consisting of two interlocked macrocycles have been reported using ATRP and ATRC. Inclusion complex topology possessing macrocycles threaded with block copolymers are called as rotaxanes. Feng and coworkers [226] designed a cyclodextrin based polyrotaxanes using ATRP and click reactions. The varying amount of γ-cyclodextrin was threaded with a linear PNIPAM-*b*-Pluronic (F68)-*b*-PNIPAM triblock copolymer and was end capped by click reaction with propargyl-terminated cyclodextrin. The LCST of resulted polyrotaxane was increased due to the cyclodextrin-covered PNIPAM, which may be ascribed to the hindered thermal aggregation of PNIPAM blocks.

3.5.3 Synthesis of block and core cross linked star polymers for emulsion stabilization

Most of the industrial applications of materials such as paints, varnishes, and so on are based on the emulsions. The formation and stabilization of emulsion is desired for the long-term storage of such materials. The formation of thin and smooth paint layer on the applied surfaces requires the coalescence of emulsions. The coalescence of such emulsion droplets needs some stimulus such as temperature, pH, or light. Block and core cross-linked star polymers stabilize the emulsions by adsorbing on the interface of oil and water leading to the minimization of interfacial tension. PDMAEMA-*b*-PEGMA-*b*-PLMA triblock copolymer [10] with hydrophilic and hydrophobic parts has been designed and utilized for stabilization of different types of emulsions. pH-dependent coalescence of oil drops has been studied due to the presence of pH-responsive PDMAEMA block. The obtained triblock copolymer stabilized the dodecane in water emulsions at lower concentration of 0.5 mg mL^{-1} polymer for about three months. Furthermore, core cross-linked star copolymers

[228] have also been used for the stabilization of oil/water and water/oil emulsions for longer time.

3.5.4 Synthesis of polymers for rheology modification of lubricating greases/oils, viscosity modifications, and compatibilization of blends

Telechelic polymers and derived architectures such as graft and hyperbranched copolymers have been used to modify the rheological properties of materials such as paints, greases, oils, and so on. Hyperbranched polymer was investigated by Jagtap and coworkers [229] from poly(ethyl acrylate) macroinitiator and trimethylol propane triacrylate by ATRP. The significant decrease in viscosity of water based paint was observed using 1 wt% of this hyperbranched polymer in the formulation.

The compatibilization is a process of modification of surface properties of immiscible polymer blends by reducing the interfacial tension and by improving the interactions between the different immiscible solid phases. Polymers prepared by ATRP, mainly the block copolymers, can be used for compatibilizations of immiscible blends. Kim et al. [230] examined the use of polycarbonate-*b*-PMMA and polycarbonate-*b*-poly(styrene-*co*-acrylonitrile) as compatibilizers for the immiscible physical blend of polycarbonate (PC) with poly(styrene-*co*-acrylonitrile) copolymer. PC-*b*-PMMA was found to be more effective compatibilizer as compared to the PC-*b*-P(SAN) by reducing the diameter and surface tension of dispersed particles.

3.5.5 Macromonomers for thermoplastics, cross-linked polymers, and gel formation

Telechelic polymers obtained by ATRP with similar functionalities at their chain ends have been utilized for the synthesis of thermoplastics as they act as macromonomers for condensation polymerization reactions. Chemical and physical properties of condensation polymers such as polyesters, polyamides, and so on can be tuned using macromonomers prepared by ATRP. For example, diamino-telechelic polystyrene (Figure 3.9) on reaction with diacid chlorides yields the chain-extended polystyrenes with amide linkages at the backbone with different properties than the precursor polystyrene [231]. However, dihydroxyl macromonomers (Figure 3.9) are widely applicable in designing segmented polyurethanes and polyesters. Macromonomers such as dihydroxyl-terminated PLMA are widely used as steric stabilizers and dispersants. The other method to obtain thermoplastic polymers is click polymerization of dialkyne- and diazido-terminated polymers or the self-click polymerization of α-azido ω-alkyne-terminated polymers (Figure 3.9) [232].

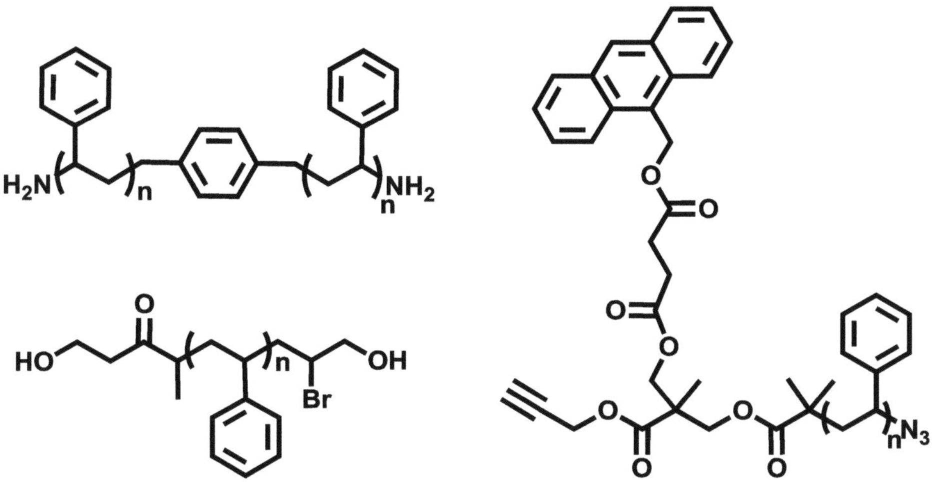

Figure 3.9: Different macromonomers for condensation reactions.

Cross-linked polymers show better thermal and mechanical properties compared to the noncross-linked thermoplastic polymers. To study the properties of cross-linked polymers, polyester-*b*-poly(meth)acrylate block copolymers were synthesized using saturated and unsaturated polyesters by combining polycondensation and ATRP [233]. The double bonds in the main chain of the obtained block copolymers were unaffected during the ATRP and the polymer was further cross-linked by photoirradiation. These cross-linked polymers have been investigated as better candidates for coating applications with overall better properties than the neat polyesters without unsaturations in their backbone (Scheme 3.20).

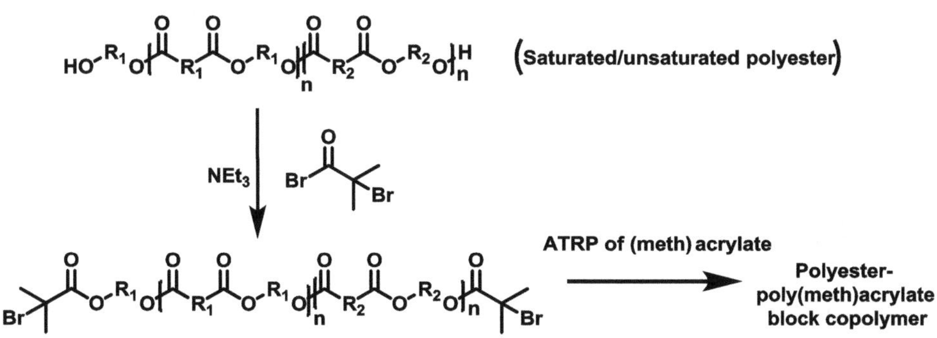

Scheme 3.20: Polyester–poly(meth)acrylate block copolymers using polycondensation and ATRP reactions. Reproduced with permission from U. Chatterjee, X. Wang, S.K. Jewrajka and M.D. Soucek, Macromolecular Chemistry and Physics, 2011, 212, 17, 1879. ©2011, John Wiley & Sons [233].

Temperature-responsive gel has been widely used for drug delivery purposes and other biological applications. Propargyl telechelic PNIPAM has been used as cross-linking agents for the formation of thermoresponsive gel as they show LCST behavior [5].

3.5.6 Synthesis of graft copolymers and surface modifying agents

The range of graft copolymers by ATRP have been reported with different phase separation behaviors in bulk and in solution [31, 234]. Phase separations of polymers in different domains affect the properties of polymers. Jiang et al. [235] reported cellulose-*g*-poly(*n*-butyl acrylate-*co*-methyl methacrylate) by activators regenerated by electron transfer ATRP (ARGET ATRP). These graft copolymer exhibited the desired properties of thermoplastic elastomers due to phase separation of rigid cellulose and rubbery poly(BA-*co*-MMA) domains. Densely grafted surfaces using ATRP macroinitiators have been used to modify the most planar substrates such as silicon, textile, wood and carbon-based surfaces such as graphene oxide and CNTs [13].

3.5.7 Synthesis of fluorescent polymers for sensing and detection applications

Fluorescent polymers prepared by ATRP are widely used in different fields such as bioimaging, sensing, labeling antibodies, acid detection, and so on. Pyrene and fullerene end functional poly(*N*-isopropylacrylamide)s were used for the temperature-sensing applications in different systems [236]. Matyjaszewski and coworkers synthesized a star polymer with pyrene functionalities at the periphery [237]. These star polymers are potentially useful as amplified fluorescent probes in the field of bioimaging. A range of fluorescent monomers have been polymerized by ATRP and are summarized in a review by Schubert and coworkers [238].

3.6 Summary and outlook

ATRP allows the synthesis of polymers with well-defined structures, controlled molecular weight, and molecular weight distribution using wide range of monomers by providing the choice to introduce variety of functionalities at the terminals. A range of ATRP initiators have been prepared bearing different reactive functional groups such as hydroxyl, carboxyl, amino, ketone, aldehyde, and so on and clickable functional groups such as alkyne, azido, maleimido, allyl, and so on. Functionally terminated polymers can be prepared by ATRP using the following approaches, namely, 1) functional initiator approach by using appropriate initiator, 2) the use of functional initiator followed by the postpolymerization transformation of terminal halide into other useful functional groups, 3) chemical modification of a functional group incorporated via functional initiator into other useful functionalities employing the strategies such as deprotection/selective hydrolysis, and 4) radical–radical coupling.

ATRP is advantageous due to its less sensitivity to the environmental conditions, more tolerance to the wide range of functional groups present in the initiators and monomers, and works at wider temperature range. However, the use of toxic metal, namely, copper decreases the applicability of polymers prepared by ATRP in biomedical sciences. A large number of initiators with different functionalities have been employed in ATRP to yield telechelic polymers that could not easily be accessible by other RDRP polymerization techniques. Marrying click reactions with ATRP is a fruitful way for exploring the complex architectures by simplifying the synthetic aspects. Due to the expanding scope of click chemistry, a range of functionalities can be introduced and variety of monomers can be polymerized to obtain enhanced properties of the resulting polymers compared to the existing materials.

Functionally terminated polymers prepared by ATRP have been used for the synthesis of different macromolecular architectures such as block, star, cyclic, and graft copolymers. Homo-telechelic polymers are applicable as prepolymers in condensation polymerization to mimic and to enhance the properties of existing materials and to modify rheology and surface properties for gel formation and for sensing applications.

Although the new versions of ATRP such as activators-generated electron transfer, ARGET, and initiators for continuous activators generation ATRP have been developed, which overcome the limitations of ATRP by enabling the reactions with use of small amount of metal catalyst (ppm level), there is still a need of a method that allows polymerization reactions to be performed in aqueous media without the metal catalysts (metal-free).

References

[1] C.A. Uraneck, H.L. Hsieh and O.G. Buck, Journal of Polymer Science Part A: Polymer Chemistry, 1960, 46(148), 535.

[2] O. Bayer, Angewandte Chemie International Edition, 1947, 59(9), 257.

[3] I. Goodman, British Polymer Journal, 1990, 22(3), 261.

[4] L.T. Nguyen, H.T. Nguyen and T.T. Truong, Journal of Polymer Research, 2015, 22(9), 186.

[5] J. Wang, Z. Kang, B. Qi, Q. Zhou, S. Xiao and Z. Shao, RSC Advances, 2014, 4(93), 51510.

[6] A.V. Ruzette and L. Leibler, Nature Materials, 2005, 4(1), 19.

[7] G. Moreno, C. Valencia, M.V.D. Paz, J.M. Franco and C. Gallegos, Industrial & Engineering Chemistry Research, 2006, 45(11), 4001.

[8] N. Hameed, N.V. Salim, T.R. Walsh, J.S. Wiggins, P.M. Ajayanc and B.L. Foxa, Chemical Communications, 2015, 51(48), 9903.
[9] A.K. Singh, R. Prakash and D. Pandey, RSC Advances, 2012, 2(27), 10316.
[10] J. Huang, J. Xu, K. Chen, T. Wang, C. Cui, X. Wei, R. Zhang, L. Li and X. Guo, Industrial & Engineering Chemistry Research, 2015, 54(5), 1564.
[11] B. Dufour, C. Tang, K. Koynov, Y. Zhang, T. Pakula and K. Matyjaszewski, Macromolecules, 2008, 41(7), 2451.
[12] H. Fengab, N.A.L. Verstappena, A.J.C. Kuehnec and J. Sprakel, Polymer Chemistry, 2013, 4(6), 1842.
[13] Q. Wei, X. Wang and F. Zhou, Polymer Chemistry, 2012, 3(8), 2129.
[14] A. Nomura, K. Ohno, T. Fukuda, T. Satobc and Y. Tsujii, Polymer Chemistry, 2012, 3(1), 148.
[15] S. Binauld and M.H. Stenzel, Chemical Communications, 2013, 49(21), 2082.
[16] T.K. Georgiou, Polymer International, 2014, 63(7), 1130.
[17] M. Xie, W. Wang, L. Ding, J. Liu, D. Yang, L. Wei and Y. Zhang, Journal of Polymer Science Part A: Polymer Chemistry, 2010, 48(2), 380.
[18] J. Couthouis, H. Keul and M. Möller, Macromolecular Chemistry and Physics, 2016, 217(1), 72.
[19] N.A.C. Lemus, R.S. Rodriguez and A.L. Claverie, Journal of Polymer Science Part A: Polymer Chemistry, 2010, 48(14), 3033.
[20] M. Nagy, M. Zsuga, D. Racz and S. Keki, Journal of Polymer Science Part A: Polymer Chemistry, 2010, 48(12), 2709.
[21] J.P. Kennedy, V.S.C. Chang and W.P. Francik, Journal of Polymer Science Part A: Polymer Chemistry, 1982, 20(10), 2809.
[22] T. Dedeoglu, H. Durmaz, G. Hizal and U. Tunca, Journal of Polymer Science Part A: Polymer Chemistry, 2012, 50(10), 1917.
[23] H. Durmaz, A. Dag, G. Hizal and U. Tunca, Journal of Polymer Science Part A: Polymer Chemistry, 2010, 48(22), 5083.
[24] O. Altintas, T. Rudolph and C.B. Kowollik, Journal of Polymer Science Part A: Polymer Chemistry, 2011, 49(12), 2566.
[25] S. Wallyn, Z. Zhang, F. Driessen, J. Pietrasik, B.G.D. Geest, R. Hoogenboom and F.E. DuPrez, Macromolecular Rapid Communications, 2014, 35(4), 405.
[26] N. Rocha, P.V. Mendonça, J.P. Mendes, P.N. Simões, A.V. Popov, T. Guliashvili, A. C. Serra and J.F.J. Coelho, Macromolecular Chemistry and Physics, 2013, 214(1), 76.
[27] N.V. Tsarevsky, B.S. Sumerlin and K. Matyjaszewski, Macromolecules, 2005, 38(9), 3558.
[28] E. Bays, L. Tao, C. Chango and H.D. Maynard, Biomacromolecules, 2009, 10(7), 1777.
[29] M.M. Stamenovi, P. Espeel, W.V. Camp and F.E. DuPrez, Macromolecules, 2011, 44(14), 5619.
[30] A. Hirao, R. Goseki and T. Ishizone, Macromolecules, 2014, 47(6), 1883.
[31] S. Banerjee, T. Maji and T.K. Mandal, Colloid and Polymer Science, 2014, 292(9), 2217.
[32] R. Gnaneshwar and S. Sivaram, Journal of Polymer Science Part A: Polymer Chemistry, 2007, 45(12), 2514.
[33] M.R. Hill, R.N. Carmean and B.S. Sumerlin, Macromolecules, 2015, 48(16), 5459.
[34] M.A. Tasdelen, M.U. Kahveci and Y. Yagci, Progress in Polymer Science, 2011, 36(4), 455.
[35] A. Anastasaki, V. Nikolaoua and D.M. Haddleton, Polymer Chemistry, 2016, 7(5), 1002.
[36] A. Anastasaki, V. Nikolaou, G. Nurumbetov, P. Wilson, K. Kempe, J.F. Quinn, T.P. Davis, M.R. Whittaker and D.M. Haddleton, Chemical Reviews, 2016, 116(3), 835.
[37] K. Matyjaszewski, Y. Gnanou and L. Leibler, Macromolecular Engineering. Precise Synthesis, Materials Properties, Applications. Edited by K. Matyjaszewski, Y. Gnanou, and L. Leibler Copyright © 2007 WILEY-VCH Verlag GmbH & Co. KGaA, Weinheim ISBN: 978-3-527-31446-1, 1–2982.

[38] K. Matyjaszewski and T.P. Davis, Handbook of Radical Polymerization, Wiley Interscience: , Hoboken, 2002, ISBN:047139274X.
[39] K. Matyjaszewski and J. Xia, Chemical Reviews, 2001, 101(9), 2921.
[40] J.S. Wang and K. Matyjaszewski, Journal of the American Chemical Society, 1995, 117(20), 5614.
[41] M. Kato, M. Kamigaito, M. Sawamoto and T. Higashimura, Macromolecules, 1995, 28(5), 1721.
[42] K. Matyjaszewski and N.V. Tsarevsky, Journal of the American Chemical Society, 2014, 136 (18), 6513.
[43] K. Matyjaszewski, Macromolecules, 1998, 31(15), 4710.
[44] K. Matyjaszewski and N.V. Tsarevsky, Nature Chemistry, 2009, 1(4), 276.
[45] V. Coessens, T. Pintauer and K. Matyjaszewski, Progress in Polymer Science, 2001, 26(3), 337.
[46] A. K. Shakya and A. Kumar, Journal of Bioscience and Biotechnology, 2013, 2(1), 1.
[47] M. Ouchi, T. Terashima and M. Sawamoto, Chemical Reviews, 2009, 109(11), 4963.
[48] K. Matyjaszewski and P. Sigwalt, Polymer International, 1994, 35(1), 1.
[49] W. A. Braunecker and K. Matyjaszewski, Progress in Polymer Science, 2007, 32(1), 93.
[50] A. Goto and T. Fukuda, Macromolecular Rapid Communications, 1999, 20(12), 633.
[51] K. Matyjaszewski, B. Goebelt, H.J. Paik and C.P. Horwitz, Macromolecules, 2001, 34(3), 430.
[52] K. Matyjaszewski, H.J. Paik, P. Zhou and S.J. Diamanti, Macromolecules, 2001, 34(15), 5125.
[53] M. Oka and M. Tatemoto, Contemporary Topics in Polymer Science, 1984, 4, 763.
[54] K. Matyjaszewski, S.G. Gaynor and J.S. Wang, Macromolecules, 1995, 28(6), 2093.
[55] K. Matyjaszewski, D.A. Shipp, J.L. Wang, T. Grimaud and T.E. Patten, Macromolecules, 1998, 31(20), 6836.
[56] Q.F. Xu, J.M. Lu, Z. Yang, X.W. Xia and L.H. Wang, Polymer Journal, 2007, 39(3), 213.
[57] A. Marsh, A. Khan, D.M. Haddleton and M.J. Hannon, Macromolecules, 1999, 32(26), 8725.
[58] K. Matyjaszewski and J. Spanswick, Materials Today, 2005, 8(3), 26.
[59] L.C. Mei, B. Rui, Q.J. Jun, H. Fen, X. Yan, Z. Chen and Z. Yun, Polymer Bulletin, 2006, 57(2), 139.
[60] D.P. Chatterjee and B.M. Mandal, Polymer, 2006, 47(6), 1812.
[61] S.S. Patil, S.K. Menon and P.P. Wadgaonkar, Polymer International, 2015, 64(3), 413.
[62] J. Opsteen and J.C.M. VanHest, Chemical Communications, 2005, 1(1), 57.
[63] W. Jakubowski, N.V. Tsarevsky, P. McCarthy and K. Matyjaszewski, Material Matters, 2010, 5(1), 16.
[64] C. Barner-Kowollik, F. E. DuPrez, P. Espeel, C. J. Hawker, T. Junkers, H. Schlaad and W. V. Camp, Angewandte Chemie International Edition, 2011, 50(1), 60.
[65] P.S. Sane, D.V. Palaskar and P.P. Wadgaonkar, European Polymer Journal, 2011, 47(8), 1621.
[66] P.S. Sane, B.V. Tawade, D.V. Palaskar, S.K. Menon and P.P. Wadgaonkar, Reactive and Functional Polymers, 2012, 72(10), 713.
[67] P.S. Sane, B.V. Tawade, I. Parmar, S. Kumari, S. Nagane and P.P. Wadgaonkar, Journal of Polymer Science Part A: Polymer Chemistry, 2013, 51(9), 2091.
[68] Y.L. Cai and S.P. Armes, Macromolecules, 2005, 38(2), 271.
[69] G. Deng, L. Zhang, C. Liu, L. He and Y. Chen, European Polymer Journal, 2005, 41(6), 1177.
[70] V. Deimede and J.K. Kallitsis, Chemistry European Journal, 2002, 8(2), 467.
[71] Y. Yamazaki, N. Ajioka, A. Yokoyama and T. Yokozawa, Macromolecules, 2009, 42(3), 606.
[72] S. Yurteri, I. Cianga, A.L. Demirel and Y. Yagci, Journal of Polymer Science Part A: Polymer Chemistry, 2005, 43(4), 879.
[73] G.J. Summers, M.P. Ndawuni and C.A. Summers, Polymer International, 2012, 61(9), 1353.
[74] K. Matyjaszewski, J.L. Wang, T. Grimaud and D.A. Shipp, Macromolecules, 1998, 31(5), 1527.
[75] T. Ando, M. Kamigaito and M. Sawamoto, Tetrahedron, 1997, 53(45), 15445.

[76] X. Zhang and K. Matyjaszewski, Macromolecules, 1999, 32(22), 7349.
[77] D.M. Haddleton, C. Waterson and P.J. Derrick, Chemical Communications, 1997, 7(7), 683.
[78] A. Likhitsup, A. Parthiban and C.L.L. Chai, Journal of Polymer Science Part A: Polymer Chemistry, 2008, 46(1), 102.
[79] H. Tai, V.K. Popov, K.M. Shakesheff and S.M. Howdle, Biochemical Society Transactions, 2007, 35(3), 516.
[80] O. Glaied, C. Delaite and P. Dumas, Journal of Polymer Science Part A: Polymer Chemistry, 2007, 45(5), 968.
[81] S. Ohno, H. Gao, B. Cusick, T. Kowalewski and K. Matyjaszewski, Macromolecular Chemistry and Physics, 2009, 210(6), 421.
[82] W. Lin, S. Nie, Q. Zhong, Y. Yang, C. Cai, J. Wang and L. Zhang, Journal of Materials Chemistry B, 2014, 2(25), 4008.
[83] H. Matsumoto, T. Nakano, K. Ohkawa and Y. Nagai, Chemistry Letters, 1978, 7(4), 363.
[84] J.C. Pelps, D.E. Bergbreiter, G.M. Lee, R. Villani and S.M. Weinreb, Tetrahedron Letter, 1989, 30(30), 3915.
[85] K.L. Heredia, Z.P. Tolstyka and H.D. Maynard, Macromolecules, 2007, 40(14), 4772.
[86] G. Mantovani, F. Lecolley, L. Tao, D.M. Haddleton, J. Clerx, J.J.L.M. Cornelissen and K. Velonia, Journal of the American Chemical Society, 2005, 127(9), 2966.
[87] M. Licciardi, Y. Tang, N.C. Billingham, S.P. Armes and A.L. Lewis, Biomacromolecules, 2005, 6(2), 1085.
[88] V. Percec and B. Barboiu, Macromolecules, 1995, 28(23), 7970.
[89] R.M. Broyer, G.M. Quaker and H.D. Maynard, Journal of the American Chemical Society, 2008, 130(3), 1041.
[90] S. Pfeifer and J.F. Lutz, Macromolecular Chemistry and Physics, 2010, 211(8), 940.
[91] J. Hegewald, J. Pionteck, L. Haussler, H. Komber and B. Voit, Journal of Polymer Science Part A: Polymer Chemistry, 2009, 47(15), 3845.
[92] V.B. Sadhu, J. Pionteck, D. Voigt, H. Komber and B. Voit, Macromolecular Symposium, 2004, 210(1), 147.
[93] V.B. Sadhu, J. Pionteck, D. Voigt, H. Komber, D. Fischer and B. Voit, Macromolecular Chemistry and Physics, 2004, 205(17), 2356.
[94] G. Toquer, S. Monge, K. Antonova, C. Blanc, M. Nobili and J.J. Robin, Macromolecular Chemistry and Physics, 2007, 208(1), 94.
[95] V. Percec and C. Grigoras, Journal of Polymer Science Part A: Polymer Chemistry, 2005, 43(21), 5283.
[96] B. Sun, C.M. Jewell, N.J. Fredin and D.M. Lynn, Langmuir, 2007, 23(16), 8452.
[97] J.T. Kopping, Z.P. Tolstyka and H.D. Maynard, Macromolecules, 2007, 40(24), 8593.
[98] F. Lecolley, C. Waterson, A.J. Carmichael, G. Mantovani, S. Harrisson and H. Chappell, Journal of Materials Chemistry, 2003, 13(11), 2689.
[99] A. Postma, T.P. Davis, G. Moad and M. Shea, Reactive and Functional Polymers, 2006, 66(1), 137.
[100] M. Matsuyama, M. Kamigaito and M. Sawamoto, Journal of Polymer Science Part A: Polymer Chemistry, 1996, 34(17), 3585.
[101] C. Boyer, B. Otazaghine, B. Boutevin, C. Joly-Duhamel and J.J. Robin, Journal of Polymer Science, Part A: Polymer Chemistry, 2005, 43(18), 4303.
[102] L. Mespouille, M. Vachaudez, F. Suriano, P. Gerbaux, W. VanCamp, O. Coulembier, P. Degee, R. Flammang, F.E. DuPrez and P. Dubois, Reactive and Functional Polymers, 2008, 68(5), 990.
[103] L. Mespouille, M. Vachaudez, F. Suriano, P. Gerbaux, O. Coulembier and P. Degee, Macromolecular Rapid Communications, 2007, 28(22), 2151.
[104] W. Agut, D. Taton and S. Lecommandoux, Macromolecules, 2007, 40(16), 5653.

[105] V.L.G. Mantovani, V. Ladmiral, L. Tao and D.M. Haddleton, Chemical Communications, 2005, 16, 2089.
[106] B. Grignard, C. Calberg, C. Jerome and C. Detrembleur, Journal of Supercritical Fluids, 2010, 53(1-3), 151.
[107] C.H. Li, J.M. Hu, J. Yin and S.Y. Liu, Macromolecules, 2009, 42(14), 5007.
[108] G. J. Chen, L. Tao, G. Mantovani, V. Ladmiral, D.P. Burt and J.V. Macpherson, Soft Matter, 2007, 3(6), 732.
[109] A. Narumi, K. Fuchise, R. Kakuchi, A. Toda, T. Satoh and S. Kawaguchi, Macromolecular Rapid Communications, 2008, 29(12–13), 1126.
[110] K. Fuchise, R. Kakuchi, S.T. Lin, R. Sakai, S.I. Sato and T. Satoh, Journal of Polymer Science Part A: Polymer Chemistry, 2009, 47(22), 6259.
[111] P.D. Topham, N. Sandon, E.S. Read, J. Madsen, A.J. Ryan and S.P. Armes, Macromolecules, 2008, 41(24), 9542.
[112] D. Haddleton and C. Waterson, Macromolecules, 1999, 32(26), 8732.
[113] J.A. Blazquez, J. Areizaga, J.J. Iruin, O. Miguel, D. Mecerreyes and J. Jouanneau, Reactive and Functional Polymers, 2006, 66(10), 1073.
[114] C. Granel, P. Dubois, R. Jerome and P. Teyssie, Macromolecules, 1996, 29(27), 8576.
[115] H. Uegaki, M. Kamigato and M. Sawamoto, Journal of Polymer Science Part A: Polymer Chemistry, 1999, 37(15), 3003.
[116] K. Matyjaszewski, S.M. Jo, H.J. Paik and S.G. Gaynor, Macromolecules, 1997, 30(20), 6398.
[117] K. Matyjaszewski, M. Wei, J. Xia and N.E. McDermott, Macromolecules, 1997, 30(26), 8161.
[118] V. Percec, B. Barboiu and H.J. Kim, Journal of the American Chemical Society, 1998, 120(2), 305.
[119] V. Percec, H.J. Kim and B. Barboiu, Macromolecules, 1997, 30(26), 8526.
[120] B. Dervaux, F. Meyer, J.M. Raquez, A. Olivier, F.E. DuPrez and P. Dubois, Macromolecular Chemistry and Physics, 2012, 213(12), 1259.
[121] S. Gong, H. Ma and X. Wan, Polymer International, 2006, 55(12), 1420.
[122] M.A. Bennett, Chemical Reviews, 1962, 62(6), 611.
[123] R. Nast, Coordination Chemistry Reviews, 1982, 47(1–2), 89.
[124] C.J. Duxbury, D. Cummins and A. Heise, Journal of Polymer Science Part A: Polymer Chemistry, 2009, 47(15), 3795.
[125] A. Hasneen, H.S. Han and H.J. Paik, Reactive and Functional Polymers, 2009, 69(9), 681.
[126] B.S. Sumerlin, N.V. Tsarevsky, G. Louche, R.Y. Lee and K. Matyjaszewski, Macromolecules, 2005, 38(18), 7540.
[127] J. Opsteen and J.C.M. VanHest, Journal of Polymer Science Part A: Polymer Chemistry, 2007, 45(14), 2913.
[128] I. Singh, Z. Zarafshani, J.F. Lutz and F. Heaney, Macromolecules, 2009, 42(15), 5411.
[129] Z. Zarafshani, O. Akdemir and J.F. Lutz, Macromolecular Rapid Communications, 2008, 29(12–13), 1161.
[130] C. Urbani, C. Bell, D. Lonsdale, M. Whittaker and M. Monteiro, Macromolecules, 2008, 41(1), 76.
[131] G.D. Fu, L.Q. Xu, F. Yao, K. Zhang, X.F. Wang, M.F. Zhu and S.Z. Nie, ACS Applied Materials & Interfaces, 2009, 1(2), 239.
[132] G.J. Chen, L. Tao, G. Mantovani, V. Ladmiral, D.P. Burt, J.V. Macpherson and D.M. Haddleton, Soft Matter, 2007, 3(6), 732.
[133] L.Q. Xu, F. Yao, G.D. Fu and L. Shen, Macromolecules, 2009, 42(17), 6385.
[134] C.H. Li, Z.S. Ge, H.W. Liu and S.Y. Liu, Journal of Polymer Science Part A: Polymer Chemistry, 2009, 47(16), 4001.
[135] W. Lin, Q. Fu, Y. Zhang and J. Huang, Macromolecules, 2008, 41(12), 4127.
[136] W. K. S. Miller and C. Pugh, Macromolecules, 2015, 48(12), 3803.
[137] J. Xu, J. Ye and S. Liu, Macromolecules, 2007, 40(25), 9103.

[138] V. Ladmiral, T.M. Legge, Y.L. Zhao and S. Perrier, Macromolecules, 2008, 41(18), 6728.
[139] S. Pfeifer, Z. Zarafshani, N. Badi and J.F. Lutz, Journal of the American Chemical Society, 2009, 131(26), 9195.
[140] Y. Li, J.W. Yang and B.C. Benicewicz, Journal of Polymer Science Part A: Polymer Chemistry, 2007, 45(18), 4300.
[141] A. Dag, H. Durmaz, G. Hizal and U. Tunca, Journal of Polymer Science Part A: Polymer Chemistry, 2008, 46(1), 302.
[142] M. Li, P. De, S.R. Gondi and B.S. Sumerlin, Journal of Polymer Science Part A: Polymer Chemistry, 2008, 46(115), 5093.
[143] A. Dag, H. Durmaz, E. Demir, G. Hizal and U. Tunca, Journal of Polymer Science Part A: Polymer Chemistry, 2008, 46(20), 6969.
[144] O. Altintas, G. Hizal and U. Tunca, Designed Monomers & Polymers, 2009, 12(1), 83.
[145] H. Durmaz, F. Karatas, U. Tunca and G. Hizal, Journal of Polymer Science Part A: Polymer Chemistry, 2006, 44(13), 3947.
[146] B. Gacal, H. Durmaz, M. A. Tasdelen, G. Hizal, U. Tunca, Y. Yagci and A.L. Demirel, Macromolecules, 2006, 39(16), 5330.
[147] H. Durmaz, A. Dag, A. Hizal, G. Hizal and U. Tunca, Journal of Polymer Science Part A: Polymer Chemistry, 2008, 46(21), 7091.
[148] M. Erdogan, G. Hizal, U. Tunca, D. Hayrabetyan and O. Pekcan, Polymer, 2002, 43(6), 1925.
[149] V. Vazquez-Dorbatt, Z.P. Tolstyka, C.W. Chang and H.D. Maynard, Biomacromolecules, 2009, 10(8), 2207.
[150] A. Klaikherd, S. Ghosh and S. Thayumanavan, Macromolecules, 2007, 40(24), 8518.
[151] A. Klaikherd, C. Nagamani and S. Thayumanavan, Journal of the American Chemical Society, 2009, 131(13), 4830.
[152] H.D. Maynard, K.L. Heredia, R.C. Li, D.P. Parra and V. Vazquez-Dorbatt, Journal of Materials Chemistry, 2007, 17(38), 4015.
[153] A.W. Jackson and D.A. Fulton, Macromolecules, 2010, 43(2), 1069.
[154] Y. Jin, L. Song, Y. Su, L. Zhu, Y. Pang, F. Qiu, G. Tong, D. Yan, B. Zhu and X. Zhu, Biomacromolecules, 2011, 12(10), 3460.
[155] V.V. Dorbatt, Z.P. Tolstyka and H.D. Maynard, Macromolecules, 2009, 42(20), 7650.
[156] Y. Shen, S. Zhu, F. Zeng and R. Pelton, Macromolecules, 2000, 33(15), 5399.
[157] F. Zeng, Y. Shen, S. Zhu and R. Pelton, Journal of Polymer Science Part A: Polymer Chemistry, 2000, 38(20), 3821.
[158] S.P.S. Koo, M.M. Stamenovic, R.A. Prasath, A.J. Inglis, F.E. DuPrez, C. Barner-Kowollik, W.V. Camp and T. Junkers, Journal of Polymer Science Part A: Polymer Chemistry, 2010, 48(8), 1699.
[159] M.M. Stamenovic, P. Espeel, W.V. Camp and F.E. DuPrez, Macromolecules, 2011, 44(14), 5619.
[160] D.J. Liaw, C.C. Huang and C.H. Tsai, Tamkang Journal of Science and Engineering, 2003, 6(3), 133.
[161] J.A. Carioscia, L. Schneidewind, C. Obrien, R. Ely, C. Feeser, N. Cramer and C.N. Bowman, Journal of Polymer Science Part A: Polymer Chemistry, 2007, 45(23), 5686.
[162] C.F. Hansell, P. Espeel, M.M. Stamenovic, I.A. Barker, A.P. Dove, F.E. DuPrez and R.K. Oreilly, Journal of the American Chemical Society, 2011, 133(35), 13828.
[163] I. Gadwal and A. Khan, Polymer Chemistry, 2013, 4(8), 2440.
[164] L.P. Yang, X.H. Dong and C.Y. Pan, Journal of Polymer Science Part A: Polymer Chemistry, 2008, 46(23), 7757.
[165] J.M. Lu, X.W. Xia, X. Guo, Q.F. Xu, F. Yan and L.H. Wang, Journal of Applied Polymer Science, 2008, 108(5), 3430.

[166] S. Yurteri, I. Cianga and Y. Yagci, Macromolecular Chemistry and Physics, 2003, 204(14), 1771.
[167] Z. Yang, J. Lu, S. Yao and L. Wang, Journal of Macromolecular Science Part A: Pure and Applied Chemistry, 2004, 41(10), 1105.
[168] G. Summers, M. Ndawuni and C. Summers, Polymer International, 2003, 52(1), 158.
[169] C.N. Urbani, C.A. Bell, D. Lonsdale, M.R. Whittaker and M.J. Monteiro, Macromolecules, 2008, 41(1), 76.
[170] C. Hou, S. Lin, F. Liu, J. Hu, G. Zhang, G. Liu, Y. Tu, H. Zouab and H. Luo, New Journal of Chemistry, 2014, 38(6), 2538.
[171] M. Degirmenci, O. Izgin, A. Acikses and N. Genli, Reactive and Functional Polymers, 2010, 70(1), 28.
[172] Y. Sun, W. Liu and Z. Ma, Polymer Bulletin, 2013, 70(5), 1531.
[173] G. Carrot, J. Hilborn, J. Hedrick and M. Trollsas, Macromolecules, 1999, 32(15), 5171.
[174] M. Urien, H. Erothu, E. Cloutet, C.R. Hiorns and L. Vignau, Macromolecules, 2008, 41(19), 7033.
[175] S.S. Patil, B.V. Tawade and P.P. Wadgaonkar, Journal of Polymer Science Part A: Polymer Chemistry, 2016, 54(6), 844.
[176] F. Zeng, Y. Shen, S. Zhu and R. Pelton, Macromolecules, 2000, 33(5), 1628.
[177] L. Campos, K. Killops, R. Sakai, J. Paulusse, D. Damiron, E. Drockenmuller, B. Messmore and C. Hawker, Macromolecules, 2008, 41(19), 7063.
[178] H. Malz, H. Komber, D. Voigt, I. Hopfe and J. Pionteck, Macromolecular Chemistry and Physics, 1999, 200(3), 642.
[179] D. Mecerreyes, B. Atthoff, K. Boduch, M. Trollsas and J. Hedrick, Macromolecules, 1999, 32(16), 5175.
[180] Y. Shen, S. Zhu, F. Zeng and R. Pelton, Macromolecular Chemistry and Physics, 2000, 201 (13), 1387.
[181] H. Malz, J. Pionteck, P. Potschke, H. Komber, D. Voigt, J. Luston and F. Bohme, Macromolecular Chemistry and Physics, 2001, 202(11), 2148.
[182] G.J. Summers, R.B. Maseko, B.M.P. Beebeejaun and C.A. Summers, Journal of Polymer Science Part A: Polymer Chemistry, 2011, 49(12), 2601.
[183] D.V. Palaskar, P.S. Sane and P.P. Wadgaonkar, Reactive and Functional Polymers, 2010, 70(12), 931.
[184] X. Yang, S. Wang, Y. Yan, Y. Wu, K. Zheng and Y. Chen, Polymer, 2014, 55(5), 1128.
[185] H. Durmaz, B. Colakoglu, U. Tunca and G. Hizal, Journal of Polymer Science Part A: Polymer Chemistry, 2006, 44(5), 1667.
[186] O.S. Taskin, B.A. Temel, M.A. Tasdelen and Y. Yagci, European Polymer Journal, 2015, 62, 304.
[187] V. Ladmiral, L. Monaghan, G. Mantovani and D.M. Haddleton, Polymer, 2005, 46(19), 8536.
[188] F. Lecolley, L. Tao, G. Mantovani, I. Durkin, S. Lautru and D.M. Haddleton, Chemical Communications, 2004, 18, 2026.
[189] A.A. Kavitha and N.K. Singha, Macromolecular Chemistry and Physics, 2007, 208(23), 2569.
[190] S.O. Sanchez, F. Marra, A. Dibenedetto, M. Aresta and A. Grassi, Macromolecules, 2014, 47 (20), 7129.
[191] P. Sun, G. Yan, Q. Tang, Y. Chen and K. Zhang, Polymer, 2015, 64, 202.
[192] Y. Chen and Z. Guan, Chemical Communications, 2014, 50(74), 10868.
[193] A. Bunha, M.C. Tria and R. Advincula, Chemical Communications, 2011, 47(32), 9173.
[194] K. Pangilinan and R. Advincula, Polymer International, 2014, 63(5), 803.
[195] J. Lyngsø, N.A. Manasir, M.A. Behrens, K. Zhu, A.L. Kjøniksen, B. Nyström and J.S. Pedersen, Macromolecules, 2015, 48(7), 2235.

[196] K. Matyjaszewski, P.J. Miller, D.C. Pyun, G. Kickelbick and S. Diamanti, Macromolecules, 1999, 32(20), 6526.
[197] D. Han, X. Tong and Y. Zhao, Journal of Polymer Science Part A: Polymer Chemistry, 2012, 50(20), 4198.
[198] J.S. Wang, D. Greszta and K. Matyjaszewski, Polymer Materials Science and Engineering, 1995, 73, 416.
[199] A.P. Narrainen, L.R. Hutchings, I. Ansari, R.L. Thompson and N. Clarke, Macromolecules, 2007, 40(6), 1969.
[200] S.P. Rwei, Y.Y. Chuang, T.F. Way, W.Y. Chiang and S.P. Hsu, Colloid and Polymer Science, 2015, 293(2), 493.
[201] G.J. Summers, M.P. Ndawuni and C.A. Summers, Polymer International, 2014, 63(5), 876.
[202] L.Z. Kong, M. Sun, H.M. Qiao and C.Y. Pan, Journal of Polymer Science Part A: Polymer Chemistry, 2010, 48(2), 454.
[203] O. Altintas, P.K. Sidenstein, H. Gliemann and C. Barner-Kowollik, Macromolecules, 2014, 47(17), 5877.
[204] C. Boyer, A.H. Soeriyadi, P.J. Roth, M.R. Whittaker and T.P. Davis, Chemical Communications, 2011, 47(4), 1318.
[205] G. Volet, T.X. Lav, J. Babinot and C. Amiel, Macromolecular Chemistry and Physics, 2011, 212 (2), 118.
[206] R. Francis, B. Lepoittevin, D. Taton and Y. Gnanou, Macromolecules, 2002, 35(24), 9001.
[207] R. Matmour, B. Lepoittevin, T.J. Joncheray, R.J. El-Khouri, D. Taton, R.S. Duran and Y. Gnanou, Macromolecules, 2005, 38(13), 5459.
[208] J. Babin, C. Leroy, S. Lecommandoux, R. Borsali, Y. Gnanou and D. Taton, Chemical Communications, 2005, 15(15), 1993.
[209] C.N. Urbani, C.A. Bell, M.R. Whittaker and M.J. Monteiro, Macromolecules, 2008, 41(4), 1057.
[210] Y. Zhang, H. Liu, H. Dong, C. Li and S. Liu, Journal of Polymer Science Part A: Polymer Chemistry, 2009, 47(6), 1636.
[211] V. Coessens, Y. Nakagawa and K. Matyjaszewski, Polymer Bulletin, 1998, 40(2–3), 135.
[212] V. Coessens and K. Matyjaszewski, Journal of Macromolecular Science Part A: Pure & Applied Chemistry, 1999, 36(5–6), 653.
[213] Y. Nakagawa, S.G. Gaynor and K. Matyjaszewski, Polymer Preprint (American Chemical Society, Division of Polymer Chemistry), 1996, 37(1), 577.
[214] J. Hegewald, J. Pionteck, L. Haubler, H. Komber and B. Voit, Journal of Polymer Science Part A: Polymer Chemistry, 2009, 47(15), 3845.
[215] M.W. Jones, R.A. Strickland, F.F. Schumacher, S. Caddick, J.R. Baker, M.I. Gibson and D.M. Haddleton, Journal of the American Chemical Society, 2012, 134(3), 1847.
[216] Y. Wang, L. Lu, H. Wang, D. Lu, K. Tao and R. Bai, Macromolecular Rapid Communications, 2009, 30(22), 1922.
[217] H. Yang, Q. Zhang, B. Lin, G. Fu, X. Zhang and L. Guo, Journal of Polymer Science Part A: Polymer Chemistry, 2012, 50(20), 4182.
[218] V. Coessens and K. Matyjaszewski, Macromolecular Rapid Communications, 1999, 20(2), 66.
[219] V. Coessens, J. Pyun, P.J. Miller, S. Gaynor and K. Matyjaszewski, Macromolecular Rapid Communications, 2000, 21(2), 103.
[220] S.A.F. Bon, A.G. Steward and D.M. Haddleton, Journal of Polymer Science Part A: Polymer Chemistry, 2000, 38(4), 678.
[221] E.G. Koulouri and J.K. Kallitsis, Macromolecules, 1999, 32(19), 6242.
[222] B. Iskin, G. Yilmaz and Y. Yagci, Macromolecular Chemistry and Physics, 2013, 214(1), 94.
[223] Y.Y. Durmaz, I. Cianga and Y. Yagci, e-Polymers, 2006, 6(1), 644.

[224] Z. Wang, L. Yuan, N.M. Trenor, L. Vlaminck, S. Billiet, A. Sarkar, F.E. DuPrez, M. Stefik and C. Tang, Green Chemistry, 2015, 17(7), 3806.

[225] J. Liu, X. Lian, F. Zhao and H. Zhao, Journal of Polymer Science Part A: Polymer Chemistry, 2013, 51(17), 3567.

[226] L. Jiang, L. Ye, A. Zhang and Z. Feng, Macromolecular Chemistry and Physics, 2014, 215(10), 1022.

[227] A.F. Voter and E.S. Tillman, Macromolecules, 2010, 43(24), 10304.

[228] W. Li, Y. Yu, M. Lamson, M.S. Silverstein, R.D. Tilton and K. Matyjaszewski, Macromolecules, 2012, 45(23), 9411.

[229] Y.T. Chimankar, Y.S. Ahire and R.N. Jagtap, Der Chemica Sinica, 2013, 4(2), 82.

[230] J.H. Kim, M.J. Kim, C.K. Kim and J.W. Lee, Korea-Australia Rheology Journal, 2001, 13(3), 125.

[231] K. Matyjaszewski, V. Coessens, Y. Nakagawa, J. Xia, J. Qiu, S. Gaynor, S. Coca and C. Jasieczek, ACS Symposium Series, 1998, 704, 16.

[232] H. Durmaz, A. Dag, G. Hizal and U. Tunca, Journal of Polymer Science Part A: Polymer Chemistry, 2011, 49(5), 1195.

[233] U. Chatterjee, X. Wang, S.K. Jewrajka and M.D. Soucek, Macromolecular Chemistry and Physics, 2011, 212(17), 1879.

[234] M. Li, G. Shan, Y. Bao and P. Pan, Journal of Applied Polymer Science, 2014, 131(22), 41115.

[235] F. Jiang, Z. Wang, Y. Qiao, Z. Wang and C. Tang, Macromolecules, 2013, 46(12), 4772.

[236] J. Hu and S. Liu, Macromolecules, 2010, 43(20), 8315.

[237] H. Gao and K. Matyjaszewski, Macromolecules, 2007, 40(6), 1789.

[238] A.M. Breul, M.D. Hager and U.S. Schubert, Chemical Society Reviews, 2013, 42(12), 5366.

Andrew B. Lowe

4 Functional (co)polymers via a combination of reversible deactivation radical polymerization techniques and thiol-based "click"/conjugation chemistries

4.1 Introduction

In the broadest sense, and regardless of polymerization mechanism, functional (co) polymers may be obtained by the direct polymerization of appropriate functional monomers or by the modification of pre-formed (co)polymers containing appropriate chemical handles facilitating modification. Prior to the discovery and development of reversible deactivation radical polymerization (RDRP) processes, *vide infra*, the latter, post-polymerization modification approach, was relatively common, and in many instances a necessity, due to a fundamental incompatibility of available synthetic techniques with certain, desirable, functionality. For example, group transfer polymerization (GTP), an enolate-mediate polymerization process based on sequential Michael addition reactions, is an extremely useful technique for the controlled polymerization of methacrylic monomers yielding (co)polymers with predetermined molecular weights and low dispersities ($Đ = \bar{M}_w/\bar{M}_n$). However, since GTP is a pseudo-living anionic process it is fundamentally incompatible with, for example, methacrylic acid (MAA). As such the introduction of MAA repeat units in (co)polymers requires the use of protecting group chemistry, that is, the direct (co)polymerization of a GTP-tolerant monomer bearing a suitable functional group that can be converted, in a facile manner, to a carboxylic acid after (co)polymerization, see Scheme 4.1. In the specific case of introducing MAA residues, the use of benzyl methacrylate (BzMA), trimethylsilyl methacrylate (TMSMA), *tert*-butyl methacrylate (TBMA), and 2-tetrahydropyranyl methacrylate (THPMA) have all been evaluated with THPMA proving to be particularly useful [1–4].

Similar approaches were, and still may be, required for the introduction of other incompatible functionality with other examples of (pseudo)living polymerization techniques. However, the discovery and development of the RDRP processes and specifically stable-free radical polymerization, best exemplified by nitroxide-mediated polymerization (NMP), atom transfer radical polymerization (ATRP), reversible addition-fragmentation chain transfer polymerization (RAFT), and tellurium-mediated radical polymerization (TERP) resulted in a decline in the

Andrew B. Lowe, Curtin Institute for Functional Molecules and Interfaces (CIFMI) and School of Molecular and Life Sciences, Curtin University, Bentley, Perth, WA, Australia

https://doi.org/10.1515/9783110643695-004

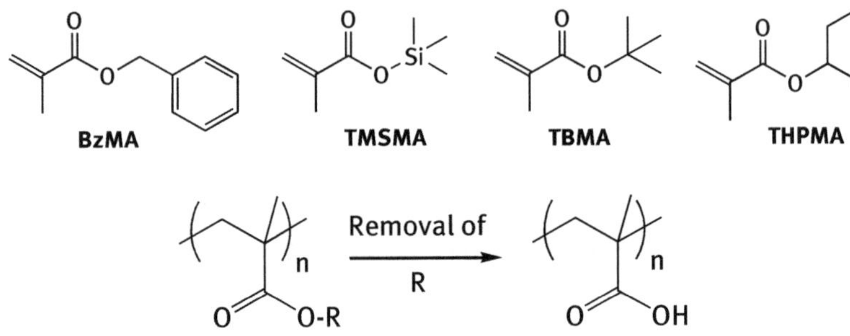

Scheme 4.1: Chemical structures of protected methacrylic acid monomers susceptible to polymerization by GTP and general approach for the synthesis of poly(methacrylic acid) via GTP.

use, or even need, for protecting group chemistry since all these processes exhibit a high functional group tolerance and few functional monomers are not amenable to direct (co)polymerization. However, performing chemistry on polymers still presented challenges and while many small molecule reactions are highly efficient it does not necessarily follow that the same is true in polymer analogous reactions. Fortunately, the development of the so-called "click" concept for (macro) molecular synthesis and modification has, in many ways, transformed the way in which synthetic polymer chemists plan and execute specific synthetic goals. The remainder of this chapter will highlight the "click" concept for macromolecular synthesis and modification with an emphasis on the combination of the suite of thiol-X chemistries with RDRP processes.

4.1.1 The "click" concept to synthesis and modification

In 2001, Kolb, Finn, and Sharpless published a paper entitled "Click Chemistry: Diverse Chemical Function from a Few Good Reactions" [5] and outlined the concept of simplifying organic syntheses via the use of a limited group of highly efficient, modular, and selective chemistries. Additionally, the authors noted a set of criteria that should be met for a chemical reaction to be accurately classified as a "click" reaction. They highlighted that such reactions "*must be modular, wide in scope, give very high yields, generate only inoffensive by-products that can be removed by nonchromatographic methods and be stereospecific.*" Also from a process perspective the authors listed certain desirable characteristics, which included "*simple reaction conditions (with the process ideally being insensitive to water and or oxygen), readily available starting materials and reagents, the use of no solvent or a solvent that is benign, i.e. water, or easily removed, and simple product isolation.*" Of all available chemistries it is those that result, most commonly, in carbon-heteroatom bond formation that meet such criteria and thus can be accurately classified as "click" reactions. Highlighted examples included 1,3-dipolar cycloadditions as well as the general Diels-Alder chemical reactions; nucleophilic substitution chemistry and especially those involving ring-opening reactions of strained heterocycles; non-Aldol carbonyl chemistry including the formation of (thio)ureas, oxime ethers, and hydrazones; and finally, additions to C–C multiple bonds in processes such as epoxidation, dihydroxylation, sulfenyl halide additions and Michael additions [5]. However, it should be noted that not every reaction that falls within these general categories necessarily meet the criteria to be "click" reactions, but certainly many do.

Following this seminal report, synthetic polymer chemists quickly recognized the potential of the "click" concept as a means of both synthesizing new materials with interesting properties but also of tackling the age-old problem of facile, quantitative chemical modification of preformed (co)polymers. So, while the concept was originally intended to motivate the simplification of chemical syntheses in traditional small molecule organic chemistry it has, arguably, had the biggest impact in the synthetic polymer/materials chemistry fields.

Since the introduction of this concept, the archetypal "click" reaction to emerge has been the Cu-catalyzed reaction between an alkyne and an azide – a modification of the well-established Huisgen 1,3-dipolar cycloaddition reaction. In fact, and confusingly, the term "click chemistry" is often used to refer specifically to this reaction. In the traditional process, the reaction of an azide and alkyne under thermal conditions, via a concerted process results in a mixture of 1,4- and 1,5-triazole derivatives, Scheme 4.2.

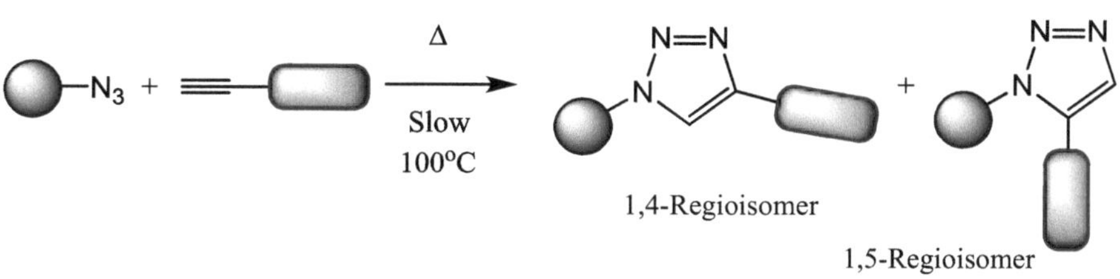

Scheme 4.2: The Huisgen 1,3-dipolar cycloaddition between an azide and terminal alkyne under thermal conditions to give an ~1:1 mixture of the 1,4- and 1,5-regioisomeric 1,2,3-triazoles. (Source: L. Liang and D. Astruc, Coordination Chemistry Reviews, 2011, 255, 23–24, 2933) [6].

In contrast, the Cu(I) catalyzed process yields, exclusively, the 1,4-regioisomer [6]. Since its original disclosure the Cu-catalyzed process has been the most widely adopted of the "click" reactions and its impact in synthetic chemistry is evidenced by the large number of reviews and articles published on this topic. See as representative examples [7–13].

While the popularity of the Cu(I) catalyzed azide-alkyne reaction remains high, its general success did spur researchers to re-evaluate many other chemistries in an effort to identify additional "click" reactions. As a result there are now significant numbers of chemical processes that can, under appropriate conditions, be accurately classified as "click" reactions. Additionally, there are certain highly efficient reactions that meet most of the required criteria to be classified as "click" chemistries but are more generally referred to as "highly efficient conjugation, or coupling, chemistries." In the case of the former group of reactions notable examples include heteroatom Diels-Alder processes, certain inverse electron-demand Diels-Alder reactions, and oxime ligation [14], while in the case of the latter the use of pentafluorophenylester [15–23] functionality and (co)polymers derived from 2-vinyl-4,4-dimethylazlactone [24–27], for example, have both proven to be versatile platforms for the preparation of libraries of interesting materials with novel properties via post-polymerization modification routes.

However, many of these alternative "click" and "click-like" reactions require the use of rather specific reagents which may not be commercially available or may require pre-"click" syntheses of new initiators, monomers, polymers, or other building blocks to enable the preparation of the target product. However, one group of such reactions for which additional syntheses are largely not required are those based on the chemistry of the thiol functional group. These reactions are drawn from a number of the general types noted by Finn, Kolb, and Sharpless and include the thiol-ene (radical addition), the intimately related thiol-Michael (conjugate addition), the sister thiol-yne reaction (radical or conjugate addition), the thiol-isocyanate (non-Aldol carbonyl chemistry), thiol-halo (nucleophilic substitution), thiol-epoxide (nucleophilic substitution), and thiol-methanethiosulfonate (nucleophilic substitution). Perhaps most significantly though is that many applications of this tool-box of thiol-based reactions can be accomplished with readily available and cheap reagents (thiols, alkenes, alkynes, isocyanates, halo compounds, and more) and thus can be readily adopted and implemented with few prior synthetic demands. Each of these thiol-based reactions will be highlighted in more detail below with an emphasis on their use in RDRP processes.

4.1.2 Reversible deactivation radical polymerization

Reversible deactivation radical polymerization (RDRP) refers to a group of radical-based chain growth polymerization techniques that possess the key characteristics associated with a traditional living (or controlled) ionic polymerization. This includes the ability to prepare (co)polymers with pre-determined molecular weights and low dispersities, targeted molar compositions in the case of copolymers, the synthesis of polymers with advanced architectures such as block copolymers (via sequential monomer addition) and the ability to control the chemical nature of the end groups either by selective termination or post-polymerization modification. In order of their open literature disclosure, the common RDRP techniques are stable-free radical polymerization (SRFP), best exemplified by nitroxide-mediated polymerizations (NMP) [28, 29], atom transfer radical polymerization (ATRP) [30, 31], reversible addition-fragmentation chain transfer (RAFT) [32–35], and tellurium-mediated radical polymerization (TERP) [36, 37]. Scheme 4.3 shows the general reaction schemes for these four different polymerization methods in the context of styrene homopolymerization.

A detailed explanation of these different techniques is beyond the scope of this chapter and readers are referred to Chapter 1 of this book and the above citations for a thorough review of these RDRP processes. However, it is worth noting that all these techniques work to suppress undesirable chain termination reactions during polymerization either by control of the radical concentration (reversible termination), and hence relative number of active chains at any given time, or by highly efficient chain transfer reactions. Successful control results in the ability to prepare (co)polymers with pre-determined average degrees of polymerization ($\bar{X}_n$), materials with advanced architectures (block copolymers for example) and in most instances gives (co)polymers with low dispersities. Additionally, given the radical nature of these techniques all of them exhibit excellent functional group tolerance with very few chemical functionalities unable to be incorporated into (co)polymers directly.

Of these four techniques, ATRP and RAFT have become the favored synthetic choices when preparing new and interesting polymers, although all techniques are highly versatile with their own associated benefits. However, in the context of this chapter we will see that ATRP and RAFT are particularly useful given the aforementioned functional group tolerance of both techniques and, in the case of ATRP, the presence of a terminal bromo group that readily facilitates further chemical transformations and, for RAFT, the presence of a thiocarbonylthio group that is likewise a versatile chemical handle that can be easily transformed into groups compatible with the thiol-x toolbox.

4.2 The thiol-x toolbox

To reiterate, the current toolbox of thiol-based chemical reactions covers, in most part, examples from the general groups of chemical transformations first identified by Kolb, Finn, and Sharpless as possible “click” reactions. At present, this family includes the thiol-ene [38–43], thiol-yne [40, 44–52], thiol-isocyanate [40, 53–55], thiol-halo [56–59], thiol-epoxy [60–65], and thiol-methanethiosulfonate [66–71] chemistries. All of these have been employed across many diverse areas of polymer synthesis and modification as evidenced by many of the above references. However, the purpose of the remainder of this chapter is to highlight their use in conjunction with the family of RDRP techniques.

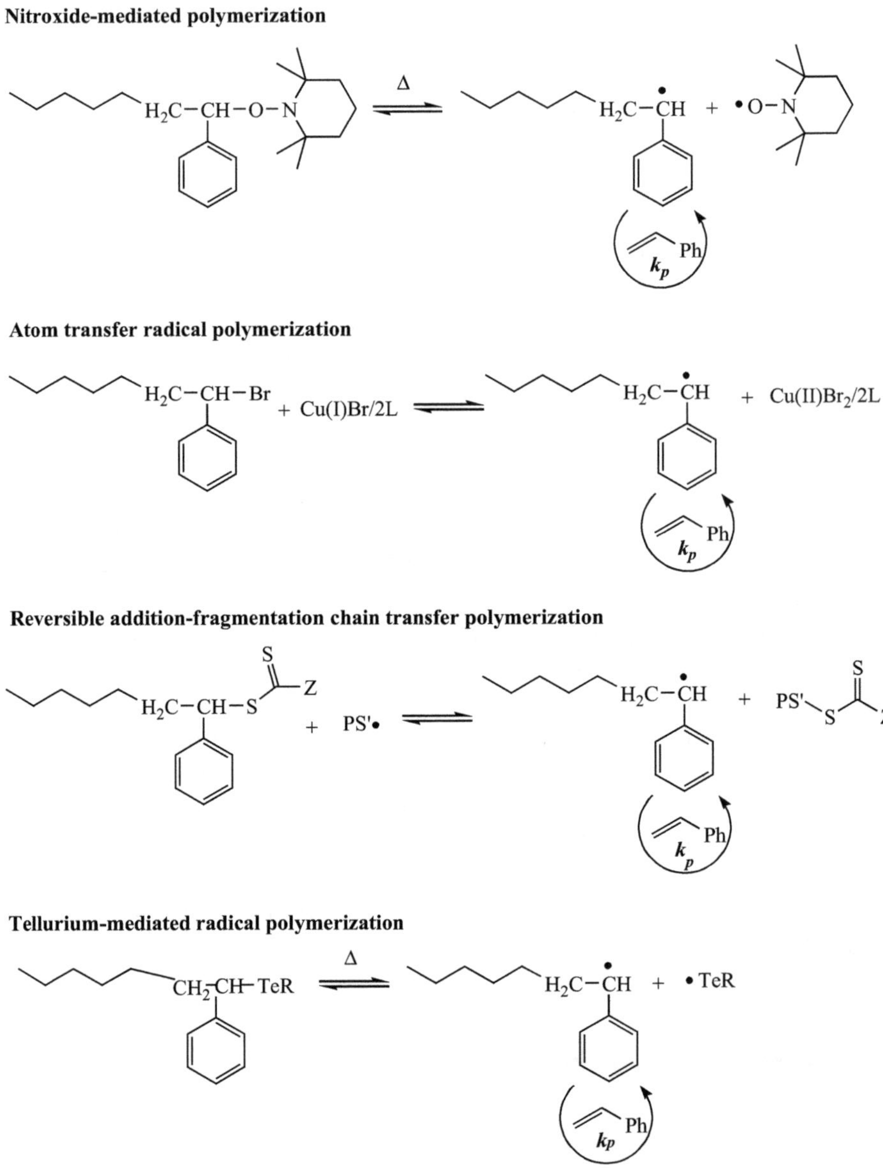

Scheme 4.3: General, simplified mechanisms for the homopolymerization of styrene via NMP, ATRP, RAFT, and TERP.

4.2.1 The thiol-ene reaction

Of the family of thiol-X chemistries the thiol-ene reaction (in its various mechanistic forms) is the most commonly utilized and best understood [38, 39, 41, 72]. The thiol-ene reaction is simply the hydrothiolation of a C = C bond yielding the corresponding thioether. There are two common variants of this general reaction: the radical mediated thiol-ene reaction and the ionic thiol-ene process, which is also commonly referred to as the thiol-Michael reaction. Several reviews are available highlighting the development, general features, and applications of these extremely versatile reactions [38, 39, 41, 42, 72], and readers are referred to these for more detailed accounts of the scope and features of these processes. For the radical version of this chemical transformation, the generally accepted mechanism is outlined in Scheme 4.4.

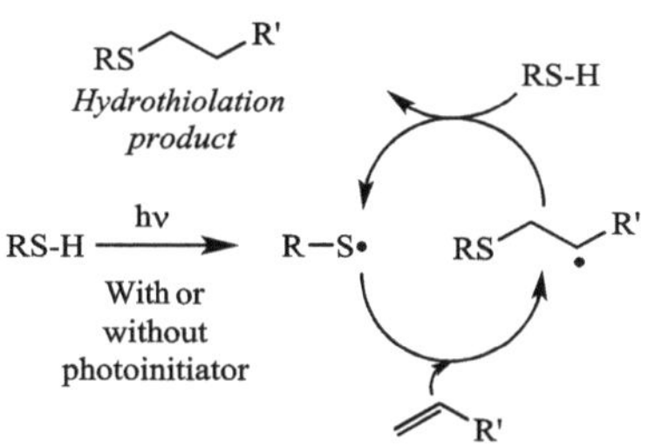

Scheme 4.4: The general mechanism for the hydrothiolation of a C = C bond under radical conditions with or without added photoinitiator.

The first step involves generation of a thiyl radical. This is most commonly achieved via photo- irradiation (with or without added photoinitiator), although any of the standard methods for radical generation can also be readily employed (heating, use of AIBN, and so on). The generated thiyl radical then adds to the C = C in the target alkene with anti-Markovnikov orientation generating an intermediate carbon-centered radical that undergoes a chain transfer reaction with addition thiol, RS-H, to give the thiol-ene product and in the process regenerating a thiyl radical. In general, such reactions are extremely rapid and, in most instances, give quantitative formation of the target thioether with essentially 100% regioselectivity. However, it is noted that the nature of the monomer, thiol, substitution pattern of the ene, method of radical generation are all known to have an effect on the overall kinetics and efficacy of the reaction, and readers are directed to the above noted reviews for specifics regarding these formulation variables.

The ionic thiol-ene, or thiol-Michael reaction, operates via a fundamentally different mechanism and is only applicable to electron deficient enes, that is, those bearing electron withdrawing groups (formally the reaction is a heteroatomic Michael addition reaction). Mechanistically, such reactions can be base catalyzed employing a weak organobase (triethylamine is common), or nucleophile catalyzed with certain tertiary phosphines being favored (dimethylphenylphosphine, DMPP, is generally the phosphine of choice but others can also be readily used). The difference between these two mechanistic pathways lies in how the thiolate species is generated. In the case of the base-catalyzed process the first step in the ionic chain reaction is abstraction of a proton from thiol by the organobase. This is followed by conjugate addition of the thiolate to the activated ene generating an intermediate carbon-centered (commonly, but not limited to, an enolate) anion (a much stronger base) that abstract a proton from addition thiol to give the product and regenerate a thiolate, Scheme 4.5.

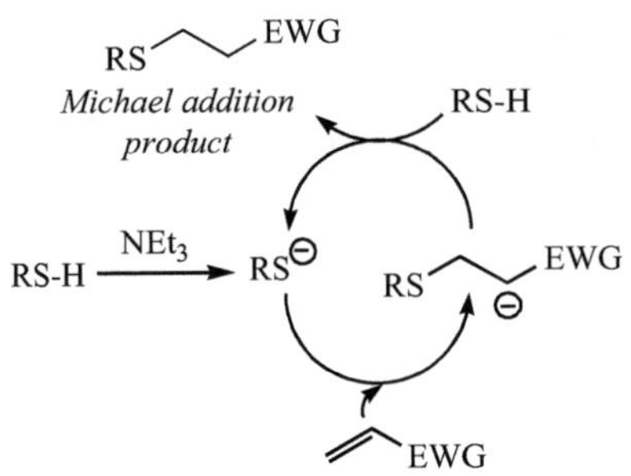

Scheme 4.5: The organobase-mediated addition of a thiol to an electron deficient ene.

In essence, a weak base, NEt_3, is utilized to generate a strong base, the carbanion that initiates the ionic chain reaction and formation of the heteroatom-Michael adduct. In contrast, the nucleophile-initiated variant proceeds as outlined in Scheme 4.6.

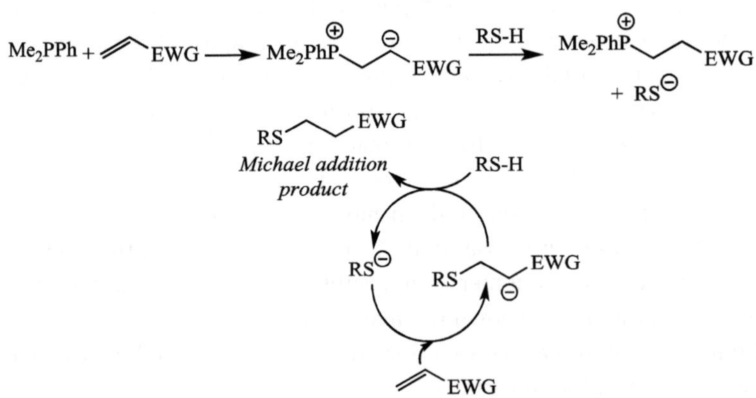

Scheme 4.6: The nucleophile-initiated hydrothiolation of an electron deficient ene.

In the first step the phosphine (DMPP, a weak base but strong nucleophile) undergoes a phospha-Michael addition reaction with the electron deficient ene to give the intermediate zwitterionic species. This carbanion abstracts a proton from RS-H generating the thiolate anion and initiating the chain reaction. So in this instance, a strong nucleophile is used to generate a strong base vs a weak base to generate a strong base as above. It is important to note, however, that the cationic phosphonium species remains as an impurity and as such phosphine levels should be kept as low as possible. Fortuitously, the phosphine-catalyzed reaction is, in general, orders of magnitude faster and more efficient than the base-mediated process even at extremely low levels of phosphine [43].

Both variants of the thiol-ene reaction have attracted significant attention in the synthetic polymer community in conjunction with RDRP chemistries and mainly as a means of modifying side or end groups or, in some instances, building more complex materials. As noted previously, the ATRP and RAFT RDRP techniques have attracted the most attention due to their high functional group tolerances and the presence of in-built functionality that facilitates the exploitation of various thiol chemistries.

4.2.1.1 Side chain modification of pre-preformed (co)polymers

The thiol-ene modification of side chains requires the presence of thiols and or enes on theses side chain groups. Both of these present particular challenges. The direct (co)polymerization of thiol-containing monomers by RAFT or ATRP is not possible due to the chain transfer properties associated with thiols and in the case of RAFT, the possibility of transthioesterification with the RAFT CTA. As such, masked thiol monomers must be employed if the goal is to introduce free thiol functionality into the (co)polymer chain. While there are in principle numerous ways this can be accomplished we will highlight a couple of the more interesting examples.

Pauloehrl et al. [73] reported the room temperature, photo-triggered formation of pendent thiol groups in a novel methacrylic substrate initially polymerized by ATRP, Scheme 4.7.

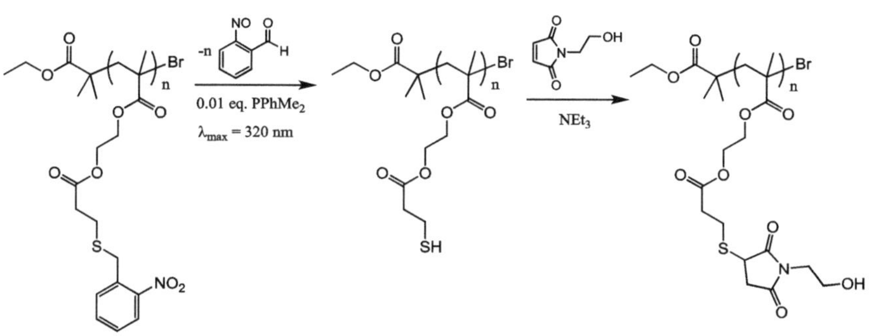

Scheme 4.7: Thiol-Michael coupling of 1-(2-hydroxyethyl)-1*H*-pyrrole-2,5-dione to pendent thiol-groups obtained via photo-sensitive protected precursor homopolymer prepared by ATRP. Reproduced with permission from T. Pauloehrl, G. Delaittre, M. Bastmeyer and C. Barner-Kowollik, Polymer Chemistry, 2012, 3, 7, 1740. ©2012, RSC [73].

The thiol-protected monomer, 2-((3-((2-nitrobenzyl)thio)propanoyl)oxy)ethyl methacrylate was homopolymerized by ATRP under standard conditions utilizing Cu(I)Br with *N,N,N′,N″,N″*-pentamethyldiethylenetriamine (PMDETA) and ethyl 2-bromoisobutyrate in anisole at 80 °C. The pendent protecting groups were then removed by photoirradiation at 320 nm in the presence of DMPP (as a reductant) to give the free-thiol containing homopolymer (nb – deprotection can be readily followed by UV analysis since the by-product, *o*-nitrosobenzaldehyde, exhibits a strong absorbance at ca. 350 nm, and the authors demonstrated quantitative cleavage after ca. 16 h irradiation). With the free thiol now available, it is, in principle, possible to perform any of the well-established thiol-x chemistries. However, given the challenges associated with quantitatively modifying pendent groups on a polymer backbone the authors utilized one of the most efficient thiol-x chemistries namely an organobase-mediated thiol-Michael reaction with the functional maleimide, 1-(2-hydroxyethyl)-1*H*-pyrrole-2,5-dione. While not-quantitative, the authors reported >90% successful modification of the pendent thiol groups although this required 60 h irradiation. The same group extended this approach to (meth)acrylamido-based copolymers prepared by RAFT copolymerization of the novel monomers *N*-(2-((2-nitrobenzyl)thio)ethyl acrylamide and the corresponding methacrylamide [74]. In this instance, sequential end-group and side group modification was accomplished via thiol-Michael reactions with *N*-benzylmaleimide with the former effected by an initial aminolysis of the thiocarbonylthio functionality with 2-aminoethanol in MeCN in the presence of tributylphosphine followed by thiol-Michael conjugation.

A complementary approach for introducing pendent thiol-functional groups has been reported by Espeel and coworkers and is based on the use of thiolactone-containing monomers [75]. For example, the trithiocarbonate-mediated RAFT copolymerization of styrene with *N*-(2-oxotetrahydrothiophen-3-yl)-4-vinylbenzenesulfonamide in dioxane yields the well-defined copolymers shown in Scheme 4.8.

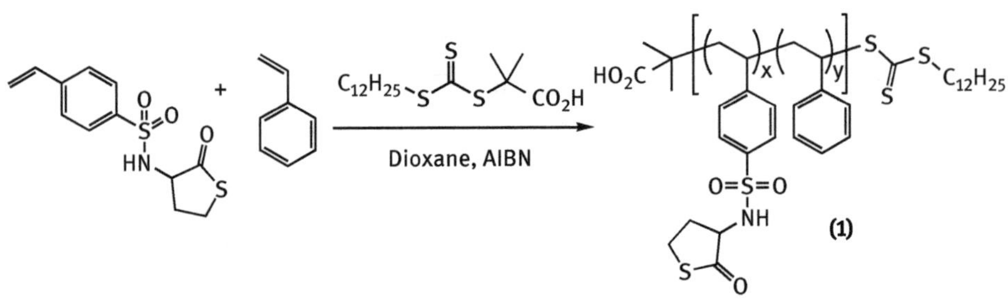

Scheme 4.8: RAFT-prepared styrenic-based copolymers containing pendent thiolactone functional groups. Reproduced with permission from P. Espeel, F. Goethals, M.M. Stamenović, L. Petton and F.E. Du Prez, Polymer Chemistry, 2012, 3, 4, 1007. ©2012, RSC [75].

Key to success of this approach is the facile reaction of thiolactones with primary and secondary amines in a ring-opening reaction that liberates a free pendent thiol. In contrast to the sequential end-group/side-group modification noted above, treatment of copolymers such as **(1)** in Scheme 4.8 with a primary amine, R-NH_2, results in *simultaneous*, thiocarbonylthio aminolysis and ring-opening of the thiolactone yielding free thiols both pendent and at the ω-chain end. The authors demonstrated that both are available for thiol-Michael type reactions with *N*-benzylmaleimide, Scheme 4.9, and reported near-quantitative modification. The same group have extended this approach to copolymers of the thiolactone monomer with methyl methacrylate as well as the preparation of statistical copolymers of poly(*N*-isopropylacrylamide) with the thiolactone-containing monomer *N*-(2-oxotetrahydrothiophen-3-yl)acrylamide [76].

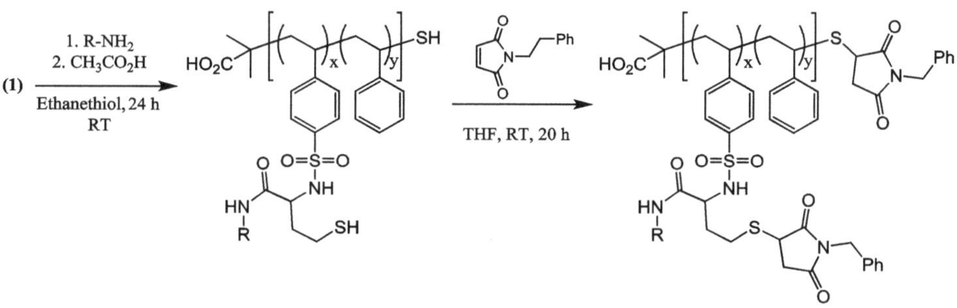

Scheme 4.9: Simultaneous end-group and side-group thiol-Michael modification of RAFT-prepared copolymers containing thiolactone functional groups.
(Source: P. Espeel, F. Goethals, M.M. Stamenović, L. Petton and F.E. Du Prez, Polymer Chemistry, 2012, 3, 4, 1007) [75].

The reverse approach, in which (co)polymers with pendent ene functionality are reacted with small molecule thiols, is likewise a viable approach for the synthesis of thioether-containing copolymers and has been more widely employed than the preparation of thiol-containing (co)polymers as described above. The direct (co)polymerization of monomers with ene groups as pendent functionality clearly presents its own problems since few, if any, such groups are completely inert under standard RDRP conditions. As such, the most common approach involves the preparation of precursor (co)polymers that can be readily modified to introduce ene functional groups susceptible to thiol-ene and or thiol-Michael chemistries. However, careful control of reagent stoichiometry can facilitate the direct incorporation of suitable ene functionality. For example, Campos et al. [77] described the direct synthesis of well-defined ene-containing (co)polymers utilizing RAFT, ATRP, and the Sn-mediated ring-opening polymerization of caprolactones, Scheme 4.10, employing appropriate ene-containing monomers.

Provided the ene-functional monomer was incorporated at no more than 10–17 mol% well defined copolymers were obtained with available pendent ene groups. The authors evaluated a series of thiols, shown in Scheme 4.10, under both thermal and photoirradiated conditions employing a 5 molar excess of thiol relative to ene groups and were able to obtain the thioether products in generally excellent yields although it is noted that some examples of the thermally initiated systems resulted in less than quantitative modification.

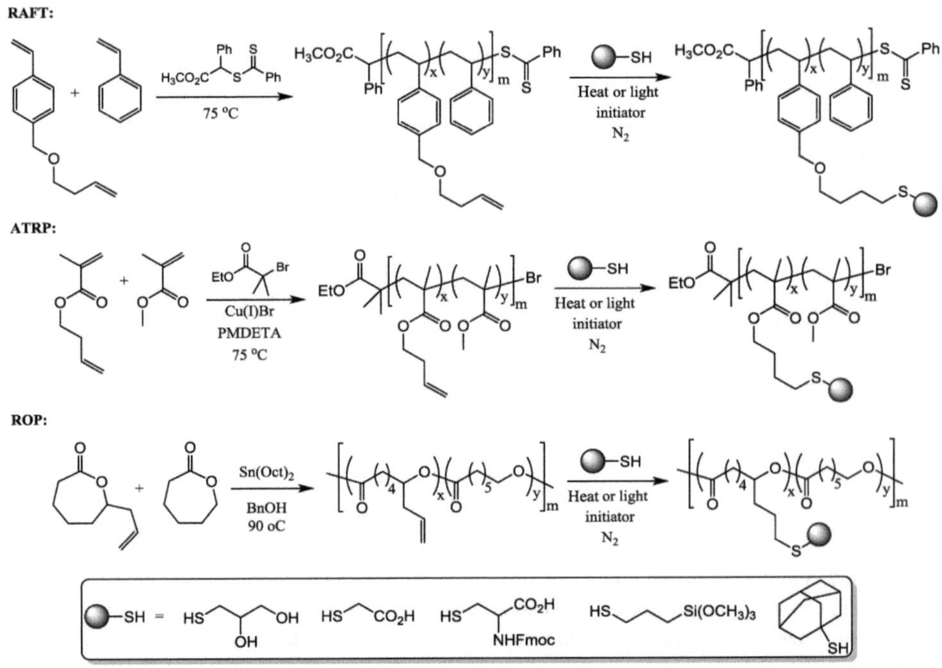

Scheme 4.10: Side chain modification of pendent ene groups in copolymers prepared by RAFT, ATRP and ring-opening polymerization (ROP).
(Source: L.M. Campos, K.L. Killops, R. Sakai, J.M.J. Paulusse, D. Damiron, E. Drockenmuller, B.W. Messmore and C.J. Hawker, Macromolecules, 2008, 41, 19, 7063) [77].

The limiting factor in the above example is clearly the relatively low incorporation of the ene monomer in the parent copolymers. While this may not preclude its use in various applications, clearly if the goal is to prepare (co)polymers with high thioether contents then alternative syntheses need to be considered. As noted, the more common approach involves the preparation of well-defined parent (co)polymers containing suitable functional groups for the facile introduction of ene species that can subsequently be employed in thiol-based reactions. There are, fortunately, many functional groups that are tolerated by RDRP processes that facilitate the incorporation of ene groups. An example of this approach was reported by Singha et al. for the preparation of polymer-protein conjugates utilizing ATRP [78]. Key to this particular approach was the use of pentafluorophenyl methacrylate (PFPMA). PFPMA, and other pentafluorophenylesters, are extremely versatile species that are susceptible to rapid and facile reaction with primary and secondary amines in nucleophilic acyl substitution reactions yielding the corresponding amide species [15]. Importantly, such reactions are commonly quantitative even in homopolymers, let alone copolymers, containing such functionality. The ATRP homopolymerization of PFPMA yielded species with targeted low molecular weights (ca. 7-8K) with corresponding

Đ's ≤ 1.30. Subsequent reaction with allylamine yielded, formally, poly(allyl methacrylamide) with NMR analysis indicating ca. 82% side group modification. The allylic homopolymers were then utilized as reactive substrates in the subsequent radical-mediated thiol-ene coupling with the peptide CVPGVG, designed to have a single cysteine residue. After heating, a water-soluble material was isolated in which ca. 50% of the allylic groups had been consumed – an apparent limiting value since all attempts to increase the degree of modification were unsuccessful. However, we should reiterate that quantitative modification of side groups, especially at such a high molar incorporation is extremely difficult and is compounded further in the radical thiol-ene modification of pendent ene groups since it has been shown that in addition to the desirable hydrothiolation product a high density of ene groups can result in undesirable radical-mediated cyclization [38].

Several additionally examples are worth mentioning since they give an indication of the wide range of chemistries that can be employed to obtain ene containing copolymers, although we note that these following examples do not represent all the possible routes to such materials via RDRP, or other polymerization, techniques [38, 39].

Chen and coworkers [79] employed this general approach for the preparation of novel glycopolymers with thermoresponsive properties. Precursor AB diblock copolymers of di(ethylene glycol)methyl ether methacrylate (DEGMA) with 2-hydroxyethyl methacrylate (HEMA) were prepared by cumyl dithiobenzoate-mediated RAFT copolymerization in DMAc with AIBN at 70 °C. The pendent -OH groups in the polyHEMA residues were subsequently esterified with 4-pentenoic anhydride to give the corresponding ene-functional block copolymer. This esterification step was judged to be quantitative by NMR spectroscopy. Finally, the ene groups were subjected to a photo initiated radical thiol-ene reaction with glucothiose in the presence of DMPA as the photoinitiator. This final step was also judged to be quantitative by NMR and FTIR analyses.

Flores and coworkers [80] detailed the side group modification of trithiocarbonate-mediated, RAFT-synthesized AB diblock copolymers of *N,N*-dimethylacrylamide (DMA) and *N*-(2-hydroxyethyl)acrylamide (HEAm) via sequential alcohol-isocyanate and hydrothiolation reactions, Scheme 4.11.

The parent poly(DMA_{96}-*block*-$HEAm_{115}$) copolymer ($\bar{M}_n^{NMR}$ = 19,000, $Đ$ = 1.11) was first reacted with two different ene-functional isocyanates (2-isocyanatoethyl acrylate and allylisocyanate) in the presence of dibutyltin dilaurate. This yielded the corresponding acrylate and allylic functional derivatives in a quantitative fashion as judged by NMR spectroscopy. Interestingly, this particular approach facilitates access to reactive substrates susceptible to both thiol-Michael and radical thiol-ene hydrothiolations. The acrylic functional block copolymer was reacted with a slight molar excess of a series of model thiols (1-propanethiol, 2-methyl-2-propanethiol, benzylmercaptan, thiophenol, 2-mercaptoethanol, thioglycerol, 3-mercaptopropionic acid, mercaptosuccinnic acid, cysteamine hydrochloride, and L-cysteine hydrochloride monohydrate) in DMSO in the presence of NEt_3 at room

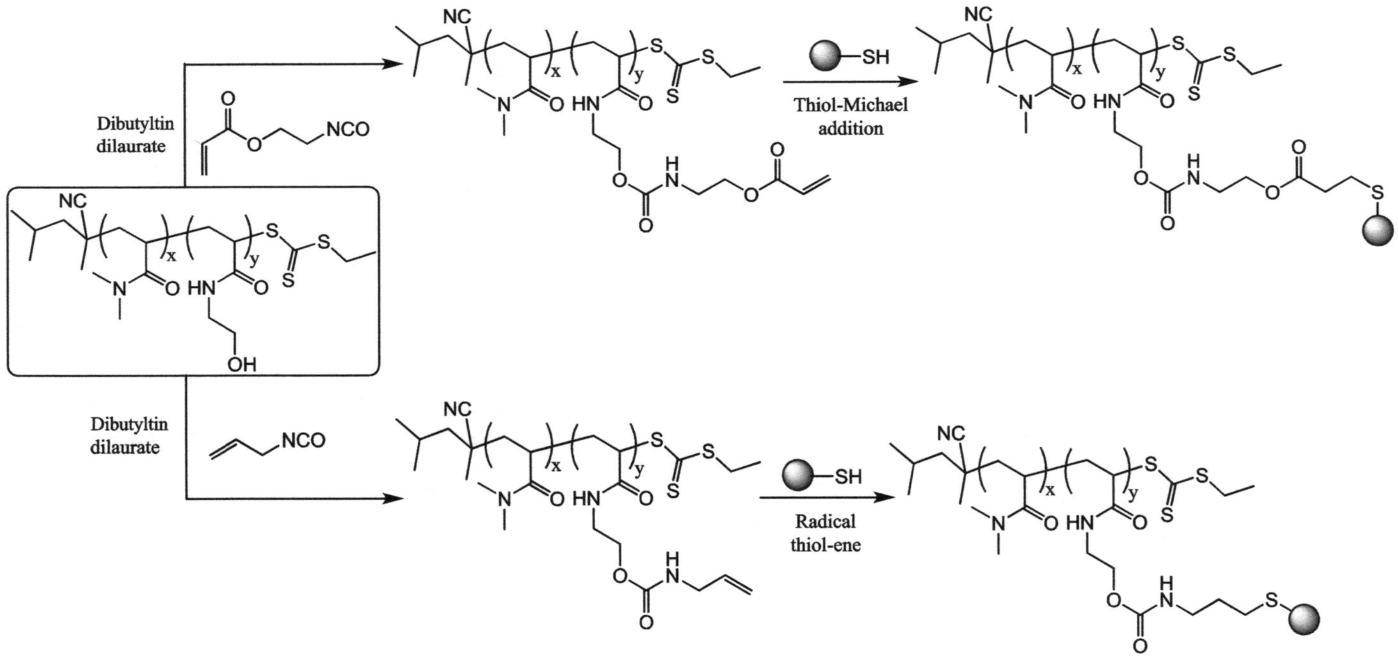

Scheme 4.11: Side chain modification of an alcohol functional AB diblock copolymer via sequential alcohol-isocyanate and hydrothiolation reactions. (Source: J.D. Flores, N.J. Treat, A.W. York and C.L. McCormick, Polymer Chemistry, 2011, 2, 9, 1976) [80].

temperature. In general, and provided reaction conditions were adjusted as necessary, essentially quantitative formation of the Michael adducts was observed as judged by NMR and FTIR spectroscopies without any observable effect on the molecular weight distribution of the parent block copolymer. In contrast, the thermally initiated radical thiol-ene reaction of the allylic functional block copolymer proved to be more sensitive to thiol structure. Where the reaction proceeded, high-to-quantitative yields were observed. However, and in contrast to the conjugate addition reactions, both thiophenol and L-cysteine hydrochloride monohydrate failed to react. While the radical-mediated reaction proved to be less versatile than the thiol-Michael process this report did nicely demonstrate that a broad range of different thiols could be readily employed in the modification reactions and clearly could be readily extended beyond the model thiol reagents noted here.

Finally, we highlight another multi-step modification route as described by Zou et al. [81] RAFT-prepared poly(glycidyl methacrylate) with an $\bar{X}_n$ of 140 and $Đ = 1.19$ was first treated with NaN_3 in a ring-opening reaction of the epoxy pendent side groups yielding a polymeric azido-alcohol. This was followed by a Cu-catalyzed alkyne azido "click" reaction with an α-ene, ω-yne functional small molecule ethylene glycol derivative to give a polymer in which the newly introduced pendent ene functional group is joined via a triazole, Scheme 4.12. The side chain ene groups were then modified employing a 10-fold excess of small molecule thiols, under UV, containing a range of functionality including -OH, $-CO_2H$, $-NH_2$, and sugar species. In all instances, conversions > 99% were reported.

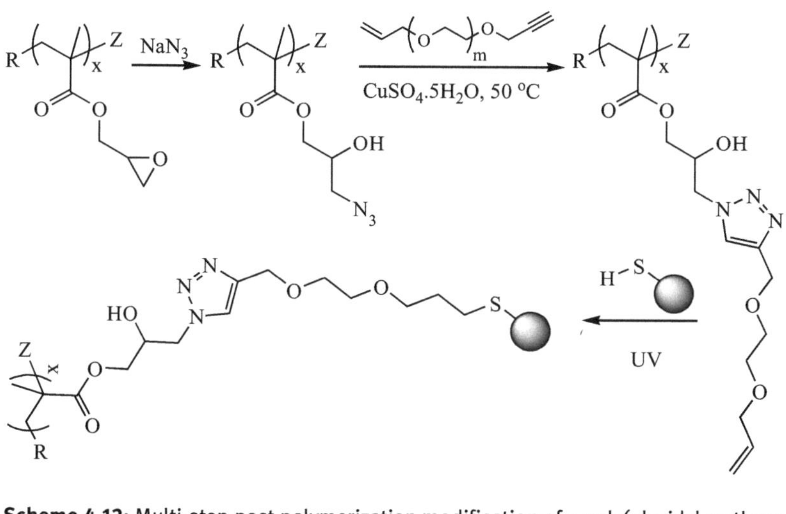

Scheme 4.12: Multi-step post polymerization modification of a poly(glycidyl methacrylate) parent homopolymer to give an ene side chain functional species susceptible to radical thiol-ene reactions. (Source: L. Zou, W. Zhu, Y. Chen and F. Xi, Polymer, 2013, 54, 2, 481) [81].

4.2.1.2 End-group modification of pre-preformed (co)polymers

End group modification of (co)polymers prepared by RDRP has become a rather facile and straightforward process for controlling polymer properties. The wide spread adoption of end group modification has been facilitated by the fact that (co)polymers prepared by ATRP and RAFT in particular by virtue of the polymerization mechanisms, contain inherent ω-end-group functionality that is readily manipulated allowing for straightforward modification as well as facilitating the incorporation of specific α-end groups via the use of appropriate functional initiators. The latter approach does, of course, require that any instilled functionality in the initiator fragment be compatible with the polymerization process.

In the context of thiol-based chemistries, the use of RAFT-prepared (co)polymers affords the most convenient route for accessing and utilizing the suite of thiol-x chemistries [82]. This is simply due to the fact that the thiocarbonylthio functional groups present at the ω-terminus of RAFT synthesized (co)polymers can be viewed as protecting groups for the corresponding macromolecular thiol. Indeed, there are a number of straightforward processes for cleaving such groups to yield, predominantly, the polymeric mercaptan [83]. The formation of a macromolecular thiol from a RAFT-prepared (co)polymer is commonly achieved via reaction with an appropriate nucleophile. The most convenient reagents to effect thiocarbonylthio cleavage are 1° or 2° amines with the former being the reagent of choice. Examples of amines used include ethylamine, piperidine, hexylamine, propylamine and butylamine. These reagents may, or may not be used in conjunction with anti-oxidants, reducing agents, or trapping agents, *vide infra*. In addition to amines, mild reducing agents such as $NaBH_4$ can be used to cleave RAFT end-groups to the corresponding thiol [84, 85]. It is perhaps also worth noting that such reactions with nucleophiles are also generally rapid with quantitative cleavage often occurring with secs-to-minutes (this is also easily observed since the end-group cleavage is associated with a colour change from the typically highly colored solutions associated with RAFT polymers to a clear solution after cleavage).

Once the RAFT end groups have been cleaved to give the corresponding polymeric thiol then, in principle, the (co)polymer can be used as a reagent in thiol-ene and thiol-Michael reactions. For example, Yu et al. [86] reported a sequential thiol-Michael/thiol-ene approach for the preparation of end functional homopolymers of poly(*N*-isopropylacrylamide) (PNIPAM), Scheme 4.13.

The parent PNIPAM homopolymer was treated with octylamine (the thiocarbonylthio end group cleavage agent) in the presence of allyl methacrylate as a Michael acceptor and DMPP as catalyst. This yielded the corresponding allyl-end functional PNIPAM that was further modified via a photoinitiated radical thiol-ene reaction with 6-mercaptohexan-1-ol, hexane-1-thiol, and 3-mercaptopropyl polyhedral oligomeric silsesquioxane. In all instances these facile reactions were quantitative, or near-quantitative. Indeed, the aminolysis of RAFT-polymer end-groups

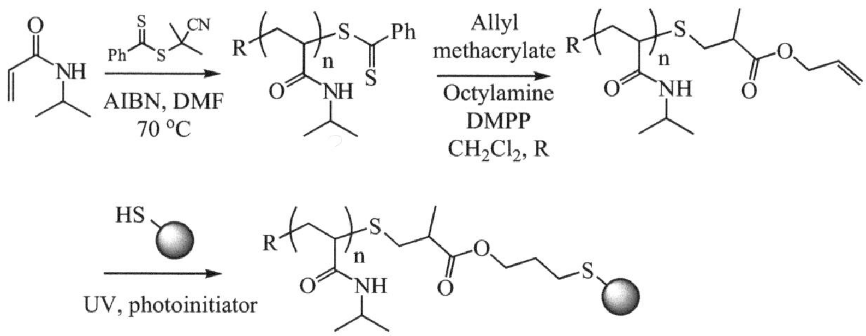

Scheme 4.13: Synthesis of end-functional poly(*N*-isopropylacrylamide) via sequential thiol-Michael and thiol-ene reactions.
(Source: B. Yu, J.W. Chan, C.E. Hoyle and A.B. Lowe, Journal of Polymer Science Part A: Polymer Chemistry, 2009, 47, 14, 3544) [86].

followed by thiol-Michael addition has become a rather ubiquitous approach for preparing specific end functionalized materials or, in a more general sense, for the trapping of the macromolecular thiol under conditions that result in end-group cleavage [19, 21, 27, 87–95].

In addition to end-functionalization, exactly the same chemistry can be utilized to build materials with more complex architectures. For example, Chan et al. [96, 97] detailed the convergent synthesis of 3-arm star polymers from homopolymers of RAFT-prepared poly(*N*,*N*-diethylacrylamide) (PDEAM) via an in situ aminolysis followed by a thiol-Michael reaction with the trifunctional acrylate trimethyloylpropane triacrylate. Such reactions were demonstrated to be highly efficient and the star structure was confirmed using a combination of ^{1}H NMR spectroscopy and matrix-assisted laser desorption/ionization time of flight mass spectrometry (MALDI-TOF MS).

Similar ω-end-group transformations can be accomplished with ATRP-prepared (co)polymers. For example, Uygun, Tasdelen, and Yagci [98] reported the polymerization of styrene under standard ATRP conditions to give low molecular weight homopolymers with narrow molecular weight distribution. The terminal –Br group was converted to a thiol *via* treatment with thiourea followed by NaOH and acidification. The thiol-terminated polystyrene was then reacted with three different enes, and specifically methyl acrylate, methyl methacrylate and allyl bromide under photoinitiated or thermal (with AIBN) conditions. In all instances the authors reported that the end group modifications proceeded cleanly and with high efficiency, Scheme 4.14.

While ω-end-group modifications are relatively common it is, of course, also possible to modify the end-groups at the α-chain end. This is most readily achieved via the use of appropriate, that is, functional initiators. This approach does however require that the ene functional group in the initiator molecule be, ideally, essentially inert under the polymerization conditions. In the example noted above from Uygun, Tasdelen, and Yagci the authors also reported, in the same publication, that α-modification could be achieved *via* the ATRP homopolymerization of styrene with

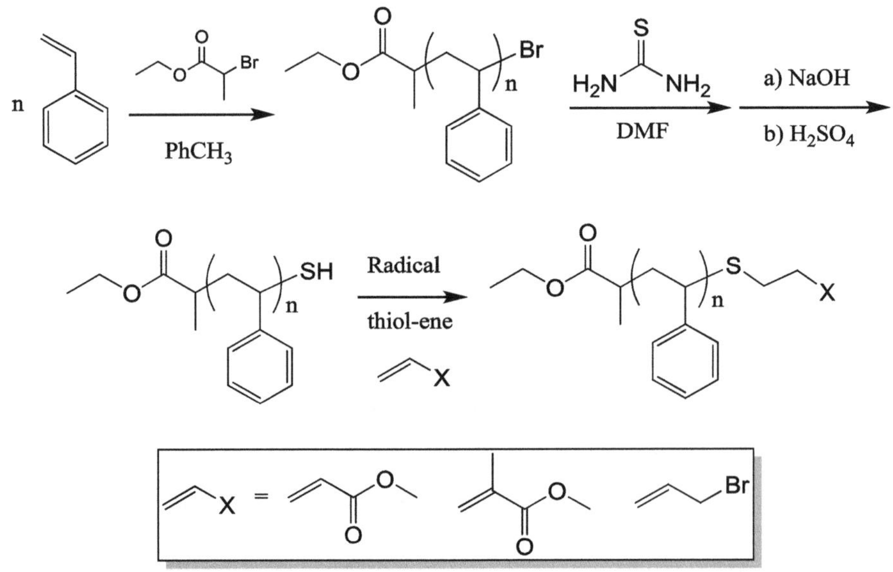

Scheme 4.14: ATRP synthesis of polystyrene, end-group transformation and radical thiol-ene modification.
(Source: M. Uygun, M.A. Tasdelen and Y. Yagci, Macromolecular Chemistry and Physics, 2010, 211, 1, 103) [98].

allylbromide followed by a radical thiol-ene reaction with 3-mercaptopropionic acid. This complementary approach was also reported to be quantitative.

Du Prez et al. [99] described the synthesis of three different norbornenyl R-group functional RAFT CTAs and their subsequent use for the polymerization of styrene, methyl acrylate, 1-ethoxyethyl acrylate, and vinyl acetate. Provided conversions were kept low (ca. < 50%) then the norbornenyl end groups remained intact and were available for subsequent reaction. For example, Scheme 4.15 shows the adopted approach in the context of a styrene homopolymer.

After homopolymerization the α-ene groups were reacted with 3-mercaptopropionic acid, 1-dodecanethiol or benzyl mercaptan in THF under photoinitiated conditions with DMPA. Due to the very high reactivity of norbornenyl functionality in radical thiol-ene reactions [72] near quantitative conversions to the thioether were observed within 15 min. The authors also noted it was possible to substitute the norbornenyl functional group for an allylic species.

4.2.2 The thiol-yne reaction

The thiol-yne reaction is intimately related to the thiol-ene reaction. Just like the thiol-ene process, the thiol-yne reaction can proceed *via* several different mechanistic

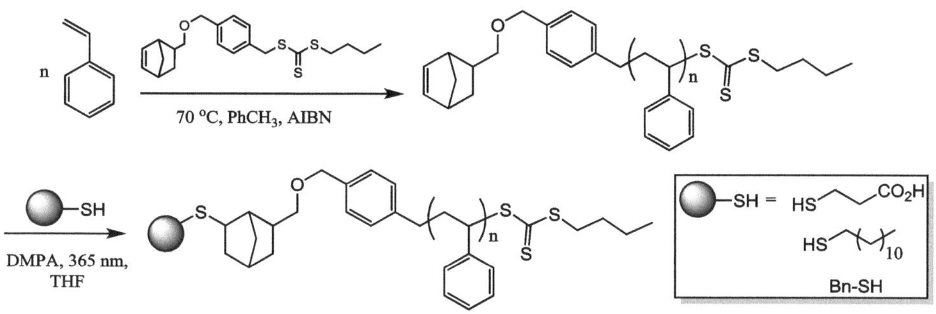

Scheme 4.15: RAFT homopolymerization of styrene with a norbornenyl-functional RAFT CTA and subsequent α-end group modification *via* a radical thiol-ene reaction.
(Source: M.M. Stamenović, P. Espeel, W.V. Camp and F.E. Du Prez, Macromolecules, 2011, 44, 14, 5619) [99].

pathways [44]. However, the ability of an alkyne group to undergo two, sequential, additions (especially in the case of terminal species) results in the possibility of preparing bisthioethers although the nature of the product (mono vs bis addition product) is dictated largely by the hydrothiolation reaction conditions. Readers are referred to the above noted reference for a more detailed discussion of the various possible products and the use of the thiol-yne reaction in a broader sense. Most commonly, however, the thiol-yne reaction is performed under radical mediated conditions, similar if not identical to those used for radical thiol-ene additions and this results in the formation of the 1,2-bisthioether adducts. The generally accepted mechanism for the radical-mediated thiol-yne process is shown in Scheme 4.16.

Thiyl radicals are first generated by any of the common approaches including thermally and via UV irradiation in the presence, or absence, or added photoinitiator. Once generated, the thiyl radical will add to an alkyne, step 1, in an anti-Markovnikov fashion, yielding the intermediate carbon-centered vinyl radical. This highly reactive species will undergo a chain transfer reaction with additional thiol, step 2, to yield an intermediate vinylthioether (A) with concurrent regeneration of a thiyl radical. The intermediate vinylthioether is *extremely* reactive toward thiyl radicals and will react in a formal thiol-ene reaction, step 3, to give another carbon-centered radical. This species undergoes a second chain transfer reaction with RS-H, step 4, to give the desired 1,2-bisthioether product, B, and additional thiyl radical. It is also worth noting that the double addition of thiols to a terminal alkyne, to give product B, also results in the generation of a chiral center. While this is generally ignored in polymer synthesis and modification it does, of course, have implications in small molecule syntheses.

The more recent interest in the thiol-yne process was inspired by research associated with the thiol-ene process for the preparation of ideal polymeric networks and was "pioneered" by the groups of Bowman, Hoyle and Lowe [45, 47–49, 100]. Recognizing that the general features and characteristics of the radical thiol-yne process also

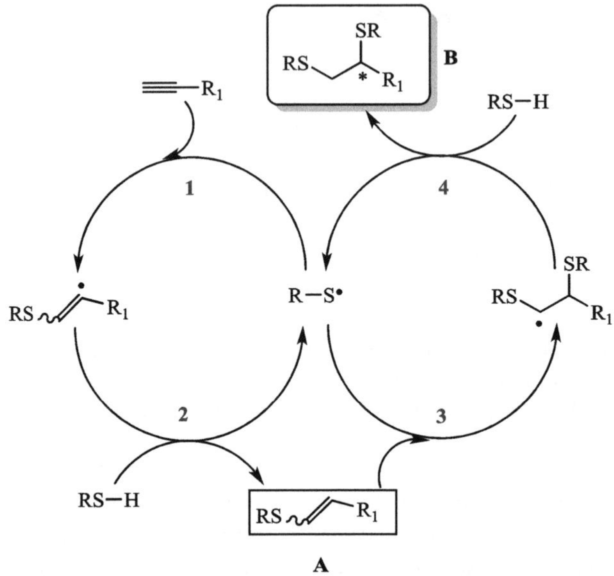

Scheme 4.16: The generally accepted, radical-mediated, thiol-yne mechanism. Reproduced with the permission from A.B. Lowe, Polymer, 2014, 55, 22, 5517. ©Elsevier, 2014 [44].

largely met the criteria to be classified as a 'click' reaction the chemistry was subsequently adapted to other chemical transformations and processes [44, 46].

Given its similarity to the thiol-ene reaction it is therefore not surprising that the thiol-yne reaction has also been examined in conjunction with RDRP processes although it has not been as widely adopted as its sister reaction.

As highlighted in Scheme 4.13, Yu et al. [86] reported end functional, RAFT-prepared, PNIPAMs *via* sequential thiol-Michael/thiol-ene reactions based on allyl methacrylate as the Michael acceptor. The group also noted that propargyl acrylate also serves as a convenient substrate for such thiol-Michael reactions and gives yne-end functional PNIPAM that can likewise be reacted with small molecule thiols to give the bisthioethers adducts.

Huynh and coworkers [101] prepared block and statistical copolymers, *via* RAFT, comprised of 2-hydroxyethyl methacrylate (HEMA) with poly[oligo(ethylene glycol) methyl ether methacrylate] (POEGMA) with the general approach shown in Scheme 4.17.

With the copolymers in hand, the pendent -OH groups, associated with the HEMA repeat units, were esterified with4-oxo-4-(prop-2-yn-1-yloxy)butanoic anhydride to yield the corresponding yne-functional copolymers. Photoinitiated radical thiol-yne reaction with 2-mercaptosuccininc acid gave the high density carboxylic acid functional material. Such highly functional copolymers were then conjugated with cisplatin (*cis*-diamminedichloroplatinum(II)) to give novel polymeric Pt-based drugs with the statistical copolymers outperforming block

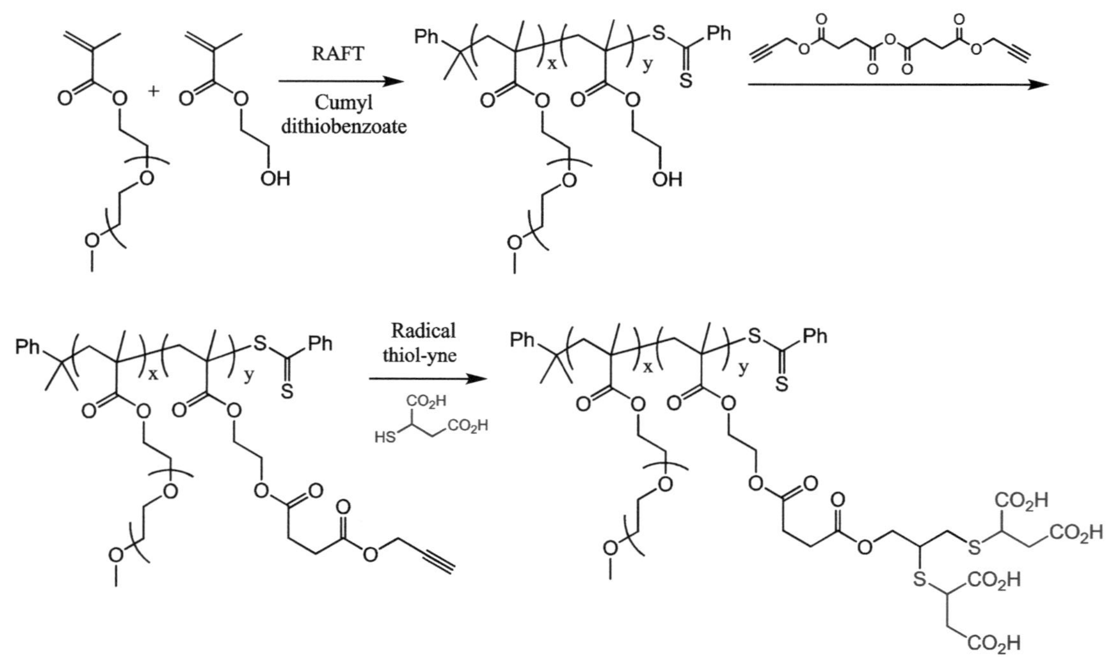

Scheme 4.17: RAFT synthesis of yne side chain functional (co)polymers and thiol-yne modification with 2-mercaptosuccinic acid.
(Source: V.T. Huynh, G. Chen, P.H. de Souza, Biomacromolecules, 2011, 12, 5, 1738) [101].

copolymers with respect to Pt drug release and hence cytotoxicity against the A549 lung cancer cell line.

The same group detailed the same general approach for the synthesis of novel high density glycopolymers with the sugar functionality instilled as pendent groups or as chain end species [102]. Again in a manner similar to the thiol-ene process, the thiol-yne reaction can also be employed to prepare (co)polymers of more advanced architecture from linear precursors. For example, Cai and co-workers [103] reported the ATRP synthesis of thiol-terminated PNIPAMs via the use of the difunctional initiator disulfanediylbis(ethane-2,1-diyl) bis(2-bromo-2-methylpropanoate) followed by disulfide reduction with dithiothreitol, Scheme 4.18. This approach yielded PNIPAM-SH's with narrow molecular weight distributions and low (ca. 3.5K) to medium (ca. 15K) $\bar{M}_n$'s as determined by SEC. The PNIPAM-SH homopolymers were subsequently conjugated to alkyne terminated poly(ε-caprolactone) in a UV radical mediated thiol-yne reaction yielding AB_2, Y-shaped, materials. The block copolymers were subsequently cast by phase inversion in an aqueous medium to give microporous membranes with uniform sizes.

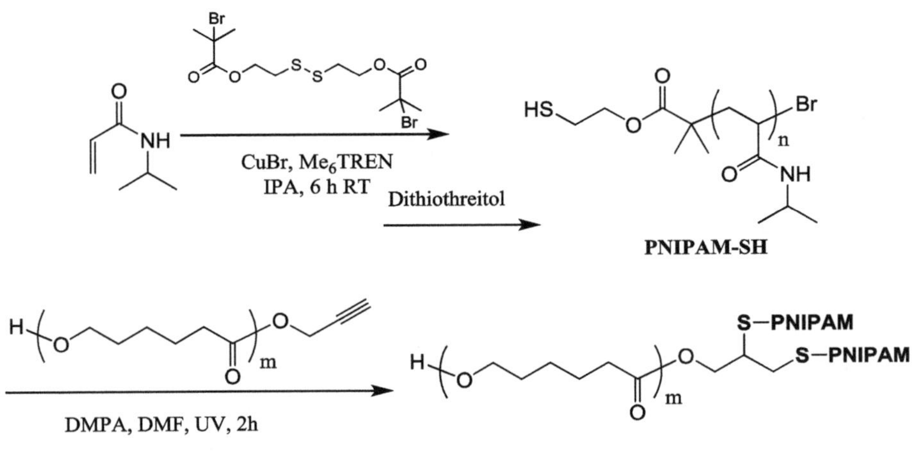

Scheme 4.18: Synthesis of AB_2 block copolymers of poly(ε-caprolactone) with poly (*N*-isopropylacrylamide) via polymer-polymer coupling thiol-yne conjugation.
(Source: T. Cai, M. Li, K.-G. Neoh and E.-T. Kang, Journal of Materials Chemistry, 2012, 22, 32, 16248) [103].

4.2.3 The thiol-isocyanate reaction

The thiol-isocyanate reaction is an example of a non-Aldol carbonyl reaction, and in general refers to the base-mediated reaction of a thiol with isocyanate functionality, Scheme 4.19, to give an *S*-thiocarbamate [40]. As with all the thiol-click reactions this results in the formation of a new C-S bond.

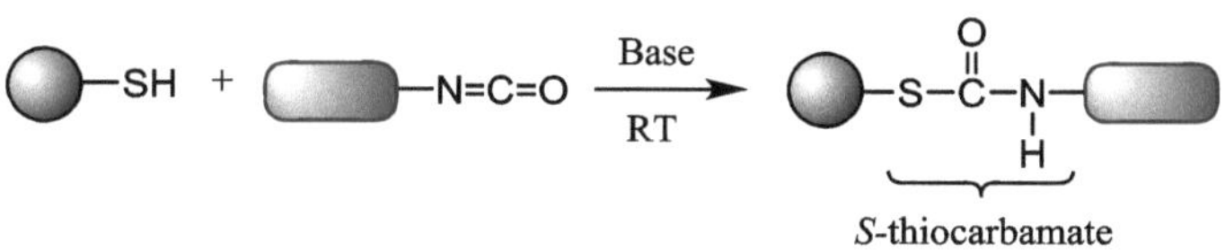

Scheme 4.19: General synthesis of an *S*-thiocarbamate from a thiol and isocyanate: t\e thiol-isocyanate 'click' reaction.
(Source: C.E. Hoyle, A.B. Lowe and C.N. Bowman, Chemical Society Reviews, 2010, 39, 4, 1355) [40].

These reactions are generally extremely rapid, with quantitative reaction possible within a few seconds in certain cases, giving extremely pure products, that is, they are generally not susceptible to the side reactions commonly observed in alcohol-isocyanate reactions. Even though such reactions are, arguably, some of the fastest and most efficient in the thiol-click toolbox there are still relatively few examples in which this reaction has been specifically combined with RDRP.

Li and co-workers [53] reported the use of the thiol-isocyanate reaction as a means of preparing a small library of ω-end functionalized homopolymers based on a RAFT-prepared poly(*N,N*-diethylacrylamide) parent species, Scheme 4.20.

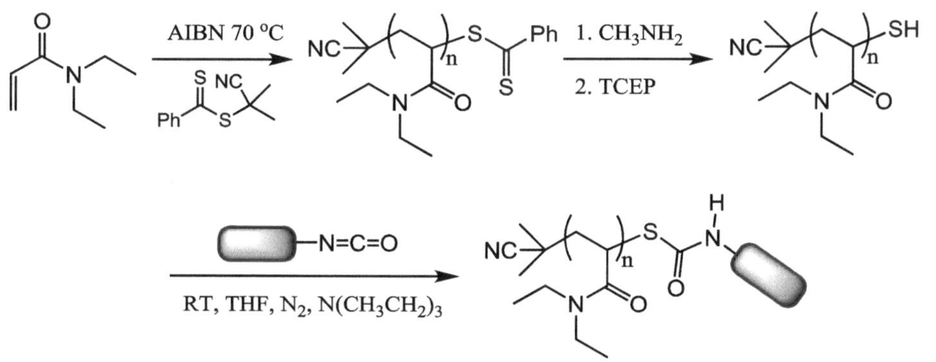

Scheme 4.20: Synthesis of ω-functional poly(*N,N*-diethylacrylamide) *via* a combination of end-group aminolysis and base-mediated thiol-isocyanate coupling.
(Source: H. Li, B. Yu, H. Matsushima, C.E. Hoyle and A.B. Lowe, Macromolecules, 2009, 42, 17, 6537) [53].

N,N-Diethylacrylamide (DEA) was initially homopolymerized under bulk conditions with AIBN as the source of primary radicals and 1-cyano-1-methylethyl dithiobenzoate (CPDB) as the RAFT chain transfer agent. Halting the polymerization at ca. 90% yielded a well-defined polyDEA (PDEA) homopolymer with a calculated average degree of polymerization ($\bar{X}_n$) of 30 which corresponds to a molecular weight of 3,800 (as determined by NMR end group analysis), and a dispersity, *Đ*, of 1.10. Subsequently the thiocarbonylthiol end-group was cleaved *via* aminolysis [82, 104] with methylamine to yield the corresponding macromolecular thiol that was then treated with tris(2-carboxyethyl)phosphine (TCEP) – a water-soluble phosphine reducing agent capable of selectively reducing disulfides to thiols [105]. The free polymeric thiol was then reacted with a library of 14 small molecule isocyanates (in the presence of triethylamine) including cyclohexylmethyl isocyanate, tetradecyl isocyanate, 2-isocyanatoethyl methacrylate, adamantly isocyanate and 9-isocyanato-9*H*-fluorene. In all cases, the reactions were reported to be essentially quantitative as judged by NMR spectroscopy and via the monitoring of isocyanate consumption by FTIR spectroscopy, while retention of the well-defined nature of the modified homopolymer after isocyanate coupling was confirmed by SEC.

Flores et al. [106] detailed the direct RAFT polymerization and subsequent modification of an isocyanate-containing acrylic monomer (Scheme 4.21). In this case, 2-isocyanatoethyl acrylate was successfully homopolymerized with 1-(dimethylamino)-1-oxopropan-2-yl benzodithioate in dioxane at 50 °C yielding materials with relatively low molecular weights and acceptable *Đ*'s although both kinetic and molecular weight vs. conversion plots suggested non-ideal behavior. The authors subsequently demonstrated that the –NCO groups were, as expected, reactive towards amines, alcohols and thiols although a detailed evaluation of the reaction with thiols was not presented.

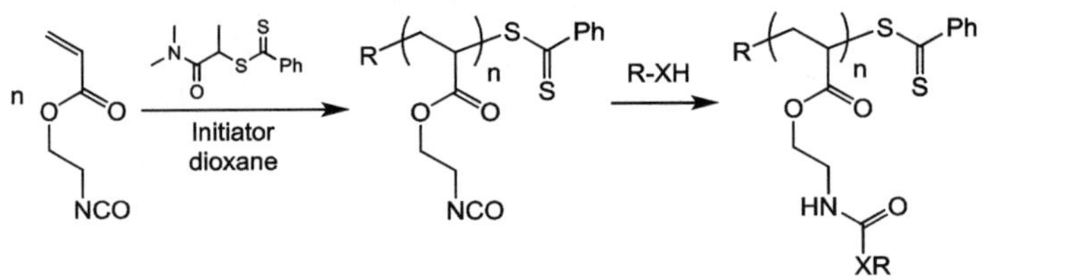

Scheme 4.21: RAFT homopolymerization of 2-isocyanatoethyl acrylate and subsequent modification with amines, thiols and alcohols roz.
(Source: J.D. Flores, J. Shin, C.E. Hoyle and C.L. McCormick, Polymer Chemistry, 2010, 1, 2, 213) [106].

Hensarling et al. [55] detailed the use of the thiol-isocyanate reaction as a means of preparing functional surfaces. An ATRP initiator was first immobilized on a SiO_2 surface. 2-Hydroxylethyl methacrylate (HEMA) was subsequently polymerized from the surface in a water/methanol mixture employing CuBr and Bpy. Esterification of the pendant OH groups with 3-(2-nitrobenzylthio)propanoic acid or 3-(2-(4-methoxyphenyl)2-oxoethylthio)propanoic acid yielded the corresponding brushes with pendant photolabile protected thiol functional groups. Free thiols were liberated by irradiation with UV light. While such pendant thiols can, in principle, be used in any of the common thiol-X reactions, the authors focussed on thiol-isocyanate chemistry and reacted the thiol-functional brushes with a selection of small molecule isocyanates including1-(isocyanatomethyl)-4-methoxybenzene, 1-isocyanato-2-nitrobenzene and 2-(isocyanatomethyl)-furan. It was demonstrated that up to 70% of the photoprotected thiol groups were available for reaction after removal of the protecting groups and that the approach lent itself to the development of sequential thiol-click reactions via the use of photomasks.

4.2.4 The thiol-halo reaction

Nucleophilic substitution (S_N) reactions were also noted as processes that might meet the criteria to be accurately defined as "click" reactions. While not all S_N reactions necessarily meet the required criteria several are generally extremely efficient, and in particular thiol-iodo/bromo reactions and nucleophilic aromatic substitution reactions based on *p*-fluoroaromatics are favoured.

Rosen et al. were the first to highlight the thiol-bromo reaction as a "click" process in their synthesis of dendrimers and dendritic macromolecules [56, 57]. Xu et al. reported the thiol-bromo reaction as a convenient method for end group modification of RAFT polymers and also for the preparation of multiblock and hyperbranched polymers [107]. Scheme 4.22 highlights an example for end group modification, while Scheme 4.23 highlights hyperbranched polymer formation.

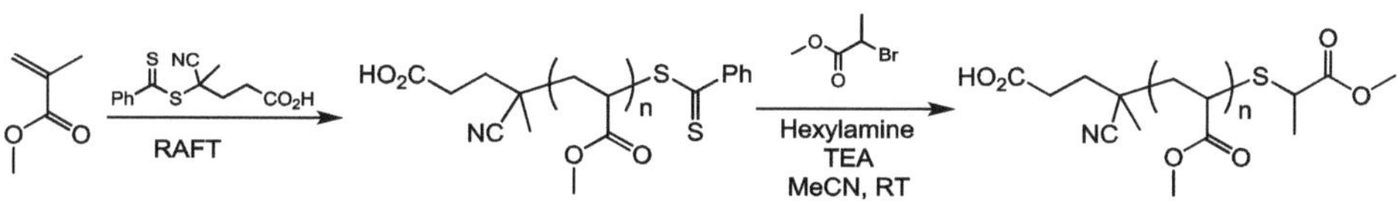

Scheme 4.22: End group modification, *via* a DS_N thiol-bromo reaction, of poly(methyl acrylate) prepared by dithiobenzoate-mediated RAFT polymerization.
(Source: J. Xu, L. Tao, C. Boyer, A.B. Lowe and T.P. Davis, Macromolecules, 2010, 43, 1, 20) [107].

Methyl acrylate was first homopolymerized by RAFT utilizing a dithiobenzoate RAFT CTA to give a low molecular weight parent homopolymer. One pot treatment of the homopolymer with hexylamine (thiocarbonylthio cleavage agent), triethylamine (base) and methyl 2-bromopropanoate resulted in end group cleavage and S_N to give the ester-end functional polymer within ca. 30 min and quantitatively. Reaction with methyl 2-bromo-2-methylpropanoate, a 3° halo species, was also successful as was replacing poly(methyl acrylate) with PNIPAM.

To give polymers with more complex architectures novel CTAs were prepared that contained the 2° and 3° bromo functional groups as part of the R-group fragment. Scheme 4.23 shows the example with the 2° bromo species, ((4-cyano-4-((phenylcarbonothioyl)thio)pentanoyl) azanediyl)bis(ethane-2,1-diyl) bis(2-bromopropanoate). RAFT homopolymerization of methyl acrylate yielded a homopolymer in which these two 2° bromo groups were part of the α-functional fragment. Treatment of the homopolymer with hexylamine, as noted for the end group modification, gives a dual reactive species able to react inter (and intra?) molecularly with similar

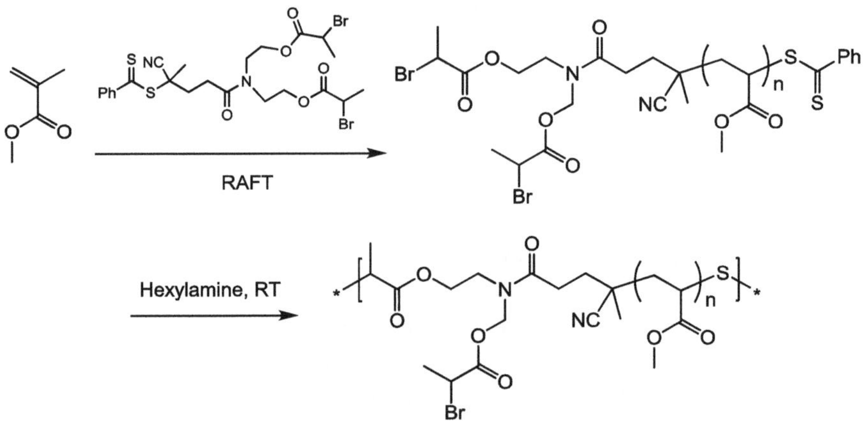

Scheme 4.23: RAFT polymerization, end group cleavage, and self-condensation via thiol-bromo S_N reactions for a methyl acrylate homopolymer.
(Source: J. Xu, L. Tao, C. Boyer, A.B. Lowe and T.P. Davis, Macromolecules, 2010, 43, 1, 20) [107].

substrates. Interestingly, in this specific example only linear multiblock species were observed. However, substituting the RAFT CTA shown in Scheme 4.23 for one containing two 3° Br groups, namely, ((4-cyano-4-((phenylcarbonothioyl)thio)pentanoyl)azanediyl)bis(ethane-2,1-diyl) bis(2-bromo-2-methylpropanoate), followed by methyl acrylate homopolymerization, end-group cleavage and in situ self-condensation resulted in the formation of hyperbranched materials.

End group modification in RAFT-prepared amphiphilic block copolymers *via* a thiol-iodo S_N reaction was detailed by Nakayama and Okano and the effect of end groups on the thermoresponsive behaviour of the materials evaluated [108]. Poly (benzyl methacrylate) (PBnMA) RAFT macro-CTAs were employed for the statistical copolymerization of NIPAM with *N,N*-dimethylacrylamide (DMA) giving the parent amphiphilic materials P[BnMA-*block*-(NIPAM-*co*-DMA], Scheme 4.24.

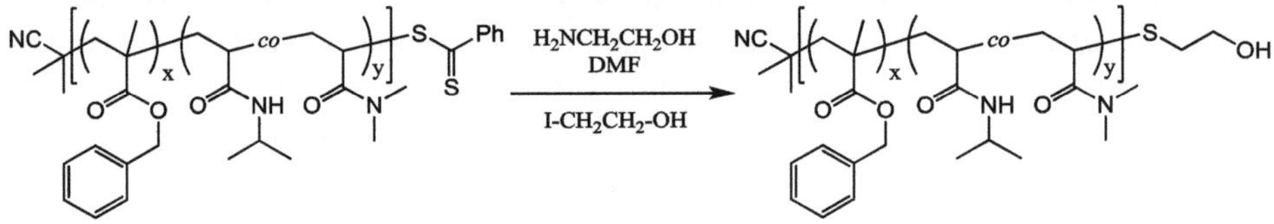

Scheme 4.24: End group modification of RAFT-synthesized amphiphilic block copolymers via in situ end group cleavage and thiol-iodo S_N modification with 2-iodoethanol.
(Source: M. Nakayama and T. Okano, Biomacromolecules, 2005, 6, 4, 2320) [108].

As a part of the solution characterization of these block copolymers, the thiocarbonylthio end-groups were transformed to alternative functionality. In one instance the copolymer was treated with ethanolamine (a thiocarbonylthio cleavage agent) in DMF in the presence of 2-iodoethanol to give the corresponding -OH terminal thioether end-group with coupling efficiencies judged to be >90% by ^{1}H NMR spectroscopy. Such block copolymers are molecularly dissolved in solvents such as d_6-DMSO but self-assemble into micellar structures in water in which the nature of the ω-end group (modified vs unmodified) were demonstrated to have a direct impact on the thermoresponsive properties.

Nucleophilic aromatic substitution reactions with pentafluorophenyl-based substrates have also emerged as a rather convenient approach for modifying preformed (co)polymers. One of the earliest examples of this type of chemistry was reported by Becer et al. for the synthesis of novel glycopolymers [59]. Pentafluorophenylstyrene (PFPS) was homopolymerized with the Bloc Builder nitroxide in THF to give a low molecular weight material with a low *Đ*. This reactive scaffold was subsequently treated with 2,3,4,6-tetra-*O*-acetyl-1-thio-β-D-glucopyranose in the presence of triethylamine to give the corresponding *para*-substituted species. Finally, the protecting acetyl groups were removed by RT treatment with sodium methoxide, Scheme 4.25. This approach was readily extended to the synthesis and modification of statistical and block copolymers with styrene.

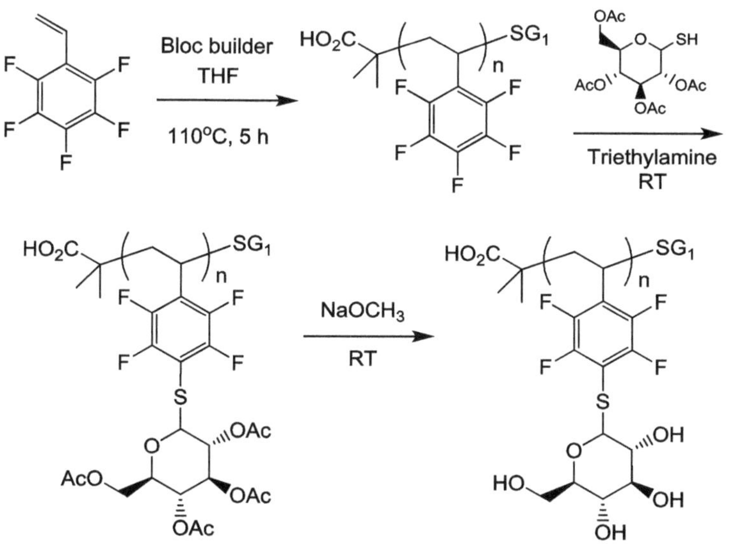

Scheme 4.25: Synthesis of well-defined glycopolymers via a combination of nitroxide-mediated (co)polymerization and nucleophilic aromatic substitution reactions.
(Source: C.R. Becer, K. Babiuch, D. Pilz, S. Hornig, T. Heinze, M. Gottschaldt and U.S. Schubert, Macromolecules, 2009, 42, 7, 2387) [59].

In more recent work, Roth et al. [109, 110] detailed the synthesis of a library of new (meth)acrylic and styrenic monomers via the Passerini reaction. This included examples of new monomers containing pentafluorophenyl functional groups, Figure 4.1.

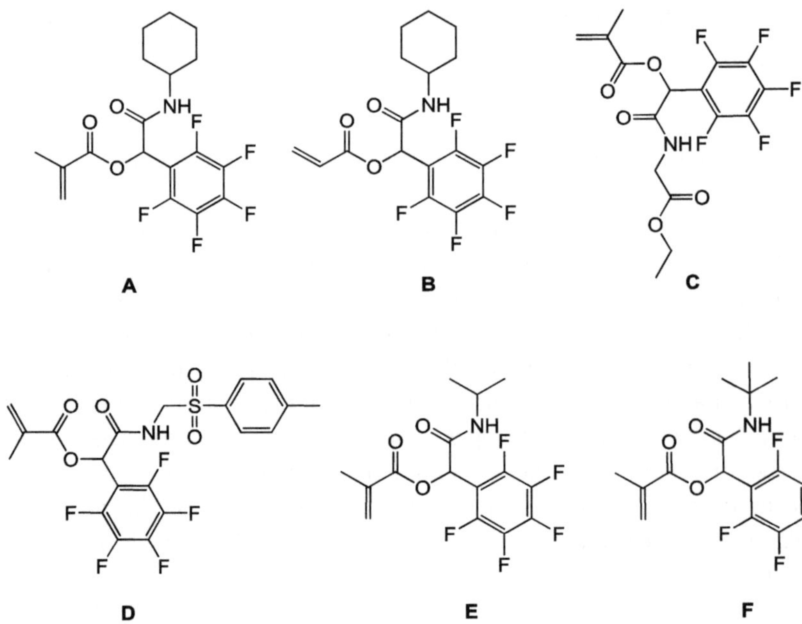

Figure 4.1: Examples of novel (meth)acrylic monomers prepared via the Passerini multicomponent process containing pentafluorophenyl functionality.
(Source: J.-M. Noy, M.Koldevitz and P.J. Roth, Polymer Chemistry, 2015, 6, 3, 436 [109].
S. Schmidt, M. Koldevitz, J.-M. Noy and P.J. Roth, Polymer Chemistry, 2015, 6, 1, 44 [110]).

All of these novel monomers could be readily homopolymerized or copolymerized with comonomers including methyl methacrylate, *tert*-butyl methacrylate, and pentafluorophenyl acrylate to give well-defined materials with low dispersities and molecular weights and compositions predetermined based on the feed ratio.

The homopolymer derived from *N*-*tert*-butyl-2-methacryloyloxy-2-(pentafluorophenyl) acetamide (“**F**” in Figure 4.1) was then used as a model substrate to evaluate the viability of conducting thiol-*para* fluoro substitution reactions. A small library of thiols were evaluated including thiophenol, 1-octanethiol, captopril, 3-mercaptopripionic acid, 1-butanethiol, 2-propanethiol and *tert*-butylthiol. Provided a suitable base (triethylamine for thiols with low pK_as and DBU for those with higher pK_as) was employed then quantitative substitution was observed within 1.5 h, and in some instances at significantly shorter reaction times. The modification approach was then extended to the statistical copolymers that had also been prepared.

The ease with which such novel Passerini monomers are prepared would suggest that this approach to new and interesting functional materials will, or should, become popular. Indeed, Pei et al. [111] recently exploited such chemistry in the modification of soft matter nanoparticles prepared by RAFT dispersion polymerization with polymerization-induced self-assembly. Macro-CTAs of 2-(dimethylamino)ethyl methacrylate (DMAEMA) with upto 5 mol% of **B, C**, or **F** (in Figure 4.1) as comonomer, were utilized for in situ formation of polymeric nanoparticles, of variable and tuneable morphology, in the dispersion copolymerization of 3-phenylpropyl methacrylate (PPMA), a convenient comonomer in such formulations in both polar and non-polar media [112–115], Scheme 4.26.

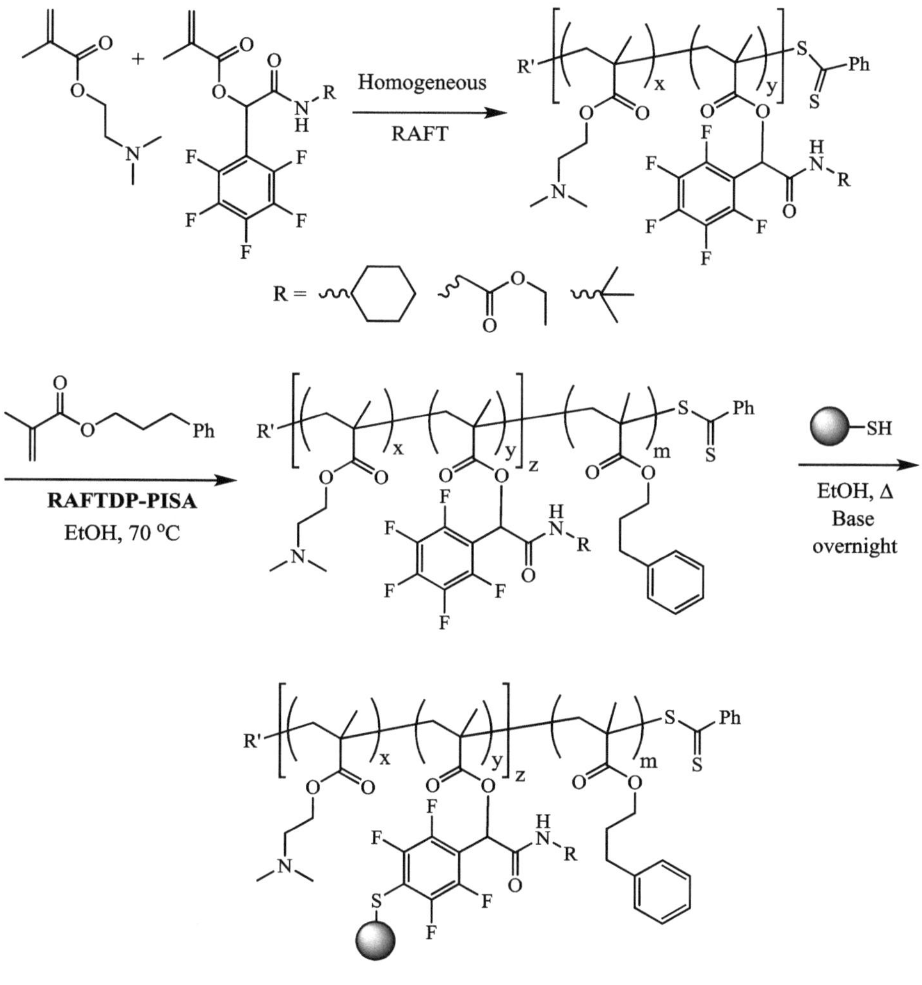

Scheme 4.26: General scheme for the RAFT dispersion polymerization of 3-phenylpropyl methacrylate with polymerization-induced self-assembly employing a poly[2-(dimethylamino) ethyl methacrylate] CTA containing a small mol% of a Passerini comonomer. Reproduced with permission from Y. Pei, J.-M. Noy, P.J. Roth and A.B. Lowe, Polymer Chemistry, 2015, 6, 11, 1928. ©2015, RSC [111].

Such formulations proceed with the in situ formation of polymeric nanoparticles that transition from spheres to worms (or cylinders) and finally to vesicles with increasing length of the PPMA block for a fixed length of the macro-CTA. In this case, the resulting nanoparticles contain the DMAEMA and Passerini-derived building blocks in the coronal structure of the nano-objects. The authors demonstrated that the spherical nanoparticles could be modified via organobase-mediated nucleophilic substitution of the *para*-fluoro groups in the Passerini repeat units with various thiols including 2-mercaptoethanol, 1-thio-β-D-glucose tetracetate, and thiophenol. In all cases the modifications were successful and yielded new examples of functional polymeric nanoparticles.

4.2.5 The thiol epoxide reaction

The nucleophilic ring-opening reaction of spring loaded heterocycles, such as epoxides or aziridines, were similarly highlighted as examples of highly efficient "click" reactions in the original article from Kolb, Finn, and Sharpless [5]. While such reactions are well known in small molecule chemistry the combination of thiol-epoxide reactions with RDRP processes are less known and, surprisingly, generally not as facile or efficient.

Harvison, Davis and Lowe [116] reported the chain end modification of homopolymers, prepared via RAFT, of polystyrene and poly(*N,N*-diethylacrylamide) with a series of small molecule oxiranes. Two different routes were examined – a sequential process and a one-pot approach. The latter proved more effective. For example, reaction of poly(*N,N*-diethylacrylamide) with a series of oxiranes including 2-(but-3-en-1-yl)oxirane, 2-butyloxirane, 2-vinyloxirane, and 3-vinyl-7-oxabicyclo[4.1.0]heptane resulted in essentially quantitative formation of the end functional polymer provided the thiocarbonylthio was cleaved with $NaBH_4$ in the presence of DMPP, Scheme 4.27.

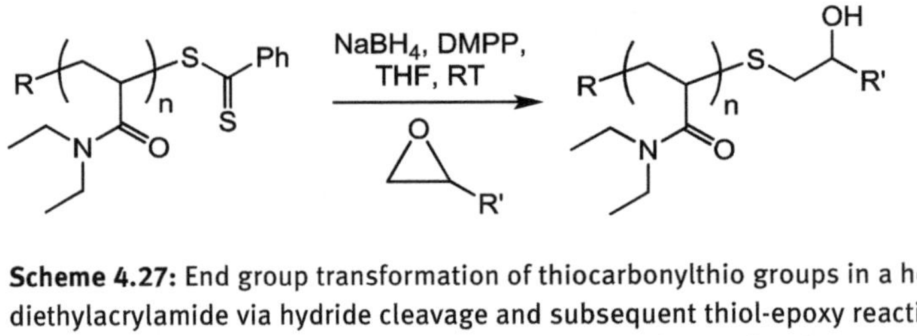

Scheme 4.27: End group transformation of thiocarbonylthio groups in a homopolymer of *N,N*-diethylacrylamide via hydride cleavage and subsequent thiol-epoxy reaction.
(Source: M.A. Harvison, T.P. Davis and A.B. Lowe, Polymer Chemistry, 2011, 2, 6, 1347) [116].

Le Neindre and Nicolaÿ [117] also reported the end-group transformation of a RAFT (co)polymer via a thiol-epoxy modification in a report covering the functionalization of a parent species via most of the common thiol-X suite of reactions.

Gadwal and Khan [118] combined ATRP and thiol-epoxy chemistry as a means of modifying the α-chain end in homopolymers of poly(methyl methacrylate), Scheme 4.28. In the first instance, the authors prepared three different epoxide-functional ATRP initiators, **I-III** in Scheme 4.28. These were then utilized for the homopolymerization of methyl methacrylate yielding well-defined homopolymers on low molecular weight and narrow molecular weight distribution. The α-epoxy functional product obtained from the use of **I** is shown in the Scheme. The α-epoxy groups were then treated with an excess of naphthalene-1-thiol in the presence of base to effect the thiol-epoxy ring-opening reaction. Finally, the liberated -OH groups were esterified with toluoyl chloride (not shown). The authors reported that both processes (thiol-epoxy and esterification) were, in all cases, quantitative.

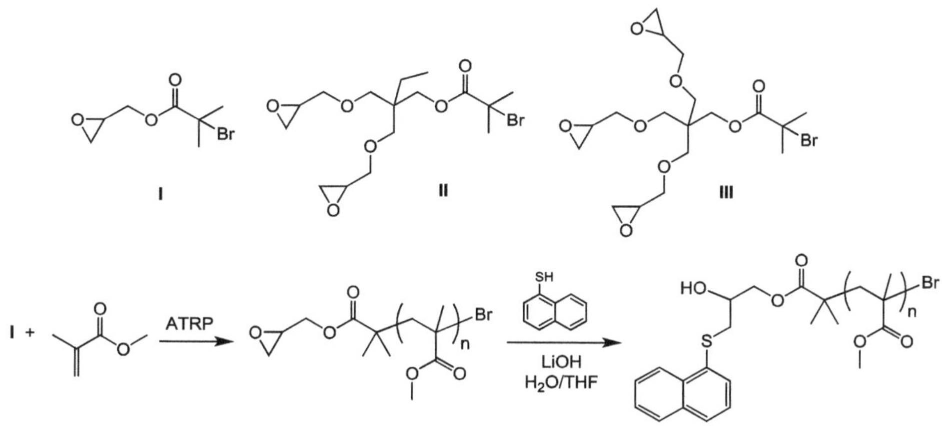

Scheme 4.28: α-End group modification of poly(methyl methacrylate) prepared by ATRP with an epoxy functional initiator and subsequent thiol-epoxy ring opening.
(Source: I. Gadwal and A. Khan, Polymer Chemistry, 2013, 4, 8, 2440) [118].

As with the other chemistries highlighted above, such chemical transformations can also be effective for side-group modification provided appropriate functionality is present. Side chain and simultaneous end-group/side chain modification of epoxy groups by thiols have been reported by Gadwal, Stuparu, and Khan [119] and by Haddleton et al. [120] in combination with ATRP.

Employing a similar approach to the end-group functionalization work noted above, glycidyl methacrylate (GMA) was first homopolymerized via ATRP according to Scheme 4.29 yielding polyGMA (PGMA) with an SEC measured $\bar{M}_n$ of ca. 20 K and dispersity of 1.3.

The authors then evaluated the ring opening of the pendant epoxy groups with *tert*-butyl (2-mercaptoethyl)carbamate and naphthalene-1-thiol under a variety of conditions with different catalysts and loadings with TBAF in THF and LiOH in aqueous THF (10% water) proving to be particularly effective. As with the

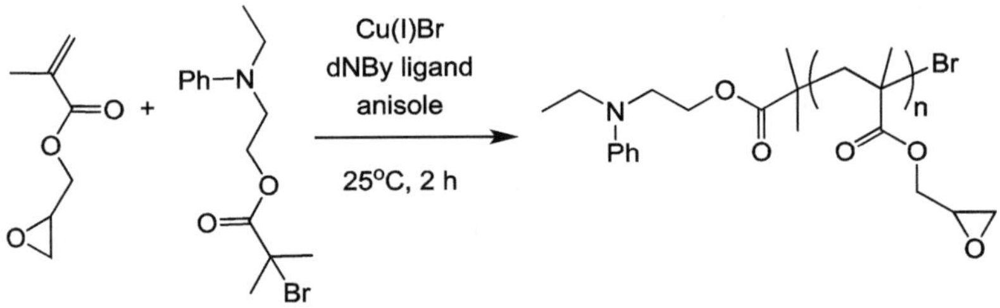

Scheme 4.29: ATRP homopolymerization of glycidyl methacrylate with 2-(ethyl(phenyl)amino)ethyl 2-bromo-2-methylpropanoate.
(Source: I. Gadwal, M.C. Stuparu and A. Khan, Polymer Chemistry, 2015, 6, 8, 1393) [119].

end-group transformations, the authors also noted that the liberated -OH groups generated during the ring-opening reaction could also be esterified although a 2 molar excess of the esterification reagent were required to achieve quantitative modification.

The approach from the group of Haddleton was very similar to that detailed above and is highlighted in Scheme 4.30. Glycidyl acrylate (GA) was homo- and copolymerized by Cu(0)-mediated ATRP to give materials with controlled molecular weights, low dispersities, and high end-group fidelity. In the case of a polyGA homopolymer (PGA) the authors treated the reactive scaffold with benzyl mercaptan (BnSH) in the presence of NEt_3 as an organobase. The authors noted that this resulted in both a thiol-bromo reaction (associated with modification of the polymer chain end) and the ring-opening thiol-epoxy reaction with the former occurring preferentially. This indicates that the possibility of selective modification is feasible. Although not noted, it is likely that exactly the same type of dual functionalization was also occurring in the work reported by Gadwal, Stuparu, and Khan.

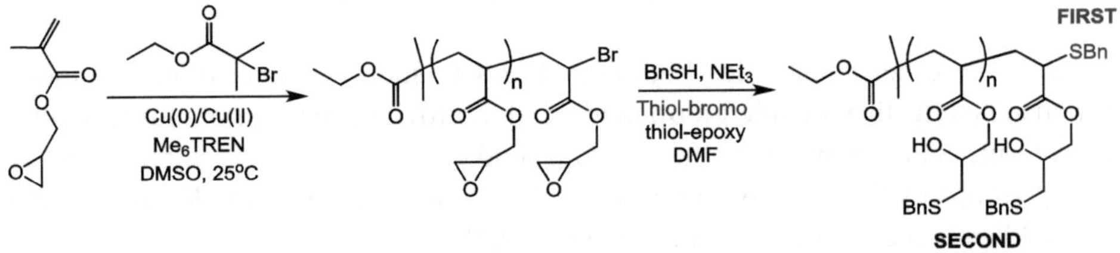

Scheme 4.30: Cu(0) RDRP of glycidyl acrylate followed by in situ, sequential thiol-bromo and thiol-epoxy reactions with benzyl mercaptan (BnSH).
(Source: Q. Zhang, A. Anastasaki, G.-Z. Li, A.J. Haddleton, P. Wilson and D.M. Haddleton, Polymer Chemistry, 2014, 5, 12, 3876) [120].

In addition to modification with BnSH the authors also reported exactly the same strategy as a means of preparing novel glycopolymers via sequential modification with 1-thio-β-D-glucose tetraacetate.

4.2.6 Thiol-methanethiosulfonate substitution reactions

Certain thiol-exchange/ nucleophilic substitution reactions have also been demonstrated to be highly efficient processes for introducing key functional groups both pendent to the main polymeric backbone as well as at chain ends. In the case of the latter, this is most usually applied to RAFT (co)polymers given that they can be formally regarded as protected macromolecular thiols. The majority of reports, certainly with methanethiosulfonate derivatives have come from the group of Theato and will not be discussed in more detail here as they are highlighted in a recent review [121].

4.3 Summary and outlook

In this chapter, we have highlighted examples in which common reversible deactivation radical polymerization processes have been utilized in conjunction with the suite of common thiol-X "click" and "click-like" reactions. An emphasis has been placed on end group and side-chain modification reactions. The thiol-ene reaction, in both its radical and ionic (thiol-Michael) forms, continues to be the most widely employed from this tool box of reactions, a trend that is likely to continue given the ready and cheap availability of both enes and thiols. However, the other reactions have also attracted significant attention. The thiol-yne reaction, a sister process to the thiol-ene reaction, still attracts significant attention and should also be considered as a viable alternative, but also complementary, reaction to the Cu-mediated alkyne-azide "click" process. The thiol-isocyanate reaction is, arguably, the fastest and most efficient of the thiol-x reaction suite. While its usefulness has been demonstrated it has not been as widely adopted as other modification reactions. The reason for this is not entirely clear but may be related to perceived handling problems of –NCO-containing reagents and or their ready availability.

The "click" approach to synthesis and modification is now an established, and highly valuable, aspect of synthetic polymer chemistry and will likely remain so for the foreseeable future. The application of now well-established chemistries continues as does the search and identification of alternative chemistries that could join this ever growing family. The future certainly appears bright in the area of polymer synthesis and modification.

References

[1] A.B. Lowe and C.L. McCormick, Synthesis and Solution Properties of Zwitterionic Polymers. Chemical Reviews, 2002, 102(11), 4177.

[2] A.B. Lowe, S.P. Armes and N.C. Billingham, Synthesis and Characterization of Zwitterionic Block Copolymers. Macromolecules, 1998, 31(18), 5991.

[3] C.S. Patrickios, A.B. Lowe, S.P. Armes and N.C. Billingham, ABC Triblock Polymethacrylates: Group Transfer Polymerization Synthesis of the ABC, ACB, and BAC Topological Isomers and Solution Characterization. Journal of Polymer Science Part A: Polymer Chemistry, 1998, 36(4), 617.

[4] V. Bütün, A.B. Lowe, N.C. Billingham and S.P. Armes, Synthesis of Zwitterionic Shell Cross-Linked Micelles. Journal of the American Chemical Society, 1999, 121(17), 4288.

[5] H.C. Kolb, M.G. Finn and K.B. Sharpless, Click Chemistry: Diverse Chemical Function from a Few Good Reactions. Angewandte Chemie, International Edition, 2001, 40(11), 2004.

[6] L. Liang and D. Astruc, The Copper(I) Catalyzed Alkyne-Azide Cycloaddition (CuAAC) "Click" Reaction and its Applications: An Overview. Coordination. Chemistry Reviews, 2011, 255(23–24), 2933.

[7] M. Meldal and C.W. Tomøe, Cu-Catalyzed Azide-Alkyne Cycloaddition. Chemical Reviews, 2008, 108(8), 2952.

[8] J.E. Hein and V.V. Fokin, Copper-catalyzed azide -alkyne cycloaddition (CuAAC) and beyond: new reactivity of copper(I) acetylides. Chemical Society Reviews, 2010, 39(4), 1302.

[9] P.L. Golas and K. Matyjaszewski, Marrying click chemistry with polymerization: expanding the scope of polymeric materials. Chemical Society Reviews, 2010, 39(4), 1338.

[10] D. Fournier, R. Hoogenboom and U.S. Schubert, Clicking polymers: a straightforward approach to novel macromolecular architectures. Chemical Society Reviews, 2007, 36(8), 1369.

[11] B.S. Sumerlin and A.P. Vogt, Macromolecular Engineering through Click Chemistry and Other Efficient Transformations. Macromolecules, 2010, 43(1), 1.

[12] H. Struthers, T.L. Mindt and R. Schibli, Metal chelating systems synthesized using the copper(I) catalyzed azide-alkyne cycloaddition. Dalton Transactions, 2010, 39(3), 675.

[13] W.H. Binder and R. Sachsenhofer, 'Click' Chemistry in Polymer and Materials Science. Macromolecular Rapid Communications, 2007, 28(1), 15.

[14] S. Ulrich, D. Boturyn, A. Marra, O. Renaudet and P. Dumy, Oxime Ligation: A Chemoselective Click Type Reaction for Accessing Multifunctional Biomolecular Constructs. Chemistry: A European Journal, 2014, 20(1), 34.

[15] P. Theato, Synthesis of Well-Defined Polymeric Activated Esters. Journal of Polymer Science Part A: Polymer Chemistry, 2008, 46(20), 6677.

[16] A.C. Pauly and P. Theato, Synthesis and Characterization of Poly(phenylacetylenes) Featuring Activated Ester Side Groups. Journal of Polymer Science Part A: Polymer Chemistry, 2011, 49 (1), 211.

[17] A.C. Pauly and P. Theato, Control of reactivity of constitutional isomers of pentafluorophenyl ethynylbenzoates for the synthesis of functional poly(phenylacetylenes). Polymer Chemistry, 2012, 3(7), 1769.

[18] Y. Zhu, J.-M. Noy, A.B. Lowe and P.J. Roth, The synthesis and aqueous solution properties of sulfobutylbetaine (co)polymers: comparison of synthetic routes and tuneable upper critical solution temperatures. Polymer Chemistry, 2015, 6(31), 5705.

[19] Y. Zhu, A.B. Lowe and P.J. Roth, Postpolymerization synthesis of (bis)amide (co)polymers: Thermoresponsive behavior and self-association. Polymer, 2014, 55(17), 4425.

[20] M. Beija, Y. Li, A.B. Lowe, T.P. Davis and C. Boyer, Factors influencing the synthesis and the post-modification of PEGylated pentafluorophenyl acrylate containing copolymers. European Polymer Journal, 2013, 49(10), 3060.

[21] P.J. Roth, T.P. Davis and A.B. Lowe, Thermoresponsive Diblock Copolymers: Insights into the Behavior of POEGMA in Alcohols. Macromolecules, 2012, 45(7), 3221.

[22] G.B.H. Chua, P.J. Roth, H.T.T. Duong, T.P. Davis and A.B. Lowe, Synthesis and Thermoresponsive Solution Properties of Poly[oligo(ethylene glycol) (meth)acrylamide]s: Biocompatible PEG Analogues. Macromolecules, 2012, 45(3), 1362.
[23] A. Das and P. Theato, Activated Ester Containing Polymers: Opportunities and Challenges for the Design of Functional Macromolecules. Chemical Reviews, 2016, 116(3), 1434.
[24] Y. Pei, O.R. Sugita, J.Y. Quek, P.J. Roth and A.B. Lowe, pH-, thermo- and electrolyte-responsive polymer gels derived from a well-defined, RAFT-synthesized, poly(2-vinyl-4,4-dimethylazlactone) homopolymer via one-pot post-polymerization modification. European Polymer Journal, 2015, 62, 204.
[25] J.Y. Quek, X. Liu, T.P. Davis, P.J. Roth and A.B. Lowe, RAFT-prepared α-difunctional poly(2-vinyl-4,4-dimethylazlactone)s and their derivatives: synthesis and effect of end-groups on aqueous inverse temperature solubility. Polymer Chemistry, 2015, 6(1), 118.
[26] J.Y. Quek, Y. Zhu, P.J. Roth, T.P. Davis and A.B. Lowe, RAFT Synthesis and Aqueous Solution Behavior of Novel pH- and Thermo-Responsive (Co)Polymers Derived from Reactive Poly(2-vinyl-4,4-dimethylazlactone) Scaffolds. Macromolecules, 2013, 46(18), 7290.
[27] Y. Zhu, J.Y. Quek, A.B. Lowe and P.J. Roth, Thermoresponsive (Co)polymers through Postpolymerization Modification of Poly(2-vinyl-4,4-dimethylazlactone). Macromolecules, 2013, 46(16), 6475.
[28] C.J. Hawker, "Living" Free Radical Polymerization: A Unique Technique for the Preparation of Controlled Macromolecular Architectures. Accounts of Chemical Research, 1997, 30(9), 373.
[29] C.J. Hawker, A.W. Bosman and E. Harth, New Polymer Synthesis by Nitroxide Mediated Living Radical Polymerizations. Chemical Reviews, 2001, 101(12), 3661.
[30] K. Matyjaszewski and J. Xia, Atom Transfer Radical Polymerization. Chemical Reviews, 2001, 101(9), 2921.
[31] M. Kamigaito, T. Ando and M. Sawamoto, Metal-Catalyzed Living Radical Polymerization. Chemical Reviews, 2001, 101(12), 3689.
[32] G. Moad, E. Rizzardo and S.H. Thang, Living Radical Polymerization by the RAFT Process. Australian Journal of Chemistry, 2005, 58(6), 379.
[33] G. Moad, E. Rizzardo and S.H. Thang, Living Radical Polymerization by the RAFT Process – A First Update. Australian Journal of Chemistry, 2006, 59(10), 669.
[34] G. Moad, E. Rizzardo and S.H. Thang, Living Radical Polymerization by the RAFT Process – A Second Update. Australian Journal of Chemistry, 2009, 62(11), 1402.
[35] G. Moad, E. Rizzardo and S.H. Thang, Living Radical Polymerization by the RAFT Process – A Third Update. Australian Journal of Chemistry, 2012, 65(8), 985.
[36] S. Yamago, Development of Organotellurium-Mediated and Organostibine-Mediated Living Radical Polymerization Reactions. Journal of Polymer Science Part A: Polymer Chemistry, 2006, 44(1), 1.
[37] Y. Nakamura and S. Yamago, Organotellurium-mediated living radical polymerization under photoirradiation by a low-intensity light-emitting diode. Beilstein Journal of Organic Chemistry, 2013, 9, 1607–1612.
[38] A.B. Lowe, Thiol-ene "click" reactions and recent applications in polymer and materials science. Polymer Chemistry, 2010, 1(1), 17.
[39] A.B. Lowe, Thiol-ene "click" reactions and recent applications in polymer and materials synthesis: a first update. Polymer Chemistry, 2014, 5(17), 4820.
[40] C.E. Hoyle, A.B. Lowe and C.N. Bowman, Thiol-click Chemistry: A Multifaceted Toolbox for Small Molecule and Polymer Synthesis. Chemical Society Reviews, 2010, 39(4), 1355.
[41] C.E. Hoyle and C.N. Bowman, Thiol-ene Click Chemistry. Angewandte Chemie, International Edition, 2010, 49(9), 1540.

[42] M.J. Kade, D.J. Burke and C.J. Hawker, The Power of Thiol-ene Chemistry. Journal of Polymer Science Part A: Polymer Chemistry, 2010, 48(4), 743.

[43] J.W. Chan, C.E. Hoyle, A.B. Lowe and M. Bowman, Nucleophile-Initiated Thiol-Michael Reactions: Effect of Organocatalyst, Thiol, and Ene. Macromolecules, 2010, 43(15), 6381.

[44] A.B. Lowe, Thiol-yne 'click'/coupling chemistry and recent applications in polymer and materials synthesis and modification. Polymer, 2014, 55(22), 5517.

[45] B.D. Fairbanks, T.F. Scott, C.J. Kloxin, K.S. Anseth and C.N. Bowman, Thiol-Yne Photopolymerizations: Novel Mechanism, Kinetics, and Step-Growth Formation of Highly Cross-Linked Networks. Macromolecules, 2009, 42(1), 211.

[46] J.W. Chan, C.E. Hoyle and A.B. Lowe, Sequential Phosphine-Catalyzed, Nucleophilic Thiol-Ene/Radical-Mediated Thiol-Yne Reactions and the Facile Orthogonal Synthesis of Polyfunctional Materials. Journal of the American Chemical Society, 2009, 131(16), 5751.

[47] J.W. Chan, H. Zhou, C.E. Hoyle and A.B. Lowe, Photopolymerization of Thiol-Alkynes: Polysulfide Networks. Chemistry of Materials, 2009, 21(8), 1579.

[48] B.D. Fairbanks, E.A. Sims, K.S. Anseth and C.N. Bowman, Reaction Rates and Mechanisms for Radical, Photoinitated Addition of Thiols to Alkynes, and Implications for Thiol-Yne Photopolymerizations and Click Reactions. Macromolecules, 2010, 43(9), 4113.

[49] J.W. Chan, J. Shin, C.E. Hoyle, C.N. Bowman and A.B. Lowe, Synthesis, Thiol-Yne "Click" Photopolymerization, and Physical Properties of Networks Derived from Novel Multifunctional Alkynes. Macromolecules, 2010, 43(11), 4937.

[50] V.X. Truong and A.P. Dove, Organocatalytic, Regioselective Nucleophilic "Click" Addition of Thiols to Propiolic Acid Esters for Polymer -Polymer Coupling. Angewandte Chemie, International Edition, 2013, 52(15), 4132.

[51] N.T. Brummelhuis and H. Schlaad, Stimuli-responsive star polymers through thiol-yne core functionalization crosslinking of block copolymer micelles. Polymer Chemistry, 2011, 2(5), 1180.

[52] S.S. Naik, J.W. Chan, C. Comer, C.E. Hoyle and D.A. Savin, Thiol-yne 'click' chemistry as a route to functional lipid mimetics. Polymer Chemistry, 2011, 2(2), 303.

[53] H. Li, B. Yu, H. Matsushima, C.E. Hoyle and A.B. Lowe, The Thiol-Isocyanate Click Reaction: Facile and Quantitative Access to w-End-Functional Poly(N,N-diethylacrylamide) Synthesized by RAFT Radical Polymerization. Macromolecules, 2009, 42(17), 6537.

[54] R.M. Hensaring, S.B. Rahane, A.P. LeBlanc, B.J. Sparks, E.W. White, J. Locklin and D.L.Patton, Thiol-isocyanate "click" reactions: rapid development of functional polymeric surfaces. Polymer Chemistry, 2011, 2(1), 88.

[55] R.M. Hensarling, E.A. Hoff, A.P. LeBlanc, W. Guo, S.B. Rahane and D.L. Patton, Photocaged Pendent Thiol Polymer Brush Surfaces for Postpolymerization Modifications via Thiol-Click Chemistry. Journal of Polymer Science Part A: Polymer Chemistry, 2013, 51(5), 1079.

[56] B.M. Rosen, G. Lligadas, C. Hahn and V. Percec, Synthesis of Dendrimers Through Divergent Iterative Thio-Bromo "Click" Chemistry. Journal of Polymer Science Part A: Polymer Chemistry, 2009, 47(15), 3931.

[57] B.M. Rosen, G. Lligadas, C. Hahn, and V. Percec, Synthesis of Dendritic Macromolecules Through Divergent Iterative Thio-Bromo "Click" Chemistry and SET-LRP. Journal of Polymer Science, Part A: Polymer Chemistry, 2009, 47(15), 3940.

[58] F. Segui, X.-P. Qiu, and F.M. Winnik, An Efficient Synthesis of Telechelic Poly (N-isopropylacrylamides) and its Application to the Preparation of a,x-Dicholesteryl and a,x-Dipyrenyl Polymers. Journal of Polymer Science Part A: Polymer Chemistry, 2007, 46(1), 314.

[59] C.R. Becer, K. Babiuch, D. Pilz, S. Hornig, T. Heinze, M. Gottschaldt and U.S. Schubert, Clicking Pentafluorostyrene Copolymers: Synthesis, Nanoprecipitation, and Glycosylation. Macromolecules, 2009, 42(7), 2387.

[60] V. Grazu, O. Abian, C. Mateo, F. Batista-Viera, R. Fernandez-Lafuente and J.M. Guisan, Novel bifunctional epoxy/thiol-reactive support to immobilize thiol containing proteins by the epoxy chemistry. Biomacromolecules, 2003, 4(6), 1495.
[61] J.A. Carioscia, J.W. Stansbury and C.N. Bowman, Evaluation and Control of Thiol-ene/Thiol-epoxy Hybrid Networks. Polymer, 2007, 48(6), 1526.
[62] S. Katogi and M. Yusa, Photobase generation from amineimide derivatives and their use for curing an epoxide/thiol system. Journal of Polymer Science Part A: Polymer Chemistry, 2002, 40(22), 4045.
[63] T. Iida, N. Yamamoto, H. Sasai and M. Shibasaki, New Asymmetric Reactions Using a Gallium Complex: A Highly Enantioselective Ring Opening of Epoxides with Thiols Catalyzed by a Gallium·Lithium·Bis(binaphthoxide) Complex. Journal of the American Chemical Society, 1997, 119(20), 4783.
[64] Y.C. Yuan, M.Z. Rong, M.Q. Zhang, J. Chen, G.C. Yang and X.M. Li, Self-Healing Polymeric Materials Using Epoxy/Mercaptan as the Healant. Macromolecules, 2008, 41(14), 5197.
[65] A. Brändle and A. Khan, Thiol-epoxy 'click' polymerization: efficient construction of reactive and functional polymers. Polymer Chemistry, 2012, 3(12), 3224.
[66] F.D. Jochum, P.J. Roth, D. Kessler and P. Theato, Double Thermoresponsive Block Copolymers Featuring a Biotin End Group. Biomacromolecules, 2010, 11(9), 2432.
[67] C. Boyer, A.H. Soeriyadi, P.J. Roth, M.R. Whittaker and T.P. Davis, Post-functionalization of ATRP polymers using both thiol/ene and thiol/disulfide exchange chemistry. Chemical Communications, 2011, 47(4), 1318.
[68] P.J. Roth, F.D. Jochum, R. Zentel and P. Theato, Synthesis of Hetero-Telechelic r,ω Bio-Functionalized Polymers. Biomacromolecules, 2010, 11(1), 238.
[69] P.J. Roth, F.D. Jochum and P. Theato, UCST-type behavior of poly[oligo(ethylene glycol) methyl ether methacrylate] (POEGMA) in aliphatic alcohols: solvent, co-solvent, molecular weight, and end group dependences. Soft Matter, 2011, 7(6), 2484.
[70] P.J. Roth, D. Kessler, R. Zentel and P. Theato, A Method for Obtaining Defined End Groups of Polymethacrylates Prepared by the RAFT Process during Aminolysis. Macromolecules, 2008, 41(22), 8316.
[71] P.J. Roth, D. Kessler, R. Zentel and P. Theato, Versatile x-End Group Functionalization of RAFT Polymers Using Functional Methane Thiosulfonates. Journal of Polymer Science Part A: Polymer Chemistry, 2009, 47(12), 3118.
[72] C.E. Hoyle, T.Y. Lee and T. Roper, Thiol-Enes: Chemistry of the Past with Promise for the Future. Journal of Polymer Science Part A: Polymer Chemistry, 2004, 42(21), 5301.
[73] T. Pauloehrl, G. Delaittre, M. Bastmeyer and C. Barner-Kowollik, Ambient temperature modification by *in situ* phototriggered deprotection and thiol-ene chemistry. Polymer Chemistry, 2012, 3(7), 1740.
[74] G. Delaittre, T. Pauloehrl, M. Bastmeyer and C. Barner-Kowollik, Acrylamide-Based Copolymers Bearing Photoreleasable Thiols for Subsequent Thiol-Ene Functionalization. Macromolecules, 2012, 45(4), 1792.
[75] P. Espeel, F. Goethals, M.M. Stamenović, L. Petton and F.E. Du Prez, Double modular modification of thiolactone-containing polymers: towards polythiols and derived structures. Polymer Chemistry, 2012, 3(4), 1007.
[76] S. Reinicke, P. Espeel, M.M. Stamenović and F.E. Du Prez, One-Pot Double Modification of p(NIPAAm): A Tool for Designing Tailor-Made Multiresponsive Polymers ACS Macro Letters, 2013, 2(6), 539.
[77] L.M. Campos, K.L. Killops, R. Sakai, J.M.J. Paulusse, D. Damiron, E. Drockenmuller, B.W. Messmore and C.J. Hawker, Development of Thermal and Photochemical Strategies for Thiol-Ene Click Polymer Functionalization. Macromolecules, 2008, 41(19), 7063.

[78] N.K. Singha, M.I. Gibson, B.P. Koiry, M. Danial and H.A. Klok, Side-chain peptide-synthetic polymer conjugates via tandem "ester-amide/thiol-ene" post-polymerization modification of poly(pentafluorophenyl methacrylate) obtained using ATRP. Biomacromolecules, 2011, 12(8), 2908.

[79] G. Chen, S. Amajjahe and M.S. Stenzel, Synthesis of thiol-linked neoglycopolymers and thermo-responsive glycomicelles as potential drug carrier. Chemical Communications, 2009, 10, 1198.

[80] J.D. Flores, N.J. Treat, A.W. York and C.L. McCormick, Facile, modular transformations of RAFT block copolymers via sequential isocyanate and thiol-ene reactions. Polymer Chemistry, 2011, 2(9), 1976.

[81] L. Zou, W. Zhu, Y. Chen and F. Xi, Modifcation of side chain terminals of PEGylated molecular bottle brushes – A toolbar of molecular nanoobjects. Polymer, 2013, 54(2), 481.

[82] M.A. Harvison, P.J. Roth, T.P. Davis and A.B. Lowe, End Group Reactions of RAFT-Prepared (Co)Polymers. Australian Journal of Chemistry, 2011, 64(8), 992.

[83] G. Moad, E. Rizzardo and S.H. Thang, End-functional polymers, thiocarbonylthio group removal/transformation and reversible addition -fragmentation -chain transfer (RAFT) polymerization. Polymer International, 2011, 60, 6.

[84] A.B. Lowe, B.S. Sumerlin, M.S. Donovan and C.L. McCormick, Facile Preparation of Transition Metal Nanoparticles Stabilized by Well-Defined (Co)polymers Synthesized via Aqueous Reversible Addition-Fragmentation Chain Transfer Polymerization. Journal of the American Chemical Society, 2002, 124(39), 11562.

[85] B.S. Sumerlin, A.B. Lowe, P.A. Stroud, P. Zhang, M.W. Urban and C.L. McCormick, Modification of Gold Surfaces with Water-Soluble (Co)polymers Prepared via Aqueous Reversible Addition-Fragmentation Chain Transfer (RAFT) Polymerization. Langmuir, 2003, 19(14), 5559.

[86] B. Yu, J.W. Chan, C.E. Hoyle and A.B. Lowe, Sequential Thiol-Ene/Thiol-Ene and Thiol-Ene/Thiol-Yne Reactions as a Route to Well-Defined Mono and Bis End-Functionalized Poly(N-isopropylacrylamide). Journal of Polymer Science Part A: Polymer Chemistry, 2009, 47(14), 3544.

[87] Y. Zhu, R. Batchelor, A.B. Lowe and P.J. Roth, Design of Thermoresponsive Polymers with Aqueous LCST, UCST, or Both: Modification of a Reactive Poly(2-vinyl-4,4-dimethylazlactone) Scaffold. Macromolecules, 2016, 49(2), 672.

[88] P.J. Roth, T.P. Davis and A.B. Lowe, Novel α,α-Bischolesteryl Functional (Co)Polymers: RAFT Radical Polymerization Synthesis and Preliminary Solution Characterization. Macromolecular Rapid Communications, 2014, 35(8), 813.

[89] P.J. Roth, T.P. Davis and A.B. Lowe, UCST-driven self-assembly and crosslinking of diblock copolymer micelles. Polymer Chemistry, 2012, 3(8), 2228.

[90] X.-P. Qiu and F.M. Winnik, Facile and Efficient One-Pot Transformation of RAFT Polymer End Groups via a Mild Aminolysis/Michael Addition Sequence. Macromolecular Rapid Communications, 2006, 27(19), 1648.

[91] J.M. Spruell, B.A. Levy, A. Sutherland, W.R. Dichtel, J.Y. Cheng, J.F. Stoddart and A. Nelson, Facile Postpolymerization End-Modification of RAFT Polymers. Journal of Polymer Science, Part A: Polymer Chemistry, 2009, 47(2), 346.

[92] G.N. Grover, S.N.S. Alconcel, M.M. Matsumoto and H.D. Maynard, Trapping of Thiol-Terminated Acrylate Polymers with Divinyl Sulfone To Generate Well-Defined Semitelechelic Michael Acceptor Polymers. Macromolecules, 2009, 42(20), 7657.

[93] T.H. Ho, M. Levere, J.-C. Soutif, V. Montembault, S. Pascual and L. Fontaine, Synthesis of thermoresponsive oxazolone end-functional polymers for reactions with amines using thiol-Michael addition "click" chemistry. Polymer Chemistry, 2011, 2(6), 1258.

[94] C.W. Scales, A.J. Convertine and C.L. McCormick, Fluorescent Labeling of RAFT-Generated Poly(N-isopropylacrylamide) via a Facile Maleimide-Thiol Coupling Reaction. Biomacromolecules, 2006, 7(5), 1389.
[95] M. Li, P. De, S.R. Gondi and B.S. Sumerlin, End Group Transformations of RAFT-Generated Polymers with Bismaleimides: Functional Telechelics and Modular Block Copolymers. Journal of Polymer Science Part A: Polymer Chemistry, 2008, 46(15), 5093.
[96] J.W. Chan, B. Yu, C.E. Hoyle and A.B. Lowe, Convergent synthesis of 3-arm star polymers from RAFT-prepared poly(N,N-diethylacrylamide) via a thiol-ene click reaction. Chemical Communications, 2008, 40, 4959–4961.
[97] J.W. Chan, B Yu, C.E. Hoyle and A.B. Lowe, The nucleophilic, phosphine-catalyzed thiol-ene click reaction and convergent star synthesis with RAFT-prepared homopolymers. Polymer, 2009, 50(14), 3158.
[98] M. Uygun, M.A. Tasdelen and Y. Yagci, Influence of Type of Initiation on Thiol-Ene "Click" Chemistry. Macromolecular Chemistry and Physics, 2010, 211(1), 103.
[99] M.M. Stamenović, P. Espeel, W.V. Camp and F.E. Du Prez, Norbornenyl-Based RAFT Agents for the Preparation of Functional Polymers via Thiol-Ene Chemistry. Macromolecules, 2011, 44(14), 5619.
[100] A.B. Lowe, C.E. Hoyle and C.N. Bowman, Thiol-yne click chemistry: A powerful and versatile methodology for materials synthesis. Journal of Materials Chemistry, 2010, 20(23), 4745.
[101] V.T. Huynh, G. Chen, P.H. de Souza, Thiol-yne and Thiol-ene "Click" Chemistry as a Tool for a Variety of Platinum Drug Delivery Carriers, from Statistical Copolymers to Crosslinked Micelles. Biomacromolecules, 2011, 12(5), 1738.
[102] J. Kumar, A. Bousquet and M.S. Stenzel, Thiol-alkyne Chemistry for the Preparation of Micelles. with Glycopolymer Corona: Dendritic Surfaces versus Linear Glycopolymer in Their Ability to Bind to Lectins. Macromolecular Rapid Communications, 2011, 32(20), 1620.
[103] T. Cai, M. Li, K.-G. Neoh and E.-T. Kang, Preparation of stimuli responsive polycaprolactone membranes of controllable porous morphology via combined atom transfer radical polymerization, ring-opening polymerization and thiol-yne click chemistry. Journal of Materials Chemistry, 2012, 22(32), 16248.
[104] H. Willcock and R. O'Reilly, End group removal and modification of RAFT polymers. Polymer Chemistry, 2010, 1(2), 149.
[105] J.A. Burns, J.C. Butler, J. Moran and G.M. Whitesides, Selective Reduction of Disulfides by Tris (2-carboxyethyl)phosphine. Journal of Organic Chemistry, 1991, 56(8), 2648.
[106] J.D. Flores, J. Shin, C.E. Hoyle and C.L. McCormick, Direct RAFT polymerization of an unprotected isocyanate-containing monomer and subsequent structipendant functionalization using "click"-type reactions. 2010, 1(2), 213.
[107] J. Xu, L. Tao, C. Boyer, A.B. Lowe and T.P. Davis, Combining Thio-Bromo "Click" Chemistry and RAFT Polymerization: A Powerful Tool for Preparing Functionalized Multiblock and Hyperbranched Polymers. Macromolecules, 2010, 43(1), 20.
[108] M. Nakayama and T. Okano, Polymer Terminal Group Effects on Properties of Thermoresponsive Polymeric Micelles with Controlled Outer-Shell Chain Lengths. Biomacromolecules, 2005, 6(4), 2320.
[109] J.-M. Noy, M.Koldevitz and P.J. Roth, Thiol-reactive functional poly(meth)acrylates: multicomponent monomer synthesis, RAFT (co)polymerization and highly efficient thiol - para-fluoro postpolymerization modification. Polymer Chemistry, 2015, 6(3), 436.
[110] S. Schmidt, M. Koldevitz, J.-M. Noy and P.J. Roth, Multicomponent isocyanide-based synthesis of reactive styrenic and (meth)acrylic monomers and their RAFT (co)polymerization. Polymer Chemistry, 2015, 6(1), 44.

[111] Y. Pei, J.-M. Noy, P.J. Roth and A.B. Lowe, Thiol-reactive Passerini-methacrylates and polymorphic surface functional soft matter nanoparticles via ethanolic RAFT dispersion polymerization and post-synthesis modification. Polymer Chemistry, 2015, 6(11), 1928.
[112] Y. Pei, N.C. Dharsana, J.A. van Hensbergen, R.P Burford, P.J. Roth and A.B. Lowe, RAFT dispersion polymerization of 3-phenylpropyl methacrylate with poly[2-(dimethylamino)ethyl methacrylate] macro-CTAs in ethanol and associated thermoreversible polymorphism. Soft Matter, 2014, 10(31), 5787.
[113] Y. Pei, L. Thurairajah, O.R. Sugita and A.B. Lowe, RAFT Dispersion Polymerization in Nonpolar Media: Polymerization of 3_Phenylpropyl Methacrylate in n_Tetradecane with Poly(stearyl methacrylate) Homopolymers as Macro Chain Transfer Agents. Macromolecules, 2015, 48(1), 236.
[114] Y. Pei, O.R. Sugita, L. Thurairajah and A.B. Lowe, Synthesis of poly(stearyl methacrylate-b-3-phenylpropyl methacrylate) nanoparticles in n-octane and associated thermoreversible polymorphism. RSC Advances, 2015, 5(23), 17636.
[115] Y. Pei, J.-M. Noy, P.J. Roth and A.B. Lowe, Soft Matter Nanoparticles with Reactive Coronal Pentafluorophenyl Methacrylate Residues via Non-Polar RAFT Dispersion Polymerization and Polymerization-Induced Self-Assembly. Journal of Polymer Science Part A: Polymer Chemistry, 2015, 53(20), 2326.
[116] M.A. Harvison, T.P. Davis and A.B. Lowe, Macromolecular thiolysis of oxiranes: end-group modification of RAFT prepared homopolymers. Polymer Chemistry, 2011, 2(6), 1347.
[117] M. Le Neindre and R. Nicolaÿ, One-pot deprotection and functionalization of polythiol copolymers via six different thiol-X reactions. Polymer International, 2014, 63(5), 887.
[118] I. Gadwal and A. Khan, Protecting-group-free synthesis of chain-end multifunctional polymers by combining ATRP with thiol-epoxy 'click' chemistry. Polymer Chemistry, 2013, 4(8), 2440.
[119] I. Gadwal, M.C. Stuparu and A. Khan, Homopolymer bifunctionalization through sequential thiol-epoxy and esterification reactions: an optimization, quantification, and structural elucidation study. Polymer Chemistry, 2015, 6(8), 1393.
[120] Q. Zhang, A. Anastasaki, G.-Z. Li, A.J. Haddleton, P. Wilson and D.M. Haddleton, Multiblock sequence-controlled glycopolymers via Cu(0)-LRP following efficient thiol-halogen, thiol-epoxy and CuAAC reactions. Polymer Chemistry, 2014, 5(12), 3876.
[121] P.J. Roth, C. Boyer, A.B. Lowe and T.P. Davis, RAFT Polymerization and Thiol Chemistry: A Complementary Pairing for Implementing Modern Macromolecular Design. Macromolecular Rapid Communications, 2011, 32(15), 1123.

Nabendu B. Pramanik, Prantik Mondal,
Dhruba J. Haloi, Nikhil K. Singha

5 Designing macromolecular architecture via reversible deactivation radical polymerization (RDRP) and Diels–Alder reaction

5.1 Introduction

The concept of "Click Chemistry" was first coined in 2001 by Sharpless. Since then the research on this particular topic has gained attention of chemists and material scientists worldwide. Now "Click Chemistry" has emerged as one of the fastest growing research areas where researchers are continuously working to develop new materials with interesting properties, using the concept of "Click Chemistry." Among the various "click" reactions known so far, the Diels–Alder (DA) cycloaddition reaction (between a diene and a dienophile) is the most prominent one, which satisfies all the criteria of "Click Chemistry" as envisioned by Sharpless [1]. This cycloaddition reaction was discovered in 1928 by Diels and Alder. They were awarded the Nobel Prize in chemistry in 1950 to acknowledge their contribution to science[2].

Synthesis of macromolecules with advanced architectures is a primary aim of polymer scientists. The advanced architectures include homopolymers, telechelic polymers, block copolymers, graft copolymers, dendronized, and star polymers [3–5]. The combination of this chemistry with reversible deactivation radical polymerization (RDRP) has emerged as an attractive and strategic tool to develop materials with interesting properties. RDRP techniques were developed in the 1990s and they are excellent tools for conjugation together with "click" chemistry.

RDRP enables the retention of some special atom or group at the end of the polymer chains. This end functionality provides enormous scope for post-polymerization modification. These groups can be utilized to reconstruct polymer architectures further via DA chemistry. RDRP and the thermo-reversible characteristic of this DA reaction has now been exploited extensively to prepare smart polymers [6], thermo-responsive polymer gel [7–11], self-healing polymers, well-defined dendrimers [12], and so on.

The DA "click" reaction in combination with living polymerization techniques including RAFT [13–16], ATRP [17–19], and NMP [20, 21] are facile routes to prepare polymers with advanced architectures.

Nabendu B. Pramanik, Department of Chemistry, Lehigh University, Bethlehem, USA
Prantik Mondal, Nikhil K. Singha, Rubber Technology Centre, Indian Institute of Technology Kharagpur, Kharagpur, West Bengal, India
Dhruba J. Haloi, Department of Chemistry, Bodoland University, Kokrajhar, Assam, India

https://doi.org/10.1515/9783110643695-005

5.2 DA reaction in homopolymers

The DA reaction is used for the preparation of linear thermoplastic [22] and thermosetting [23] polymers like polyamides [24–29], polyphenylenes [30–32], and ladder polymers [33–35]. Polymerization carried out by the DA reaction is achieved by starting with monomers containing both diene and dienophile functional groups in a single molecule (A-B type) or via the intermolecular reaction between bisdienes and bisdienophiles (A-A and A-B type of comonomers) as shown in Scheme 5.1. The DA linkage in the polymeric backbone has added advantages of adjusting the chain length, controlling processing viscosity, and improving recyclability [36, 37].

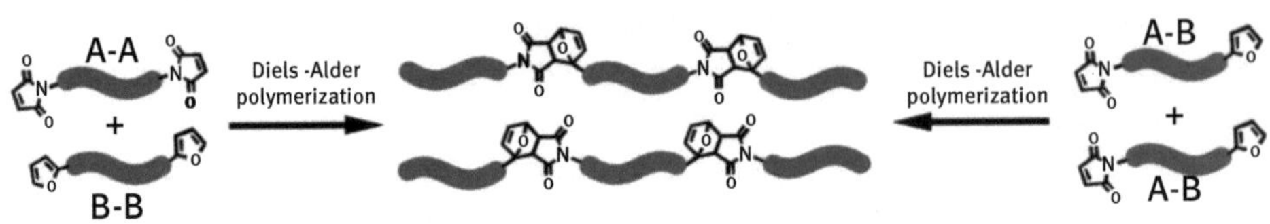

Scheme 5.1: A-A/A-B type of monomers via Diels–Alder polymerization. Reproduced with permission from M.A. Tasdelen, Polymer Chemistry, 2011, 2, 10, 2133. ©2011, RSC [35].

5.2.1 Block copolymers

The preparation of block copolymers via the DA reaction is a facile route for preparation of complex macromolecular architectures. The DA reaction based on the [4 + 2] cycloaddition coupling between anthracene and furan protected maleimide end-functionalized polymers has been used to prepare several block copolymers [38, 39]. Well-defined functional precursors of poly(ethylene oxide) (PEO), polystyrene, and poly(methyl methacrylate) (PMMA) were prepared via RDRP methods, like NMP and ATRP techniques, and later on block copolymers of PEO-*b*-PSt and PEO-*b*-PMMA were prepared via DA reactions. Tunca and co-workers [40] developed ABCD 4-Miktoarm Star Quarter polymer from nitroxide-mediated radical polymerization (NMP) of styrene and free radical polymerization of methyl methacrylate (MMA) using a macroinitiator prepared from DA "click" conjugation of polycaprolactone (PCL) with anthracyl and poly-*t*-butyl acrylate (PtBA) with furan-protected maleimide end group at 100 °C. Kimura and coworkers prepared poly(L-lactide)-*b*-poly(D-lactide) block copolymer via DA "click" adduct between the anthracene-terminated poly(L-lactide) and maleimide-terminated poly(D-lactide) [41]. The hetero-DA reaction (HDA) reaction between the imines, aldehydes, and thiocarbonyls as dienophile with suitable dienes are also powerful strategy to prepare heterocycle compounds [42–44]. Barner-Kowollik *et al.* reported the efficient HDA reaction between the diene and dithioester functionalized polymer as dieneophile to prepare block copolymers [45, 46]. They reported the formation of polystyrene-*b*-poly(ε-caprolactone) (PSt-*b*-PCL) block copolymer via DA reaction of dithioester-terminated

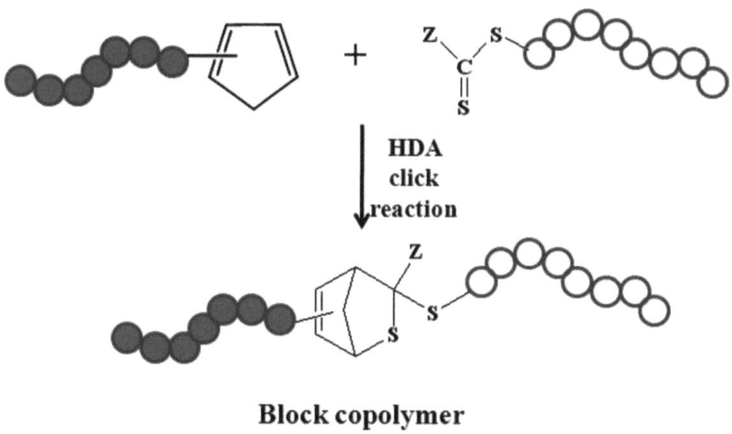

Scheme 5.2: Synthesis of block copolymers via Hetero-DA click reaction.
(Source: A.J. Inglis and C. Barner-Kowollik, Macromolecular Rapid Communications, 2010, 31, 14, 1247) [15].

polystyrene and diene-terminated PCL at 50 °C in the presence of catalyst, where the polymer precursors were prepared via RAFT polymerization as shown in Scheme 5.2 [15]. In this approach the electron withdrawing dithioester moiety played an important role in controlling the RAFT polymerization and the efficiency toward the HDA reaction. The reaction was catalyzed by a Lewis acid like zinc chloride or trifluoroacetic acid. These catalysts coordinate with sulfur atoms of dithioesters end-group and increased the electrophilicity of the thiocarbonyl bond, as well as the efficiency of the HDA reaction. They also reported that the efficiency of HDA reaction was dramatically increased when cyclopentadiene was used as a diene. In that case, the reaction was completed in few minutes at ambient temperature without addition of catalyst. This ultrafast HDA reaction was used to prepare block copolymers of molecular weight ranging from 34,000 g/mol to 100,000 g/mol with low dispersity, only by shaking at room temperature.

Barner-Kowollik and coworkers also reported the preparation of a block copolymer via a catalyst free, but extremely rapid, hetero-Diels–Alder "click" reaction [47–49]. In this process, ATRP prepared cyclopentadienyl-capped polymer was clicked at ambient temperature with the RAFT prepared polymer having an electron-deficient dithioester to prepare the block copolymer. This block copolymer preparation process uses ATRP, RAFT, and the Diels–Alder "click" reaction for the conjugation. Cyclopentadiene was introduced at the end of the ATRP prepared polymer by reacting its halide end group with the sodium salt of cyclopentadiene. Durmaz *et al.* reported the formation of triblock copolymers of PEO-*b*-PSt-*b*-PMMA and PCL-*b*-PSt-*b*-PMMA by using CuAAC and followed by a DA "click" reaction as shown in Scheme 5.3 [50]. Well-defined polymer precursors of PEO and PCL were obtained via ROP and PSt and PMMA were obtained via ATRP. The middle block PSt contained anthracene at the initiator end and azide at the terminated

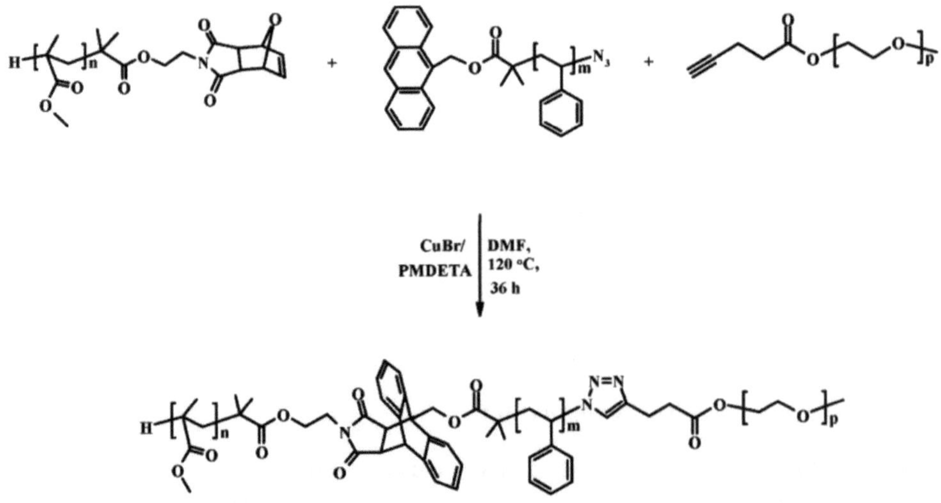

Scheme 5.3: Synthesis of ABC triblock copolymer via one-pot CuAAC and DA click reactions. Reproduced with permission from H. Durmaz, A. Dag, O. Altintas, T. Erdogan, G. Hizal and U. Tunca, Macromolecules, 2007, 40, 2, 191. ©2007, ACS [50].

end. The copper catalytic system accelerated both the CuAAC and DA "click" reactions at higher temperature.

5.2.2 DA reaction in graft copolymers

Gacal *et al.* reported the preparation of well-defined graft copolymers via DA "click" reaction between the pendent anthryl groups of polystyrene backbone and the maleimide end groups of poly(methyl methacrylate) (PMMA) and poly(ethylene glycol) (PEG) [7]. This process involves first the preparation of random copolymer of styrene (St) and 4-chloromethylstyrene (CMS) via nitroxide-mediated radical polymerization (NMP) at 125 °C. Anthracene functionality was then introduced to this copolymer by reacting the chloride functionality of the copolymer with 9-anthracenemethanol. The polymer, PMMA with protected maleimide as end group was prepared via ATRP using a specially designed functional initiator. PEG with protected maleimide as end group was also prepared by modifying commercial PEG. Finally, an in situ deprotection process via retro-DA reaction exposed the maleimide functionality of PEG and PMMA for the click reaction with the anthracenyl moieties of polystyrene copolymer to prepare PS-*g*-PEG and PS-*g*-PMMA copolymers as shown in Scheme 5.4.

The one pot CuAAC and DA "click" reaction were used to prepare hetero-graft copolymers from maleimide-terminated P*t*BA, alkyne-terminated PEO and anthracene and azide-functionalized PSt backbone. The graft copolymers were prepared with grafting efficiency of 90%–100% with low dispersity. The use of consecutive DA and CuAAC

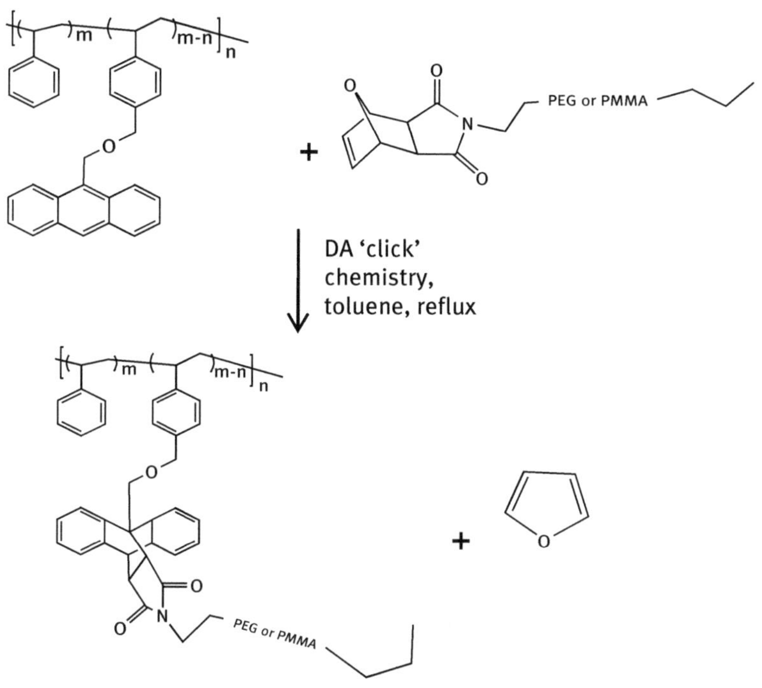

Scheme 5.4: Grafting via in situ retro-DA and DA process. Reproduced with permission from B. Gacal, H. Durmaz, M.A. Tasdelen, G. Hizal, U. Tunca, Y. Yagci and A.L. Demirel, Macromolecules, 2006, 39, 16, 5330. ©2006, ACS [7].

reaction was an easy way to prepare well-defined hetero-graft copolymers. Barner-Kowollik *et al.* demonstrated the preparation of comb polymers using a "grafting onto" approach via RAFT and HDA "click" chemistry [51, 52]. In this process, the copolymer backbone was prepared by copolymerizing trans-hexa 2, 4-dienylacrylate (tHA) with styrene via RAFT polymerization. Thus the prepared copolymer had pendant diene functionality, which was later targeted as grafting sites for the preparation of comb like graft copolymer via DA reaction. n-Butyl acrylate was polymerized separately via RAFT using electron-deficient thioester benzyl pyridin-2-yldithioformate as RAFT reagent. This thioester can acts as a dienophile for DA reaction with the diene groups present in the copolymer backbone. Later these two polymers were clicked via DA reaction at 50 °C to prepare the comb-like graft copolymer, polystyrene-*g*-poly(n-butyl acrylate) (PSt-*g*-PBA). The grafting yield for this reaction varied from 75% to 100%.

Anthracene functionalized oxanorbornene was polymerized by ROMP using 1st generation Grubbs catalyst. Poly(oxanorbornenylanthracene) or poly(oxanorbornene)-*g*-polystyrene-anthracene copolymer was coupled with maleimide end functionalized polymers like PMMA, PEO, and PBA via DA reaction to yield graft and graft-block copolymers [53]. This strategy was further used to prepare brush copolymers with AB block-brush architecture having poly-oxanorbornene in backbone and PCL, PMMA, or PtBA in side chains [54].

5.2.3 Telechelic polymer

Telechelic polymers are macromolecules having reactive end groups that can be used for further polymerization or other reactions [55]. Haddleton and coworkers have reported the synthetic route for the preparation of maleimido-functionalized telechelic polymer via a retro-DA reaction [56]. Well-defined telechelic polymethacrylates were prepared by this strategy, which reduced potential cross-linking side reactions.

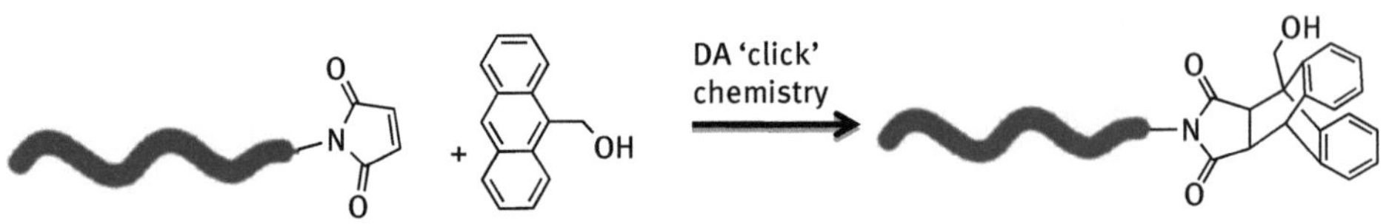

Scheme 5.5: Synthesis of hydroxyl-functionalized telechelic polymer via anthracene-maleimide type DA "click" reaction.
(Source: M. Li, P. De, S.R. Gondi and B.S. Sumerlin, Journal of Polymer Science Part A: Polymer Chemistry, 2008, 46, 15, 5093) [57].

The DA reaction in combination with RAFT polymerization was used to prepare telechelic polymers. Sumerlin *et al.* reported the strategy for the end group modification of RAFT-generated polymers via DA reaction as shown in Scheme 5.5 [57]. In this case, thiol terminated poly(N-isopropylacrylamide) (PNIPAM) was prepared via RAFT polymerization, which was reacted with 1,8 bismaleimide to yield maleimido-terminated PNIPAM. This maleimide dienophile was coupled with 9-anthracene methanol via DA reaction to yield hydroxyl monotelechic polymer. In a different study, thioxanthone end-functionalized PEO was synthesized via DA "click" chemistry strategy.

5.3 Complex macromolecular architectures

The DA reaction is a very useful synthetic tool to prepare complex macromolecular architectures, for example, formation of star polymers, due to its high efficiency, mild reaction conditions, and high tolerance toward various functional groups. Tunca and Hizal *et al.* demonstrated the formation of star polymers via a DA "click" reaction in combination with CuAAC reaction [58]. Several A_3 star polymers containing arms of PEO, PMMA, and PtBA were efficiently synthesized via DA "click" reactions. ABC type star polymers were prepared via a DA reaction by combination of ATRP and NMP processes [59]. The combination of DA and CuAAC reaction was used to prepare three-arm star-block copolymers of (PSt-*b*-PMMA)$_3$ and (PSt-*b*-PEO)$_3$ from maleimide-terminated PMMA or PEO and anthracene-azide terminated PSt [60]. Linear PMMA, PEO, and PSt precursors were prepared via ATRP techniques [61]. Barner-Kowollik and coworkers have reported the formation of star polymers via RAFT and HDA "click" reactions. In this case, first styrene was polymerized via RAFT polymerization followed by HDA reaction leading to formation of four-arms star polymers. They have shown that the coupling efficiency decreases with increasing number of arms. Barner-Kowollik *et al.* also reported the formation of three-arm star polymers of (PSt-*b*-PCL)$_3$ prepared via combination of CuAAC and HDA reaction [62]. In this case, the strategy involved the preparation of α-diene-ω-alkyne functionalized PCL via anionic ring opening polymerization (AROP), a PSt having an electron-deficient dithioester end group via RAFT and a triazide coupling agent as building blocks. In addition, twelve-arm star-block copolymer of poly(isobornyl acrylate) (P*i*BorA) as internal and PSt as external blocks was prepared via a combination of ATRP, RAFT, and HDA reactions [63]. In this case, bromine functionalized multi-functional (P*i*BorA-Br)$_{12}$ was prepared via ATRP and the end –Br group was converted into diene end group and subsequently coupled with diethioester functionality of RAFT polymerized PSt via HDA "click" chemistry.

5.3.1 Dendrimer and dendronized polymers

Coupling of anthracene and maleimide derivatives via the DA "click" reaction has given excellent results in the formation of dendrimers and dendronized polymers, as shown in Scheme 5.6. McElhanon and coworkers have reported the formation of benzyl aryl ether dendrimers up to third generation by the use of thermo-reversible DA and rDA reactions [64]. This group later on reported the formation of fourth generation benzyl aryl ether dendrimers containing bismaleimide moieties in the core and furan moieties at their focal points via DA reactions as shown in Scheme 5.7 [65].

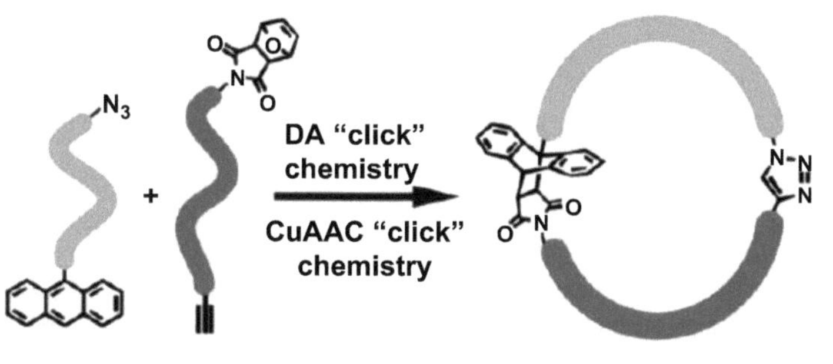

Scheme 5.6: Cyclic block copolymers synthesis via tandem CuAAC and DA "click" reactions. Reproduced with permission from M.A. Tasdelen, Polymer Chemistry, 2011, 2, 10, 2133. ©2011, RSC [6].

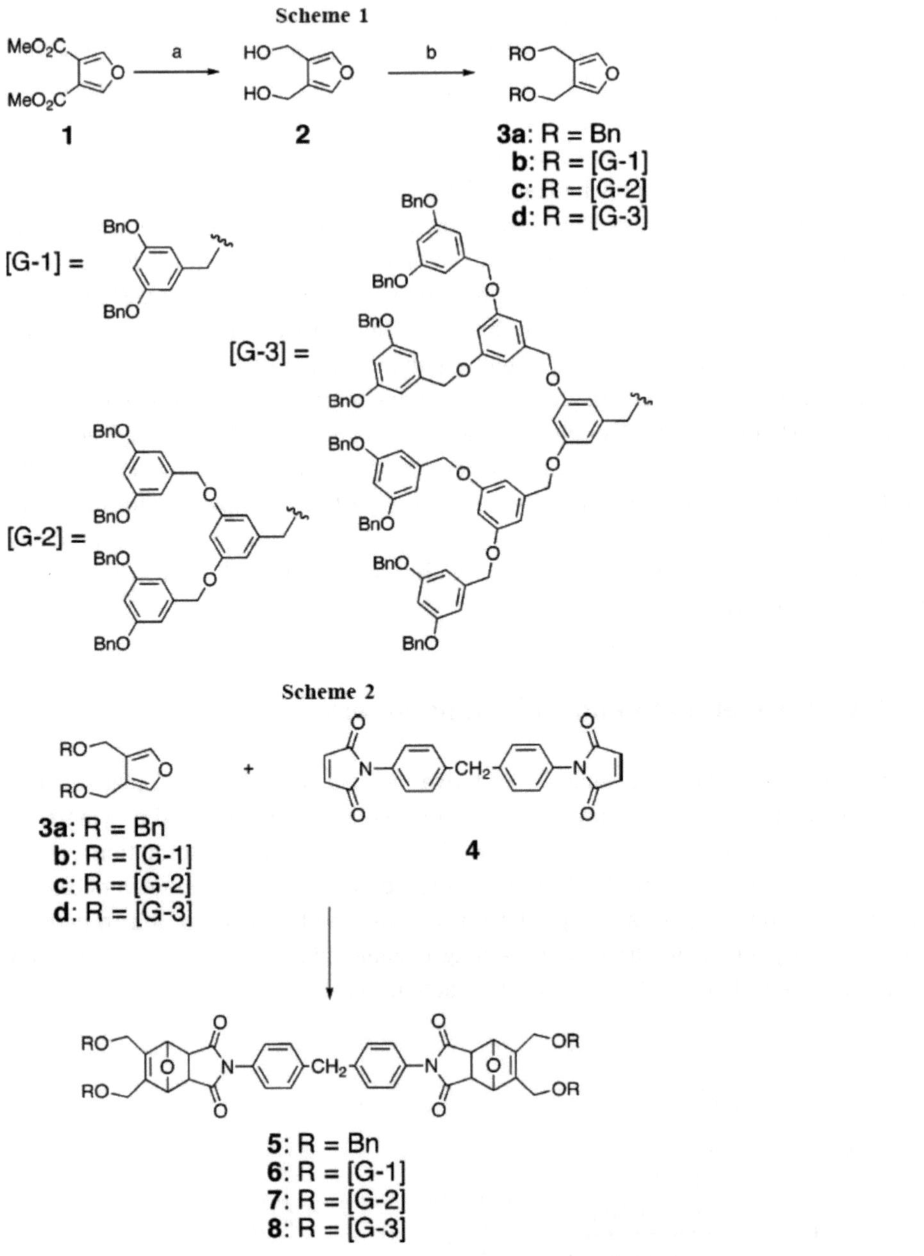

Scheme 5.7: Synthesis of DA dendrimers based on reaction between the functionalized furan dendrons and multifunctional maleimide. Developing covalently reassembling DA dendrimers was based on reacting appropriately functionalized furan dendrons with a multifunctional maleimide. Reproduced with permission from I. Kosif, E.J. Park, R. Sanyal and A. Sanyal, Macromolecules, 2010, 43, 9, 4140. ©2010, ACS [12].

Sanyal *et al.* reported the synthesis of unsymmetrical dendrimers (AB type) via DA reaction between the furan functionalized polyaryl ether dendrons and maleimide functionalized polyester dendrons. Dendronized polymers were prepared via DA "click" reaction using anthracene-bearing PSt backbone and third-generation polyester dendrons containing reactive maleimide at the focal point as shown in Scheme 5.8 [66].

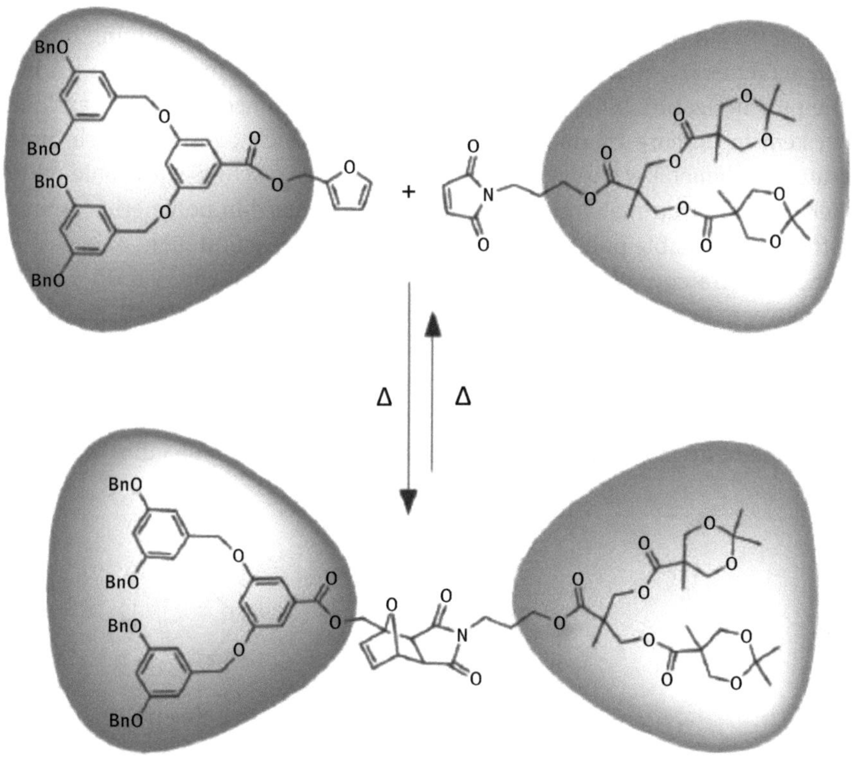

Scheme 5.8: Segment block dendrimers consisting of polyester and polyaryl ether dendrons synthesized reagent free Diels–Alder cycloaddition reactions. Reproduced with permission from M.M. Kose, G. Yesilbag and A. Sanyal, Organic Letters, 2008, 10, 12, 2353. ©2008, ACS [66].

Kakkar *et al.* have reported the synthesis of thermos-responsive dendrimers via of CuAAC and furan-maleimide type DA "click" reaction in a sequential manner [67]. MeElahanon and coworkers have reported the formation of linear dendronized step via CuAAC and DA "click" reactions [68]. First, third generation dendritic bisfuran monomers were prepared by a CuAAC reaction and followed by thermally reversible polymerization by a DA "click" reaction between the bisfuran and bismaleimide.

5.3.2 Cross-linked and self-healing polymers

The reversible DA "click" reaction is being employed for the preparation of thermo-reversible cross-linked polymer networks [69] and hydrogels [70], which are widely being used in self-healing materials and coating applications [71, 72]. Three synthetic strategies have been employed for the preparation of cross-linked polymer network via the DA reaction, namely (i) direct DA cyclo-addition reaction between multifunctional monomers, for example, di- or tri- furan derivative and bismaleimide [73–76], (ii) DA reaction between the linear polymer chain containing furan or maleimide moieties [77–83], and (iii) a cross-linker or an initiator having a DA linkage in the polymerization [84, 85]. Craven *et al.* (1969) first used the furan and maleimide system to prepare thermoreversible polymer, which consisted of saturated condensation polymer backbone bearing furan groups reacting with maleimides. Since then several furan-maleimide systems have been used in polymeric system to prepare thermo-reversible polymeric network via the DA reaction [86].

For the first strategy, Wudl *et al.* prepared self-healing polymer via DA and rDA reaction between multifunctional furan moieties and multifunctional maleimide moieties as shown in Scheme 5.9 [87]. They used small molecules containing four furan moieties and another type of small molecule containing three maleimide moieties to construct a cross-linked polymer network via DA reaction between the furan and maleimide moieties [88]. The multifunctional small molecules were used

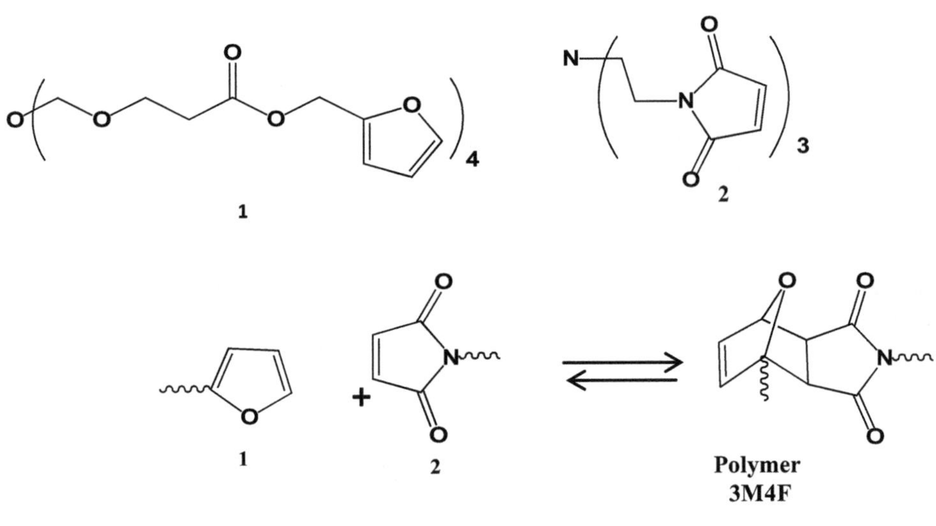

Scheme 5.9: Thermoreversible polymeric materials prepared via DA cycloaddition reaction between a multi diene (multifuran F) and multi dienophile (multimaleimide M).
(Source: X. Chen, M.A. Dam, K. Ono, A.K. Mal, H. Shen, S.R. Nutt, K. Sheran and F. Wudl, Science, 2002, 295, 5560, 1698) [88].

as a monomer for the construction of cross-linked polymers exhibiting highly efficient of self-healing behavior. The main advantage of using polymers having low-melting points is that it can show high mobility in the cross-linked matrix, to heal the crack effectively. Films of these polymers were cut and subjected to thermal healing. The self-healing property was studied by differential scanning calorimetry (DSC), solid-state ^{13}C NMR, and scanning electron microscopy (SEM) analysis.

Liu and Hsieh employed liquid monomers for the construction of self-healing epoxy polymers [89]. They prepared trifuran (TF) and trimaleimide (TMI) compounds that were liquid in room temperature. For this purpose a ring-opening epoxy structure containing furan and maleimide moieties were used as monomers. The mixture of furan and maleimide monomer formed cross-linked polymeric network via DA reaction when the mixture was heated in a low boiling solvent like acetone at 50 °C for 24 h. “To avoid the rDA reaction, the solvent removal was outrightly done at lower temperature.”

For the second approach, Goiti *et al.* and Gandini *et al.* reported the formation of cross-linked polymer network based upon homo- and copolymers of poly(furfuryl methacrylate) (PFMA) via DA reactions [90–95]. They used the free radical polymerization (FRP) technique for the preparation of PFMA, which leads to gel formation. The PFMA contains reactive furfuryl groups which were used in the DA reaction as a diene with bismaleimide as a dienophile. This DA cross-linked gel undergoes rDA reaction leading to get back the starting materials when it is heated at higher temperature.

Kavitha and Singha reported the formation of self-healing polymers based upon the homo- and block copolymers PFMA containing the furfuryl group in the polymeric backbone via ATRP and RAFT polymerization as shown in Scheme 5.10 [96–99]. Later on they used post-polymerization modification to prepare self-healing polymeric materials via a DA “click” reaction.

Wei and coworkers reported the synthesis of thermoresponsive hydrogels based upon the aqueous DA reaction of polymers having pendant furfuryl groups and PEO bismaleimide. For the last strategy, a cross-linker or an initiator containing a DA linkage was synthesized via a series of reactions containing DA cycloaddition and esterification reaction. The thermally labile cross-linker or initiator was converted to polymer networks via different polymerization techniques [100–102].

Recently, Barner-Kowollik *et al.* reported a newly designed low temperature reversible polymeric system capable of self-healing application using HDA “click” reaction. For this purpose a new cyanodithioester (CDTE) compound having an electron deficient C = S bond was employed, which acts as a dienophile in a HDA reaction with cylclopentadiene (C_p) as diene. The CDTE/C_p HDA pair was used to prepare a novel self-healing material. This system showed repetitive cyclability of bonding and debonding in a very rapid fashion (5 min) at a temperature range of

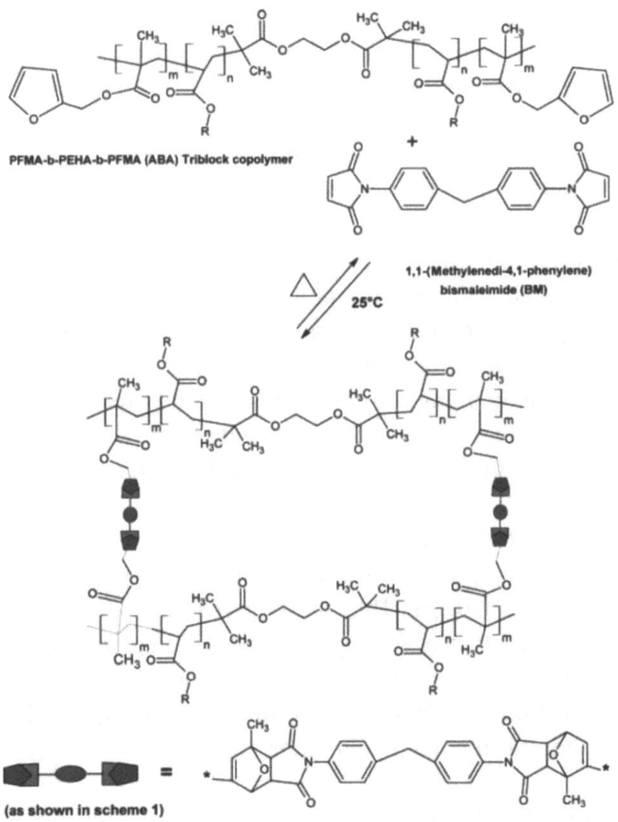

Scheme 5.10: Synthesis of cross-linked Diels–Alder smart polymer. Reproduced with permission from A.A. Kavitha and N.K. Singha, Macromolecules 2010, 43, 7, 3193. ©2010, ACS [98].

40–120 °C. There are several reports on the preparation of thermally mendable polymeric networks via Diels–Alder system based on furan-maleimide, cyclopentadiene, and anthracene-maleimide adducts [103].

Schubert *et al.* reported the synthesis of well-defined poly(furfurylglycidyl ether) homopolymer and poly(ethylene oxide-b-furfurylglycidyl ether) (PEO-*b*-PFGE) block copolymers via RDRP and their self-healing capability via DA and rDA chemistry. They used living anionic ring-opening polymerization (AROP) for the preparation of well-defined homo- and block copolymers of PFGE. The thermo-reversible network was achieved via the DA and rDA reaction between the furan group in the side-chain of the PFGE segments and a bifunctional maleimide cross-linker. This PEO-*b*-PFGE cross-linked polymer was capable of healing complex scratch patterns multiple times when heated to 155 °C. The process was studied by differential scanning calorimetry (DSC), depth-sensing indentation, and profilometry [104].

Schubert *et al.* also reported the formation of terpolymers containing both furan and maleimide moieties in the methacrylate backbone, which were reversibly cross-linked via DA reaction and used as self-healing coatings. The terpolymer of furfuryl methacrylate (FMA), 2-(1,3-Dioxo-3a,4,7,7a tetrahydro-1H-4,7-epoxyisoindol-2(3H)-yl)ethyl methacrylate (MIMA) and alkyl methacrylate (AMA) was synthesized via ATRP. No additional cross-linker was needed as the backbone had both the furan and maleimide moieties. The crack is repaired via heating the materials above retro-DA temperature, and subsequent cooling to room temperature leads to healing the crack due to coupling of two reacting functional group via a DA reaction as shown in Scheme 5.11 [105].

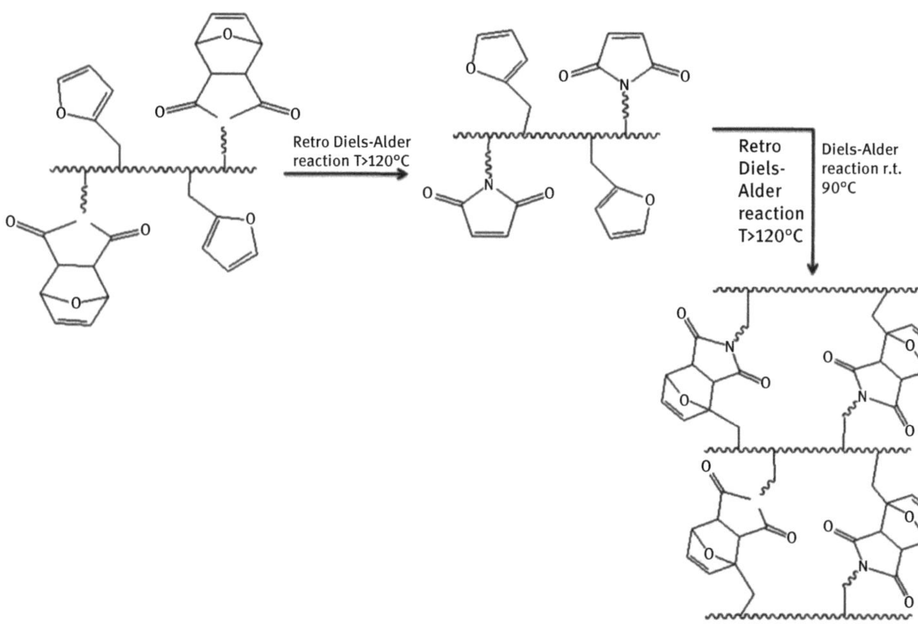

Scheme 5.11: Schematic representation of the self-healing concept via reversible DA reactions. (Source: J. Kötteritzsch, S. Stumpf, S. Hoeppener, J. Vitz, M.D. Hager and U.S. Schubert, Macromolecular Chemistry and Physics, 2013, 214, 14, 1636) [105].

Haddleton *et al.* reported the preparation of self-healing polymers via ATRP as shown in Scheme 5.12. For this purpose two polymerization initiators and one cross-linker was prepared via DA reaction. Well-defined linear and star poly(methyl methacrylate) (PMMA) was prepared using this DA initiator. Thus, PMMA bearing DA adducts in their polymeric backbone have the ability to cleave and reform under thermal stimuli to achieve self-healing characteristics [106].

Yao *et al.* demonstrated the preparation of self-healing thermoplastics via RAFT polymerization [107]. In this case, living PMMA was used as the polymer matrix and

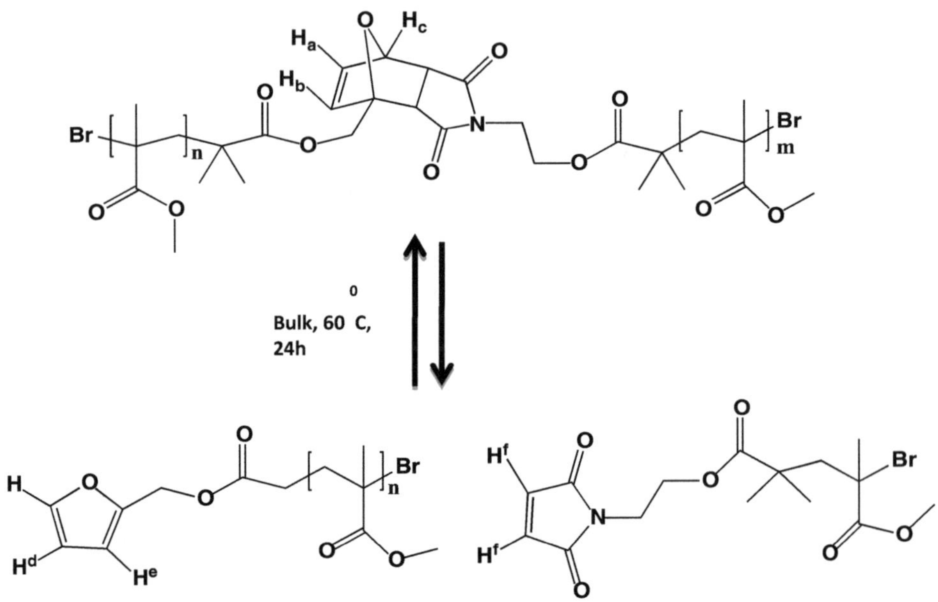

Scheme 5.12: Cleaved polymer via heating and reformed polymer. Reproduced with permission from J.A. Syrett, G. Mantovani, W.R.S. Barton, D. Price and D.M. Haddleton, Polymer Chemistry, 2010, 1, 1, 102. ©2010, RSC [85].

prepared at an ambient temperature using living RAFT polymerization with cumyl phenyl dithioacetate as a RAFT agent. GMA-loaded poly(melamine formaldehyde) (PMF) microcapsules were also added into the living PMMA to supply a fluidic healing agent. As a non- toxic and low viscosity liquid chemical at ambient temperature, GMA possesses C = C bonds that can react with the living PMMA matrix. Upon mechanical damage of the composites, the monomer released from broken microcapsules resumed polymerization with the living matrix, establishing chemical bonding between the cracked planes. As a result, full recovery of mechanical strength was achieved at room temperature without manual intervention.

Zhang *et al.* reported the preparation of self-healing linear polymers based on RAFT polymerization [108]. In this case, PSt composites were fabricated, in which glycidyl methacrylate (GMA)-loaded microcapsules were embedded. The PSt matrix was synthesized by RAFT polymerization, it kept living characteristics and was able to resume the polymerization so long as monomers were available. Upon damage of the composites, the GMA released from the broken capsules infiltrated into the cracks and was copolymerized with the matrix PSt via RAFT polymerization. As a result, the cracked planes were covalently re-bonded, offering recovery of impact strength. Compared to the self-healing system based on ATRP, which is also an approach of living polymerization, the current one possesses robust vitality in air and eliminates the possibility of acceleration of matrix degradation aroused by metal ions.

Pramanik *et al.* reported the preparation of self-healing polymer based on the RAFT polymerization and DA "click" chemistry as shown in Scheme 5.13 [109]. In this case, homopolymer and random copolymers of furfuryl methacrylate (FMA) and butyl methacrylate (BMA) were prepared via RAFT polymerization. The (PFMA-*co*-PBMA) BCP contains reactive furfuryl groups that act as diene to react with bis-maleimide as dienophile via a DA reaction to prepare a cross-linked polymer network. This cross-linked polymer showed self-healing characteristics via DA and rDA reactions. Scheme 5.14 shows the SEM images of the self-healing experiment carried out with the copolymer.

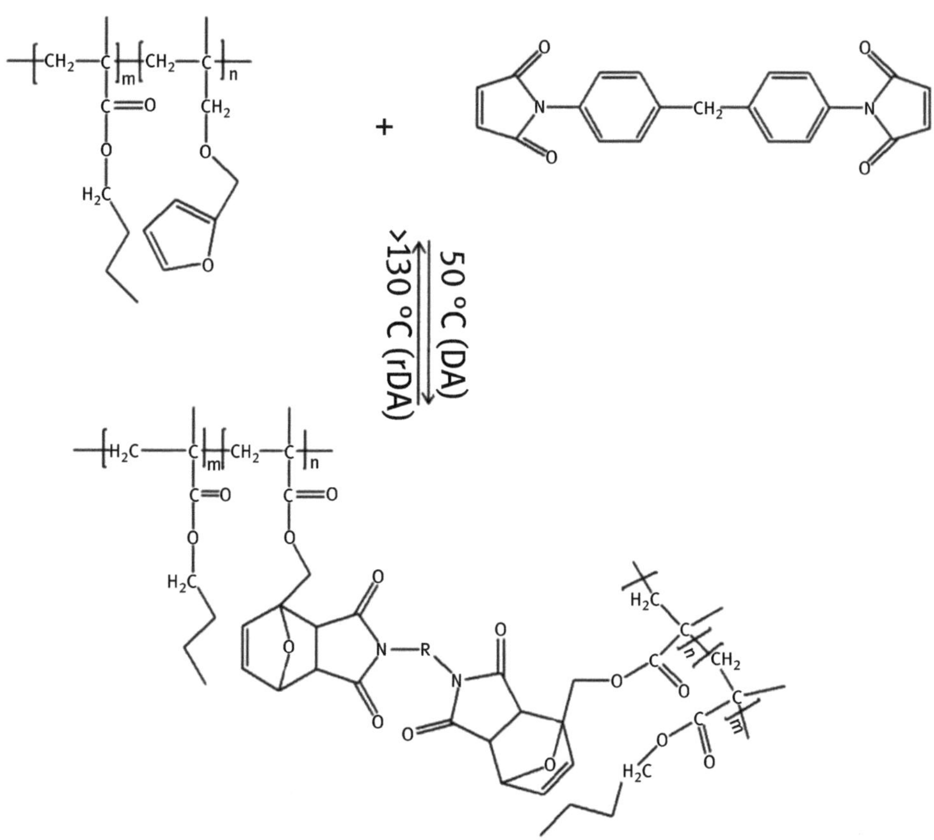

Scheme 5.13: Thermally amendable characteristics of the cross-linked DA polymer. Reproduced with permission from N.B. Pramanik, G.B. Nando and N.K. Singha, Polymer, 2015, 69, 349. ©2015, Elsevier [109].

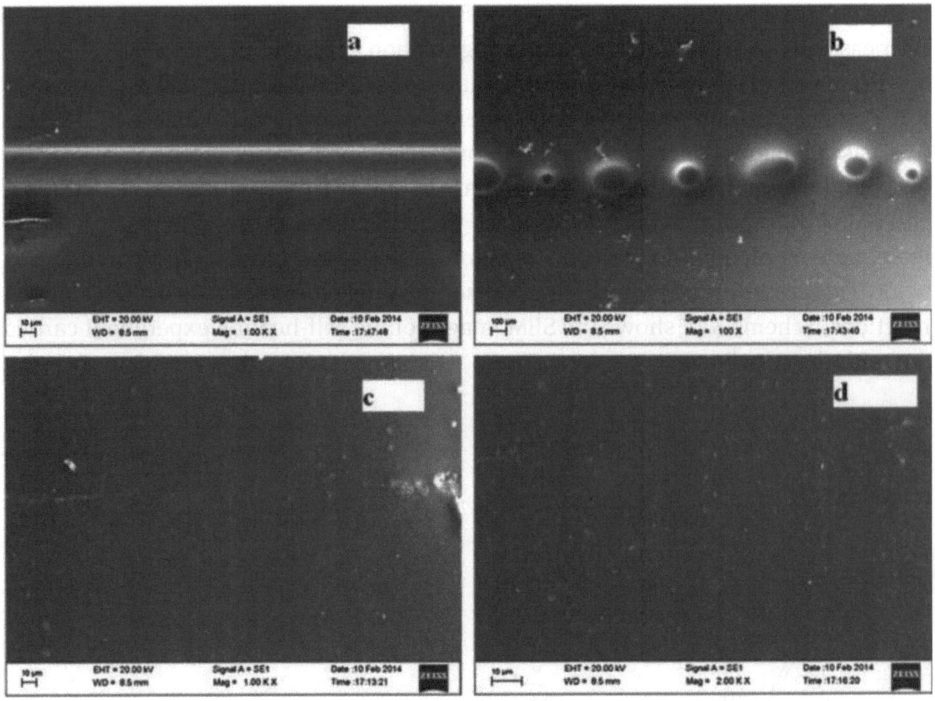

Scheme 5.14: SEM photographs of self-healing experiments of copolymer/BM DA adduct; a) notch surface, b) heated at 130 °C for 1 h, c) heated at 130 °C for 2 h, d) heated at 130 °C for 4 h, and followed by maintaining temperature at 50 °C for 24 h every time. Reproduced with permission from N.B. Pramanik, G.B. Nando and N.K. Singha, Polymer, 2015, 69, 349. ©2015, Elsevier [109].

5.4 Hybrid materials via RDRP and DA chemistry

DA cycloaddition reactions have been extensively used for the functionalization of organic/inorganic nanostructures and surfaces. The graphene planes of carbon black [110, 111], nanotubes [112,–114], fibers [115, 116] and fullerenes [117, 118] are functionalized by DA cycloaddition reactions to their π-electrons. Abetz *et al.* have shown that carbon-rich materials can act as a both diene and dienophile in DA reaction depending upon the reaction partner [119]. Single-walled carbon nanotubes (SWCNTs) are more reactive than multi-walled carbon nanotubes (MWCNTs) and carbon nanofibers are least reactive in this group due to less available surface area. Consecutive DA reactions of 1,3-butadienes with carbon nanofibers formed a polymeric layer onto the surface of nanofibers as shown in Scheme 5.15 [120, 121].

Barner-Kowollik *et al.* reported the functionalization of SWCNTs with cyclopentadiene (C_p) capped polymer as diene via DA reaction at room temperature and at 80 °C as shown in Scheme 5.16 [122]. For that purpose PMMA was prepared with

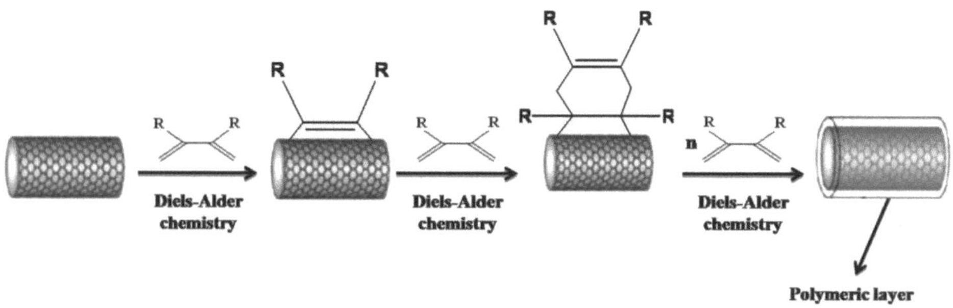

Scheme 5.15: Surface modification of carbon nanotubes with organic molecules via Diels–Alder reactions. Reproduced with permission from M.A. Tasdelen, Polymer Chemistry, 2011, 2, 10, 2133. ©2011, RSC [6].

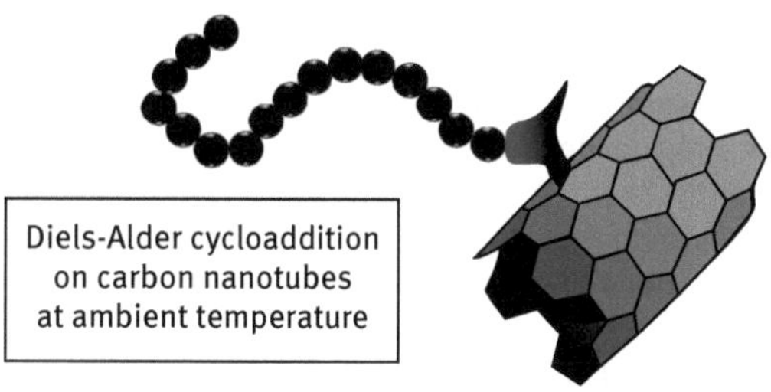

Scheme 5.16: Functionalization of SWCNT with cyclopentadienyl functionalized polymer via Diels–Alder reaction. Reproduced with permission from N. Zydziak, C. Hubner, M. Bruns, and C. Barner-Kowollik, Macromolecules, 2011, 44, 9, 3374. ©2011, ACS [122].

preselected chain length and narrow dispersity via ATRP having a –Br group in the chain end which was later converted into reactive cyclopentadiene group. Similarly, DA and HDA reactions have been used for the functionalization of fullerenes.

Nebhani and Barner-Kowollik have reported a direct and rapid functionalization of fullerence with cyclopentadienyl capped polymer, poly(ethylene glycol) [M_n = 2,000 g/mol] via DA reaction at ambient temperature without any catalyst [123]. These modification processes offer a new and powerful methodology to obtain functionalized carbon nanomaterials with better solubility for different applications.

The DA "click" reaction was also used for the synthesis of polymer hybrid materials onto silicon wafers, tetrahydrosilane, and gold nanoparticles. Pramanik *et al.* reported the functionalization of MWCNTs via grafting of PFMA onto the surface of MWCNTs by a DA click reaction [124]. In this case, PFMA was prepared via RAFT polymerization of FMA and grafted onto the surface of MWCNTs. The grafting was confirmed via SEM, XPS, and TEM analysis. At higher temperature the detachment of polymer layer took place from the MWCNTs surface via the rDA reaction and was confirmed by TEM analysis. The MWCNT grafted with PFMA was soluble in various solvents.

References

[1] a) H.C. Kolb, M.G. Finn and K.B. Sharpless, Angewandte Chemie International Edition, 2001, 113(11), 2056; (b) H.C. Kolb, M.G. Finn and K.B. Sharpless, Angewandte Chemie International Edition, 2001, 40, 11, 2004.

[2] O. Diels and K. Alder, Justus Liebigs Annalen der Chemie, 1928, 460(1), 98.

[3] M.L. Szalai, D.V. McGrath, D.R. Wheeler, T. Zifer and J.R. McElhanon, Macromolecules, 2007, 40(4), 818.

[4] B.S. Sumerlin and A.P. Vogt, Macromolecules, 2010, 43(1), 1.

[5] R.K. Iha, K.L. Wooley, A.M. Nystrom, D.J. Burke, M.J. Kade and C.J. Hawker, Chemical Reviews, 2009, 109(11), 5620.

[6] M.A. Tasdelen, Polymer Chemistry, 2011, 2(10), 2133.

[7] B. Gacal, H. Durmaz, M.A. Tasdelen, G. Hizal, U. Tunca, Y. Yagci and A.L. Demirel, Macromolecules, 2006, 39(16), 5330.

[8] Y. Chujo, K. Sada and T. Saegusa, Macromolecules, 1990, 23(10), 2636.

[9] S.A. Canary and M.P. Stevens, Journal of Polymer Science Part A: Polymer Chemistry, 1992, 30(8), 1755.

[10] J.R. Jones, C.L. Liotta, D.M. Collard and D.A. Schiraldi, Macromolecules, 1999, 32(18), 5786.

[11] J. Canadell, H. Fischer, G. De With and R.A.T.M. Van Benthem, Journal of Polymer Science Part A: Polymer Chemistry, 2010, 48(15), 3456.

[12] I. Kosif, E.J. Park, R. Sanyal and A. Sanyal, Macromolecules, 2010, 43(9), 4140.

[13] U. Mansfeld, C. Pietsch, R. Hoogenboom, C.R. Becer and U.S. Schubert, Polymer Chemistry, 2010, 1(10), 1560.

[14] A.J. Inglis and C. Barner-Kowollik, Macromolecular Rapid Communications, 2010, 31(14), 1247.

[15] A.J. Inglis, S. Sinnwell, M.H. Stenzel and C. Barner-Kowollik, Angewandte Chemie International Edition, 2009, 48(13), 2411.

[16] S. Sinnwell, A.J. Inglis, T.P. Davis, M.H. Stenzel and C. Barner-Kowollik, Chemical Communications, 2008, 17, 2052.

[17] A.J. Inglis, S. Sinnwell, T.P. Davis, C. Barner-Kowollik and M.H. Stenzel, Macromolecules, 2008, 41(12), 4120.

[18] H. Durmaz, B. Colakoclu, U. Tunca and G. Hizal, Journal of Polymer Science Part A: Polymer Chemistry, 2006, 44(5), 1667.

[19] O. Altintas, G. Hizal and U. Tunca, Journal of Polymer Science Part A: Polymer Chemistry, 2008, 46(4), 1218.

[20] O. Altintas, G. Hizal and U. Tunca, Journal of Polymer Science Part A: Polymer Chemistry, 2006, 44(19), 5699.

[21] B. Gacal, H. Durmaz, M.A. Tasdelen, G. Hizal, U. Tunca, Y. Yagci and A.L. Demirel, Macromolecules, 2006, 39(16), 5330.

[22] H. Durmaz, A. Dag, O. Altintas, T. Erdogan, G. Hizal and U. Tunca, Macromolecules, 2007, 40 (2), 191.

[23] C. Gousse and A. Gandini, Polymer International, 1999, 48(8), 723.

[24] K. Kamahori, S. Tada, K. Ito and S. Itsuno, Macromolecules, 1999, 32(3), 541.

[25] S.E. Mallakpour, A.R. Hajipour, A.R. Mahdavian and S. Khoee, Journal of Applied Polymer Science, 2000, 76(2), 240.

[26] J.A. Mikroyannidis, Journal of Polymer Science Part A: Polymer Chemistry, 1992, 30(9), 2017.

[27] G. Alhakimi, H. Gorls and E. Klemm, Macromolecular Chemistry and Physics, 1994, 195(5), 1569.

[28] U. Kumar and T.X. Neenan, ACS Symposium Series, 1995, 614, 518.

[29] H. Ben Romdhane, M. Baklouti, M.R. Chaabouni, M.F. Grenier-Loustalot, F. Delolme and B. Sillion, Polymer, 2002, 43, 2, 255.
[30] G. Alhakimi, E. Klemm and H. G€orls, Journal of Polymer Science Part A: Polymer Chemistry, 1995, 33(7), 1133.
[31] C.L. Schilling, J.A. Reed and J.K. Stille, Macromolecules, 1969, 2(1), 85.
[32] J.K. Stille, R.O. Rakutis, H. Mukamal and F.W. Harris, Macromolecules, 1968, 1(5), 431.
[33] H. Mukamal, F.W. Harris and J.K. Stille, Journal of Polymer Science Part A: Polymer Chemistry, 1967, 5(11), 2721.
[34] A.D. Schluter, Advanced Materials, 1991, 3(6), 282.
[35] U. Scherf and K. Mullen, Advanced Polymer Science, 1995, 123, 1.
[36] N. Teramoto, Y. Arai and M. Shibata, Carbohydrate Polymers, 2006, 64(1), 78.
[37] T. Brand and M. Klapper, Designed Monomers and Polymers, 1999, 2(4), 287.
[38] K.B. Wagener and L.P. Engle, Journal of Polymer Science Part A: Polymer Chemistry, 1993, 31 (4), 865.
[39] H. Durmaz, B. Colakoclu, U. Tunca and G. Hizal, Journal of Polymer Science Part A: Polymer Chemistry, 2006, 44(5), 1667.
[40] O. Altintas, G. Hizal and U. Tunca, Designed Monomers and Polymers, 2009, 12(1), 83.
[41] H. Durmaz, F. Karatas, U. Tunca and G. Hizal, Jounal of Polymer Science PartA: Polymer Chemistry, 2006, 44(13), 3947.
[42] K. Masutani, S. Kawabata, T. Aoki and Y. Kimura, Polymer International, 2010, 59(11), 1526.
[43] L.F. Tietze and G. Kettschau, Topics in Current Chemistry, 1997, 189, 1.
[44] K.A. Jorgensen, Angewandte Chemie International Edition, 2000, 39(20), 3558.
[45] H. Waldmann, Synthesis, 1994, 06, 535.
[46] S. Sinnwell, C.V. Synatschke, T. Junkers, M.H. Stenzel and C. Barner-Kowollik, Macromolecules, 2008, 41, 21, 7904.
[47] A.J. Inglis and C. Barner-Kowollik, Polymer Chemistry, 2011, 2(1), 126.
[48] K.K. Oehlenschlaeger, N.K. Guimard, J. Brandt, J.O. Mueller, C.Y. Lin, S. Hilf, A. Lederer, M.L. Coote, F.G. Schmidt and C. Barner-Kowollik, Polymer Chemistry, 2013, 4(16), 4348.
[49] A.J. Inglis, M.H. Stenzel and C. Barner-Kowollik, Macromolecular Rapid Communication, 2009, 30(21), 1792.
[50] H. Durmaz, A. Dag, O. Altintas, T. Erdogan, G. Hizal and U. Tunca, Macromolecules, 2007, 40 (2), 191.
[51] A. Bousquet, C. Barner-Kowollik and M.H. Stenzel, Journal of Polymer Science Part A: Polymer Chemistry, 2010, 48(8), 1773.
[52] A. Bousquet, C. Boyer, T.P. Davis and M.H. Stenzel, Polymer Chemistry, 2010, 1(8), 1186.
[53] H. Durmaz, A. Dag, N. Cerit, O. Sirkecioglu, G. Hizal and U. Tunca, Journal of Polymer Science Part A: Polymer Chemistry, 2010, 48(24), 5982.
[54] A. Dag, H. Sahin, H. Durmaz, G. Hizal and U. Tunca, Journal of Polymer Science Part A: Polymer Chemistry, 2011, 49(4), 886.
[55] M.A. Tasdelen, M.U. Kahveci and Y. Yagci, Progress in Polymer Science, 2011, 36(4), 455.
56. G. Mantovani, F. Lecolley, L. Tao, D. M. Haddleton, J. Clerx, J. Cornelissen and K. Velonia, Journal of the American Chemical Society, 2005, 127(9), 2966.
[57] M. Li, P. De, S.R. Gondi and B.S. Sumerlin, Journal of Polymer Science Part A: Polymer Chemistry, 2008, 46(15), 5093.
[58] A. Dag, H. Durmaz, G. Hizal and U. Tunca, Journal of Polymer Science Part A: Polymer Chemistry, 2008, 46(1), 302.
[59] H. Durmaz, F. Karatas, U. Tunca and G. Hizal, Journal of Polymer Science Part A: Polymer Chemistry, 2006, 44(1), 499.

[60] H. Durmaz, A. Dag, A. Hizal, G. Hizal and U. Tunca, Journal of Polymer Science Part A: Polymer Chemistry, 2008, 46(21), 7091.

[61] E. Gungor, G. Hizal and U. Tunca, Journal of Polymer Science Part A: Polymer Chemistry, 2009, 47(13), 3409.

[62] S. Sinnwell, A.J. Inglis, M.H. Stenzel and C. Barner-Kowollik, Macromolecular Rapid Communications, 2008, 29(12–13), 1090.

[63] S. Sinnwell, M. Lammens, M.H. Stenzel, F.E. Du Prez and C. Barner-Kowollik, Journal of Polymer Science Part A: Polymer Chemistry, 2009, 47(8, 2207.

[64] J.R. McElhanon and D.R. Wheeler, Organic Letters, 2001, 3(17), 2681.

[65] M. L. Szalai, D. V. McGrath, D. R. Wheeler, T. Zifer and J. R. McElhanon, Macromolecules, 2007, 40(4), 818.

[66] M.M. Kose, G. Yesilbag and A. Sanyal, Organic Letters, 2008, 10(12), 2353.

[67] A. Vieyres, T. Lam, R. Gillet, G. Franc, A. Castonguay and A. Kakkar, Chemical Communications, 2010, 46(11), 1875.

[68] N.W. Polaske, D.V. McGrath and J.R. McElhanon, Macromolecules, 2010, 43(3), 1270.

[69] A.A. Kavitha and N.K. Singha, ACS Applied Material Interfaces, 2009, 1(7), 1427.

[70] H.L. Wei, Z. Yang, L.M. Zheng and Y.M. Shen, Polymer, 2009, 50(13), 2836.

[71] S.A. Canary and M.P. Stevens, Journal of Polymer Science Part A: Polymer Chemistry, 1992, 30(8), 1755.

[72] E. Goiti, M.B. Huglin and J.M. Rego, European Polymer Journal, 2004, 40(2), 219.

[73] C. Gousse and A. Gandini, Polymer Bulletin, 1998, 40(4–5), 389.

[74] F. Ghezzo, D.R. Smith, T.N. Starr, T. Perram, A.F. Starr, T.K. Darlington, R.K. Baldwin and S.J. Oldenburg, Journal of Composite Materials, 2010, 44(13), 1587.

[75] K. Inoue, M. Yamashiro and M. Iji, Journal of Applied Polymer Science, 2009, 112(2), 876.

[76] B.J. Adzima, H.A. Aguirre, C.J. Kloxin, T.F. Scott and C.N. Bowman, Macromolecules, 2008, 41 (23), 9112.

[77] H. Laita, S. Boufi and A. Gandini, European Polymer Journal, 1997, 33(8), 1203.

[78] C. Gousse, A. Gandini and P. Hodge, Macromolecules, 1998, 31(2), 314.

[79] S. Magana, A. Zerroukhi, C. Jegat and N. Mignard, Reactive and Functional Polymers, 2010, 70(7), 442.

[80] Y. L. Liu, C.Y. Hsieh and Y.W. Chen, Polymer, 2006, 47(8), 2581.

[81] Y. Imai, H. Itoh, K. Naka and Y. Chujo, Macromolecules, 2000, 33(12), 4343.

[82] A.A. Kavitha and N.K. Singha, Macromolecules, 2010, 43(7), 3193.

[83] R. Gheneim, C. Perez-Berumen and A. Gandini, Macromolecules, 2002, 35(19), 7246.

[84] W.H. Heath, F. Palmieri, J.R. Adams, B.K. Long, J. Chute, T.W. Holcombe, S. Zieren, M.J. Truitt, J.L. White and C.G. Willson, Macromolecules, 2008, 41(3), 719.

[85] J.A. Syrett, G. Mantovani, W.R.S. Barton, D. Price and D.M. Haddleton, Polymer Chemistry, 2010, 1(1), 102.

[86] J.M. Craven, inventor; US Patent no. 3435003, 1969.

[87] X. Chen, F. Wudl, A.K. Mal, H. Shen and S.R. Nutt, Macromolecules, 2003, 36(6), 1802.

[88] X. Chen, M.A. Dam, K. Ono, A.K. Mal, H. Shen, S.R. Nutt, K. Sheran and F. Wudl, Science, 2002, 295(5560), 1698.

[89] Y.L. Liu and C.Y. Hsieh, Journal of Polymer Science Part A: Polymer Chemistry, 2006, 44(2), 905.

[90] C. Gousśe, A. Gandini and P. Hodge, Macromolecules, 1998, 31(2), 314.

[91] R. Gheneim, C. Perez-Berumen and A. Gandini, Macromolecules, 2002, 35(19), 7246.

[92] E. Goiti, M.B. Huglin and J.M. Rego, European Polymer Journal, 2004, 40(2), 219.

[93] A. Gandini, D. Coelho and A.J.D. Silvestre, European Polymer Journal, 2008, 44(12), 4029.

[94] A. Gandini, A. Silvestre and D. Coelho, Polymer Chemistry 2013, 4(5), 1364.

[95] E. Goiti, F. Heatley, M.B. Huglin and J.M. Rego, European Polymer Journal, 2004(40), 7, 1451.
[96] A.A. Kavitha and N.K. Singha, Journal of Polymer Science Part A: Polymer Chemistry, 2007, 45 (19), 4441.
[97] A.A. Kavitha and N.K. Singha, ACS Applied Material Interfaces, 2009, 1(7), 1427.
[98] A.A. Kavitha and N.K. Singha, Macromolecules, 2010, 43(7), 3193.
[99] N.B. Pramanik, D.S. Bag, S. Alam, G.B. Nando and N.K. Singha, Journal of Polymer Science Part A: Polymer Chemistry, 2013, 51(16), 3365.
[100] H.L. Wei, Z. Yang, Y. Chen, H.J. Chu, J. Zhu and Z.C. Li, European Polymer Journal, 2010, 46(5), 1032.
[101] H.L. Wei, Z. Yang, L.M. Zheng and Y.M. Shen, Polymer, 2009, 50(13), 2836.
[102] H.L. Wei, Z. Yang, H.J. Chu, J. Zhu, Z.C. Li and J.S. Cui, Polymer, 2010, 51(8), 1694.
[103] K.K. Oehlenschlaeger, J.O. Mueller, J. Brandt, S. Hilf, A. Lederer, M. Wilhelm, R. Graf, M.L. Coote, F.G. Schmidt and C. Barner-Kowollik. Advanced Materials 2014, 26(21), 3561.
[104] M.J. Barthel, T. Rudolph, A. Teichler, R.M. Paulus, J. Vitz, S. Hoeppener, M.D. Hager, F.H. Schacher and U.S. Schubert, Advanced Functional Materials, 2013, 23(39), 4921.
[105] J. Kötteritzsch, S. Stumpf, S. Hoeppener, J. Vitz, M.D. Hager and U.S. Schubert, Macromolecular Chemistry and Physics, 2013, 214(14), 1636.
[106] J.A. Syrett, G. Mantovani, W.R.S. Barton, D. Pricecand and D.M. Haddleton. Polymer Chemistry, 2010, 1(1), 102.
[107] L. Yao, M.Z. Rong, M.Q. Zhang and Y.C. Yuan, Journal of Materials Chemistry, 2011, 21(25), 9060.
[108] L. Yao, Y.C. Yuan, M.Z. Rong and M.Q. Zhang, Polymer, 2011, 52(14), 3137.
[109] N.B. Pramanik, G.B. Nando and N.K. Singha, Polymer, 2015, 69, 349.
[110] X.J. Zhou, Q.Y. Li and C.F. Wu, Applied Organometallic Chemistry, 2008, 22(2), 78.
[111] M.T. Beck, G. Mandy, S. Papp and I. Dekany, Colloid Polymer Science, 2004, 283(3), 237.
[112] C. Menard-Moyon, F. Dumas, E. Doris and C. Mioskowski, Journal of the American Chemical Society, 2006, 128(46), 14764.
[113] X. Lu, F. Tian, N.Q. Wang and Q.N. Zhang, Organic Letters, 2002, 4(24), 4313.
[114] C.M. Chang and Y. L. Liu, Carbon, 2009, 47(13), 3041.
[115] B. Xu, X.S. Wang and Y. Lu, Applied Surface Science, 2006, 253(5), 2695.
[116] F.M. Fernandes, R. Araujo, M.F. Proenca, C.J.R. Silva and M.C. Paiva, Journal of Nanoscience and Nanotechnology, 2007, 7(10), 3514.
[117] Y. Vida, R. Suau, J. Casado, A. Berlin, J. T. L. Navarrete and E. Pérez-Inestrosa, Macromolecular rapid communications, 2007, 28(12), 1345.
[118] A. Kraus and K. Mullen, Macromolecules, 1999, 32(13), 4214.
[119] S. Munirasu, J. Albuerne, A. Boschetti-de-Fierro and V. Abetz, Macromolecular Rapid Communications, 2010, 31(6), 574.
[120] M.F. Proenca, R.F. Araujo, M.C. Paiva and C.J.R. Silva, Journal of Nanoscience and Nanotechnology, 2009, 9(10), 6234.
[121] M.C. Paiva, R.M. Novais, R.F. Araujo, K.K. Pederson, M.F. Proenca, C.J.R. Silva, C.M. Costa and S. Lanceros-Mendez, Polymer Composites, 2010, 31(3), 369.
[122] N. Zydziak, C. Hubner, M. Bruns, and C. Barner-Kowollik, Macromolecules, 2011, 44(9), 3374.
[123] L. Nebhani and C. Barner-Kowollik, Macromolecular Rapid Communications, 2010, 31(14), 1298.
[124] N.B. Pramanik and N.K. Singha, RSC Advances, 2015, 5(114), 94321.

Sanjib Banerjee, Arindam Chakrabarty, Nikhil K. Singha and Bruno Ameduri

6 Recent advances in the reversible deactivation radical (co)polymerization of fluorinated alkenes/acrylates/methacrylates/styrenes

6.1 Introduction

Due to their (i) unique combination of extraordinary physical properties (originating from their low polarizability, strong electronegativity, small van der Waals radius of the F atom (1.32 Å), and strong C-F bond (bond energy dissociation = 485 kJ mol^{-1}) and (ii) great diversity of morphologies, ranging from semicrystalline to completely amorphous structure to thermoplastic elastomers, fluorinated (co)polymers have emerged as remarkable speciality macromolecules [1–5]. Hence, fluoropolymers bearing highly F-rich backbones exhibit excellent thermal, chemical, aging/weather resistance, as well as high inertness to solvents and acids, and low flammability/refractive index/dielectric constants/moisture absorption. Polymers bearing a dangling perfluorinated group are highly repellent toward oil, soils, and water. Furthermore, the presence of strong C-F bonds in fluoropolymers improves the resistance to oxidative and hydrolytic degradation, and imparts low surface energy.

Due to the exceptional properties of fluoropolymers as described above, these specialty macromolecules have been extensively used in many high-tech applications such as the automobile industry (~ 300 g of fluoropolymers per car) [5], in components of fuel cells and lithium-ion batteries, in aerospace and aeronautics (as seals/gaskets/O-rings for use at extreme temperatures for tanks of liquid hydrogen or hydrazine in boosters of space shuttles [5, 6]), in the petrochemical industry (pipes and coatings as liners), microelectronics, chemical engineering (for production of high performance membranes [7]), textile treatment, building UV-resistant paints and coatings and for graffiti and stone protection, mainly as coating materials for protecting old heritage

Sanjib Banerjee, Department of Chemistry, Indian Institute of Technology, Bhilai, Raipur, Chhattisgarh, India
Arindam Chakrabarty, JSPS Post-doctoral Fellow, Division of Forest and Biomaterials Science, Graduate School of Agriculture, Kyoto University, Kyoto, Japan
Nikhil K. Singha, Rubber Technology Centre, Indian Institute of Technology Kharagpur, Kharagpur, India
Bruno Ameduri, Engineering and Macromolecular Architectures, Institut Charles Gerhardt UMR (CNRS) 5253, Ecole Nationale Supérieure de Chimie de Montpellier, Montpellier (France)
Sanjib Banerjee and Arindam Chakrabarty contributed equally to this book chapter.

https://doi.org/10.1515/9783110643695-006

monuments [8], and for producing waveguides, core, and cladding of optical fibers [9]. In spite of their high production cost (arising from the small scale of fabrication, the production/purification of the gaseous monomers, requirements of unique polymerization processes/instrumental set up), extensive academic/industrial research is focussed on developing technologies involving fluoropolymers.

The journey of fluoropolymers began with the accidental discovery of polytetrafluroethylene (PTFE) in 1938 by the DuPont Company, prepared by the free radical polymerization (FRP) of tetrafluoroethylene (TFE) [6, 10]. PTFE gained immediate recognition as an efficient coating material due its excellent hydrophobicity, oleophobicity, antisticking, and thermal properties. Since then, the development of new fluorinated polymers has attracted research interest in industry as well as in academia. In spite of their immense potential in high value applications, fluoropolymers suffer from two main limitations: (i) poor solubility in common organic solvents arising from their high crystallinity (especially of homopolymers, PTFE is 95% crystalline) and (ii) difficulty in curing or cross-linking [11]. To overcome these limitations, extensive research was carried out in the 1950s and 1960s to develop a new generation of fluorinated macromolecules [4, 5, 11–13] containing comonomers bearing bulky pendant groups. These groups induce a certain degree of disorder in the macromolecule to suppress the high crystallinity of the fluorinated homopolymers and thus enhance their solubility in organic solvents. By fine-tuning the choice of the comonomer, fluorinated copolymers having exciting properties can be prepared to achieve high-value applications in specific areas.

Free radical polymerization (FRP) methods are generally employed for the synthesis of fluoropolymers based on fluorinated monomers [1–6]. However, FRP techniques suffer from certain limitations in producing polymers with uncontrolled molecular weight, broad dispersity, gel fraction, and so on. In contrast, the development of reversible deactivation radical polymerization (RDRP) techniques allows the limitations of FRP to be overcome. However, few strategies of RDRP have been developed for the polymerization of fluorinated monomers to prepare fluorinated homopolymers, random copolymers, and block copolymers, as reviewed in a recent article [14]. Additionally, RDRP techniques offer the scope to modify various substrates via surface-initiated controlled polymerization. This provides the opportunity to introduce fluoropolymers to improve the water and oil resistance in a substrate. Due to the presence of highly electronegative fluorine atoms, fluoromonomers display certain differences in reactivity as compared to hydrocarbon monomers. Moreover, the incompatibility between a fluoropolymer and its hydrocarbon analogues leads to the self- stratification of fluoropolymer segments. Thus, a polymer composed of a fluoropolymer and its hydrocarbon analogue displays the enrichment of the fluoropolymer phase at the surface as compared to the bulk. This phenomenon has lead researchers to prepare fluorinated random/block copolymers with low fluoropolymer content.

RDRP has undergone a continuous development to become a precision technique to yield polymers bearing pre-designed end groups with precisely controlled molar masses, narrow dispersities (*Đ*), and well-defined architectures (telechelic, block, graft, or star copolymers) [15]. The exceptional and inventive development of modern techniques of RDRP has been discussed in detail in many book and reviewes (more than 13,000 papers have been published on RDRP over the past 20 years) [4, 15, 16]. Thus, in this chapter we will discuss these developments briefly and, for further reading, readers may consult the above references. RDRP is beginning to translate from academic research labs to the industries for the development of adhesives, dispersants, lubricants, high-performance elastomers, for application as novel electrical [17] and biomedical materials [18], in electronics [19], and in aerospace, automotive, and medical industries [20, 21].

This chapter will first briefly review and highlight fundamental aspects of RDRP for the production of well-defined polymers. Some exciting recent examples of macromolecular architectures prepared by RDRP of fluorinated alkenes/acrylates/methacrylates/styrenes for designed applications will also be presented, considering the fact that a more exhaustive review on controlled radical (co)polymerization of fluoroalkenes was published 5 years ago [22]. We conclude with an outline of future directions in the applications of RDRP of fluorinated alkenes toward the synthesis of advanced polymer materials for many high-value applications.

6.2 Fundamentals and Developments of Controlled Radical (Co)polymerization

Since the mid-1990s, extensive research was carried out to develop controlled/living radical polymerization [15]. The origin of control in RDRP process arises from the (i) nature of the (macro)radicals having very short "active" life times) and (ii) fast exchange between living "active" macroradicals and dormant chains (Scheme 6.1). Both features reduce undesirable chain transfer and terminations.

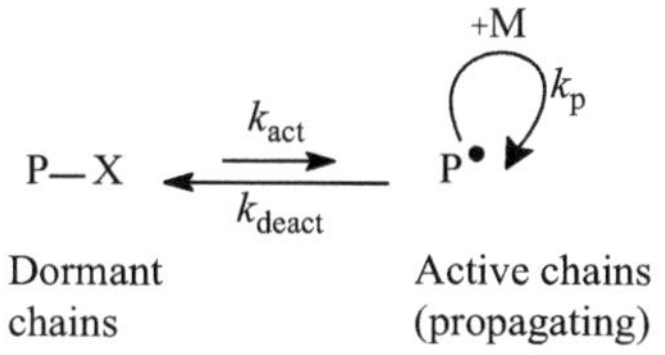

Scheme 6.1: Reversible fast exchange between active and dormant species in RDRP.

In a pioneering work, Tatemoto reported iodine transfer (co)polymerization of fluoroalkenes in 1978 [20, 23]. Subsequently, Otsu and coworkers developed the initiation-transfer-termination (INIFERTER) method [24], but not on fluorinated monomers. Recently, nitroxide-mediated radical polymerization (NMP) [25], atom transfer radical polymerisation (ATRP) [26], reverse iodine transfer polymerization [26, 27], reversible addition-fragmentation chain transfer (RAFT) [28] including macromolecular design via interchange of xanthates (MADIX) polymerization [21, 29], organoheteroatom mediated radical polymerization based on tellurium, bismuth, antimony (TERP, SBRP, and BIRP, respectively) [30], and cobalt mediated radical polymerization (CRMP) using bis(acetylacetonato)cobalt(II) ($Co(acac)_2$) [31] were successfully developed as RDRP techniques. Chung *et al.* [32] reported the synthesis of chain end functionalized fluoropolymers via RDRP using functional borane initiators. Recenlty, Goto *et al.* [33] developed reversible complexation mediated RDRP.

Development of RDRP methods for fluoromonomers has been hindered by the recombination of the growing macroradicals [34]. Thus, various combinations of fluorinated organic compounds have been used as initiators, monomers, chain transfer agents, or solvents to achieve RDRP of fluoromonomers [35]. Detailed surveys of the development of state-of-the-art strategies for RDRP of fluorinated monomers are given in two previous reviews [14, 22]. Thus, this section focuses on recent exciting discoveries in RDRP of fluoroalkenes, especially on iodine transfer polymerization (27) and RAFT/MADIX polymerization, followed by a second section devoted to the RDRP of fluorinated (meth)acrylates and styrenic monomers.

6.3 Controlled Radical Polymerization of Fluoroalkenes

6.3.1 Iodine Transfer Polymerization of Fluoroalkenes

6.3.1.1 History and Production of Copolymers Produced by ITP

Development of ITP of fluoropolymers has been inspired by radical telomerization of fluoroalkenes [36], as reviewed previously [4]. In the late 1970s, Tatemoto introduced the clever concept of RDRP as iodine transfer (co)polymerization of fluoroalkenes including vinylidene fluoride (VDF) [23a]. In ITP of fluoroalkenes, the terminal active bond (pseudo living center) of the growing polymer always contains a C-I bond which originates from the iodine-containing chain transfer agent (CTA) and the monomer, as displayed in Scheme 6.2. The reactivity of this terminal group always remains the same during the entire duration of the polymerization, and even after quenching of the polymerization [13, 37].

The high transfer rate (C_{tr} = 7 at 74 °C [38]) as compared to the termination rate provides the control during ITP. Apostolo's group [37] studied the kinetics of radical copolymerization of VDF and hexafluoropropylene (HFP) in microemulsion to

$$C_nF_{2n+1}—I + (p+1)\ H_2C{=}CF_2 \xrightarrow{R^{\bullet}\ \text{or}\ \Delta} C_nF_{2n+1}\ (C_2H_2F_2)_p—CH_2CF_2—I$$

Scheme 6.2: Iodine transfer polymerization of vinylidene fluoride in the presence of 1-iodoperfluoroalkanes as the chain transfer agent. Reproduced with permission from B. Ameduri, Macromolecules, 2010, 43, 24, 10,163. ©2010, ACS [14].

produce polymers with well-defined microstructure. In ITP termination occurs exclusively by recombination [34].

Peroxide initiators and perfluorinated or chlorofluorinated solvents originally used by Tatemoto *et al.* [23] were subsequently improved by employing diiodo-[23, 39] and polyiodo- compounds [23]. Boyer *et al.* studied ITP of VDF utilizing various 1-iodofluoroalkanes as CTAs. Among the CTAs employed, $C_6F_{13}I$ and $C_6F_{13}CH_2CF_2I$ induce a pseudo-living tendency [38], originating from the lability of CF_2-I bond and the fast decomposition of the $C_6F_{13}I$, whereas $HCF_2CF_2CH_2I$ induces uncontrolled ITP (C_{tr} = 0.7 at 74–140 °C) [38]. However, in the case of unsymmetrical fluoroolefins (such as VDF), after incorporation of ~12–15 VDF units in the growing PVDF macroradical, head-to-head chain defects start to become evident. Reactivation from these defective chain ends (-CF_2CH_2-I) is not possible, resulting in broader dispersities.

Beuermann's group developed ITP of fluoromonomers in supercritical carbon dioxide ($scCO_2$) at 120 °C and 1,500 bar, in a homogeneous phase using $C_6F_{13}I$ [40a] or $IC_6F_{12}I$ [40b] as the CTA, and produced polymers with narrow dispersities ($Đ$ = 1. 2–1.5). The use of $scCO_2$ in the polymerization was inspired by DeSimone *et al.*'s article [41] which verified that $scCO_2$ can solubilize fluoropolymers.

Tatemoto *et al.* [20] also pioneered the synthesis of Hard-*b*-Soft-*b*-Hard triblock copolymers by ITP of VDF and HFP in the presence of an α,ω-diiodoperfluoroalkane (IC_4F_8I or $IC_6F_{12}I$). Subsequent chain extension with VDF (or ethylene and TFE) (Scheme 6.3) yielded thermoplastic elastomers (TPEs) with potential application as O-rings, hot melts, pressure-sensitive adhesives, tough transparent films, and sealants for high tech applications in aeronautics or automotive industries. Daikin Company marketed these TPEs as early as 1984, under the trademark Daiel. They are composed of soft segments (e.g., low T_g blocks containing the poly(VDF-*co*-HFP) as above or composed of poly(VDF-*co*-CTFE), poly(VDF-*ter*-HFP-*ter*-TFE), poly(E-*ter*-PMVE-*ter*-TFE) elastomeric blocks, where CTFE, E, and PMVE stand for chlorotrifluoroethylene, ethylene, and pefluoromethyl vinyl ether, respectively. Hard blocks are composed of various crystalline polymeric sequences (PVDF in case of Dai-el T-630 [20, 42], PTFE, or poly(ethylene-*alt*-tetrafluoroethylene) or poly(E-*ter*-TFE-*ter* -HFP) hard block in case of Dai-el T530) thermoplastics) [42]. After a gap of several years, these TPEs were also produced by Ausimont (now Solvay Specialty Polymers) and DuPont, under the Tecnoflon and Viton tradenames, respectively.

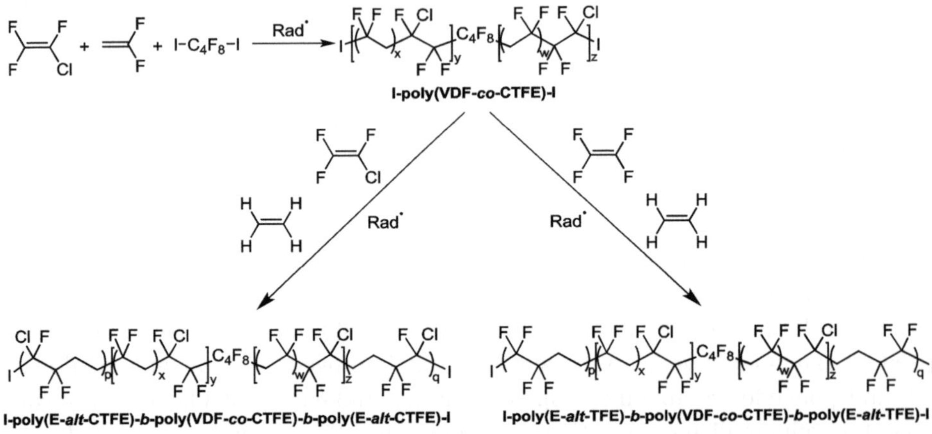

Scheme 6.3: Use of telechelic diiodo poly(CTFE-*co*-VDF) copolymers as chain transfer agent in the radical copolymerization of ethylene and tetrafluoroethylene to achieve triblock thermoplastic elastomers [42a]. Reproduced with permission from F. Boschet and B. Ameduri, Chemical Reviews, 2014, 114, 2, 927. ©2014, ACS [42b].

ITP of VDF and CTFE, first achieved by the Daikin Company [43] in the presence of IC_4F_8I as the CTA, led to I-poly(VDF-co-CTFE)-I (Scheme 6.3), where the mol.% of VDF and CTFE were 55 and 45, respectively, as an elastomeric block (T_g = −7 °C). This α,ω-bis(iodinated) functionalized soft segment was able to undergo chain extension with a poly(E-*alt*-CTFE or TFE) to yield Hard-Soft-Hard triblock copolymer TPEs for use as artificial lenses [43]. The melting points of the hard blocks containing either poly(E-*alt*-CTFE) or poly(E-*alt*-TFE) copolymers were 247 °C or 252 °C, respectively (Scheme 6.3).

Mladenov and coworkers reported the synthesis and characterization of fluorinated copolymers containing VDF and HFP, through the use of 1,6- diiodoperfluorohexan as the CTA via an ITP process [44], as well as fluorinated telomers based on VDF using $C_6F_{13}I$ or C_4F_9I [45].

Recently, Asandei and coworkers developed photomediated controlled radical polymerization and block copolymerization of VDF and other monomers using different metal carbonyls employing $Mn_2(CO)_{10}$- as a visible light photocatalyst, in conjunction with iodoperfluoroalkanes (R_FIs) and respective alkyl halides (Table 6.1). The reactions were carried out in sealed glass tubes. Their extensive research established dimethyl carbonate (DMC) as the best solvent under the reaction conditions [46].

They later employed this $Mn_2(CO)_{10}$-system to prepare a well-defined sulfonated and fluorinated block copolymers for potential application in fuel cell membranes by sequential copolymerization of neopentyl styrene sulfonate and VDF [47]. Subsequently, they employed hypervalent iodide carboxylate, $(CX_3CO_2)_2I^{III}Ph$ to develop a metal-free protocol in conjunction with external $[I(CF_2)_6I]$ and in situ

Table 6.1: Dependence of M_n and M_w/M_n on conversion in $Mn_2(CO)_{10}$-photomediated VDF-IDT.

Monomer (M)	PVDFI or I-PVDF-I		[M]/[PVDF]/ [$Mn_2(CO)_{10}$]	Conv. (%)	M/VDF	M_n	$Đ$
	M_n	$Đ$					
Styrene	2,500	1.34	60/1/2	67	70/30	14,500	2.25
Styrene	11,500	1.48	4,000/1/5	12	92/8	44,700	1.92
Butadiene	1,400	1.48	200/1/1	25	62/38	4,700	2.00
Vinyl Chloride	1,800	1.29	100/1/1	35	77/23	20,100	1.52
Vinyl Acetate	1,500	1.49	100/1/0.2	30	65/35	11,000	1.70
Methyl Acrylate	2,300	1.52	75/1/4	40	72/28	9,000	2.46
Acrylonitrile	2,100	1.31	50/1/1	25	74/26	25,800	2.33

Reproduced with permission from A.D. Asandei, O.I. Adebolu and C.P. Simpson, Journal of the American Chemical Society, 2012, 134, 14, 6,080. ©2012, ACS [46].

generated (CF_3I) iodinated CTAs to achieve RDRP of VDF and to produce poly (VDF)-*b*-poly(M) block copolymers [48] (where “M” stands for styrene, butadiene, 2,2,2-trifluoroethyl methacrylate, and surprisingly methyl-2-(trifluoromethyl) acrylate, known not to homopolymerize under radical conditions [49]. They later modified the end group of PVDF with azide to prepare various functional polymers [50]. Then, these authors studied the effects of metal and ligand employing a series of metal carbonyls in the ITP of VDF and block copolymerization of VDF [51]. Despite the advances of the studies, a major drawback of this procedure stems from the fact that it is of academic interest only and can not be scaled up for commercial applications. Nonetheless, these interesting results could open up synthesis of a wide range of poly(VDF)-*b*-poly(M) block copolymers (where “M” could be vinyl acetate (VAc), vinyl chloride, acrylonitrile, methyl acrylate, styrene, and butadiene).

6.3.1.2 Iodine Transfer Copolymerization of VDF and Perfluoroalkyl Vinyl Ethers

Non-homopolymerizable perfluoroalkyl vinyl ethers (PAVEs) (r_{PAVE} = 0) can be easily copolymerized with either TFE or VDF to yield random copolymers that are semicrystalline or amorphous (with reduced crystallinity compared to PTFE and PVDF) [52]. The key innovative aspect of these polymers is producing materials with T_gs lower than −40 °C for potential low temperature applications [53], along with good thermal stability for potential future application in aeronautics and automotive industries.

Despite extensive reports on successful conventional radical copolymerizations of VDF (or TFE) and PAVEs [52], only one patent and one publication mention the CR copolymerization of VDF with perfluoromethyl vinyl ether (PMVE) via

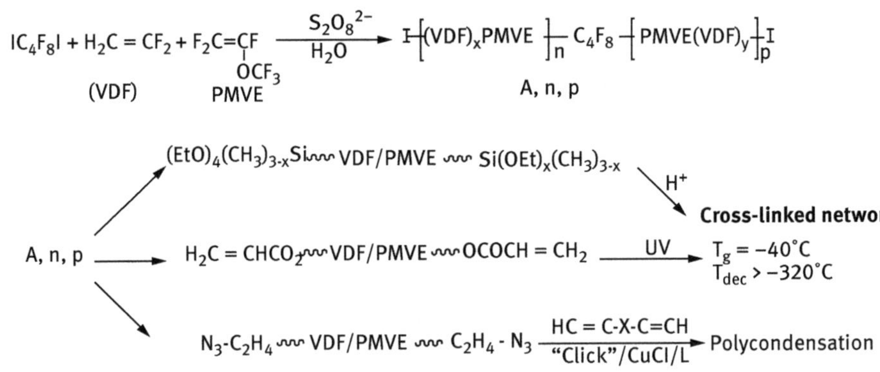

Scheme 6.4: Iodine transfer copolymerization of vinylidene fluoride (VDF) with perfluoromethyl vinyl ether (PMVE) in the presence of 1,4-diiodoperfluorobutane chain transfer agent, followed by chemical modification and cross-linking.
(Source: C. Boyer, B. Ameduri and M.H. Hung, *Macromolecules,* 2010, 43, 8, 3,652) [54].

ITP in conjunction with $C_6F_{13}I$ or α,ω-diiodoperfluoroalkanes as the CTAs [54] (Scheme 6.4) in water, without using any surfactant. The linear molar mass-monomer conversion relationship confirmed the controlled nature of the polymerization [54]. Using the signals assigned to the central -C_4F_8 group of the CTA as a marker, ^{19}F NMR spectroscopy allowed us to determine the molecular weights and to unequivocally identify the end groups (which are exclusively composed of VDF-I). Indeed, the higher the molecular weight the lower the –CH_2CF_2-I content and the higher the –CF_2CH_2-I content. The latter could not reactivate chains because the bond dissociation energy of -CH_2-I is higher than that of –CF_2-I [54]. The resulting poly(VDF-*co*-PMVE) copolymers exhibit low T_g values (ca. –60 °C) and satisfactory thermal stability upon cross-linking by the chemical modification of these diiodides into original telechelic bis(hydroxyl) [55], diazido [56], diacrylates [55], bis(triethoxysilanes) [57], bis(methyldiethoxysilanes) [57] poly(VDF-*co*-PMVE) copolymers.

Ameduri and Alaaddine developed a method to synthesize block copolymers via controlled free radical copolymerization of trifluoroethylene (TrFE) with another fluoromonomer such as VDF, or 2,3,3,3-tetrafluoroprop-1-ene (1234) in the presence of 1-iodo-PVDF-macro CTA (Scheme 6.5) [58].

Lopez *et al.* have reported the synthesis of various fluorinated macro-chain transfer agents (CTAs) based on VDF, chlorotrifluoroethylene (CTFE), or CTFE and vinylidene chloride (VDC) to synthesize block polymers by ITP (Scheme 6.6) [59].

In summary, ITP was the first RDRP that led to commercial production in the fluoropolymer industry, pioneered by the Daikin Company and followed by Dupont and Solvay Speciality Polymers. Further academic studies and improvements have been carried out.

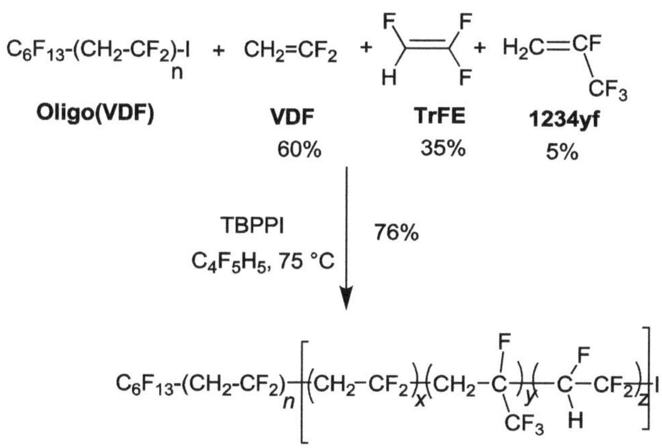

Scheme 6.5: Synthesis of PVDF-*b*-poly(VDF-*ter*-TrFE-*ter*-1234) block copolymers by ITP, where TrFE, 1,234 and TBPPI represent trifluoroethylene, 2,3,3,3-tetrafluoroprop-1-ene and *tert*-butylhydroperoxide.
(Source: B. Ameduri and A. Allaaddine, US Patent 2015/0119523 2015 (deposited by Arkema and CNRS).

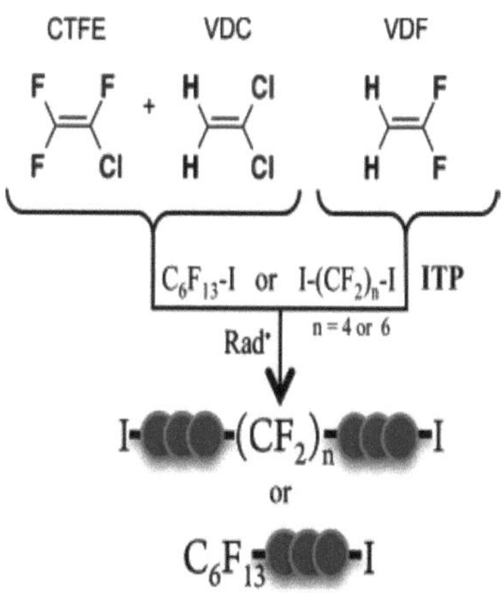

Scheme 6.6: Synthesis of fluorinated macro-chain transfer agents by ITP of various halogenated monomers. Reproduced with permission from G. Lopez, A. Thenappan and B. Ameduri, *ACS Macro Letters,* 2015, 4, 1, 16. ©2015, ACS [59].

6.3.2 Reversible Addition-Fragmentation Chain Transfer (RAFT)/ Macromolecular Design via the Interchange of Xanthates (MADIX) Polymerization of Fluoroalkenes

Reversible addition-fragmentation chain transfer (RAFT)/macromolecular design via the interchange of xanthates (MADIX) polymerization, pioneered by the Rhodia Company [21] for RDRP of VAc, has recently emerged as one of the most promising RDRP techniques because of its tolerance toward diverse functional groups, applicability to a wide range of vinyl monomers, and its less-stringent experimental conditions [28]. The fundamentals of RAFT polymeriwation were recently reviewed [28g]. Thus, the subsection below focuses on recent advances toward the synthesis of specifically designed fluorinated copolymers via MADIX.

6.3.2.1 RAFT/MADIX Homopolymerization of Fluoromonomers

Development of initiating systems to achieve RDRP of fluoromonomers, especially VDF, is still very challenging. Guerre and coworkers has developed an efficient protocol for the first MADIX polymerization of VDF in DMC using O-ethyl-S-(1-methoxycarbonyl)ethyldithiocarbonate as the CTA (Scheme 6.7) [60].

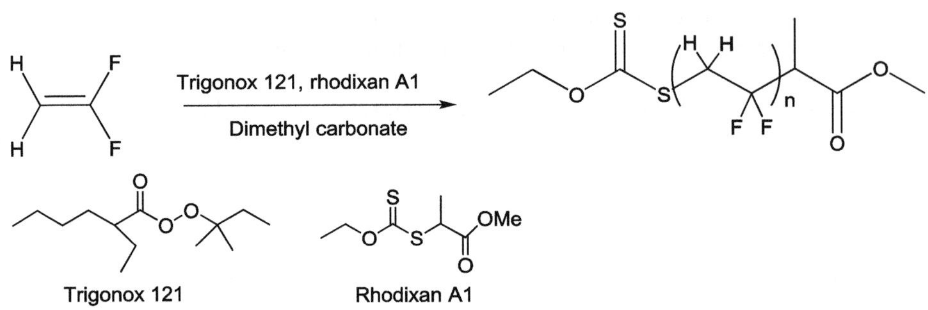

Scheme 6.7: Synthesis of PVDF by RAFT/MADIX polymerization of vinylidene fluoride. Reproduced with permission from M. Guerre, B. Campagne, O. Gimello, K. Parra, B. Ameduri and V. Ladmiral, *Macromolecules,* 2015, 48, 21, 7,810. ©2015, ACS [60].

A comprehensive NMR study was performed to determine the end-group functionality of the polymer, mainly $-CH_2CF_2$-SC(S)OEt and $-CF_2CH_2$-SC(S)OEt. This technique allowed us to prepare relatively well-defined PVDF. Detailed chain-end analysis using MALDI-TOF MS as well as ^{1}H, ^{19}F, and HETCOR ^{1}H–^{19}F NMR established that VDF reverse additions and chain transfer reactions to solvent (DMC) are responsible for a significant loss of CTA and the accumulation of unreactive polymer chains in the reaction medium, leading to a loss of control of the polymerization [60].

Subsequenty, the chemical modification of the xanthate end group of the above PVDF into thiols via two strategies was carried out (viz. aminolysis and elimination using sodium azide). The thios were then immediately added onto the acrylate functionalities of 3-(acryloyloxy)-2-hydroxypropyl methacrylate (AHPMA) via regioselective thia-Michael addition to form new PVDF-MA macromonomers. Results revealed that the aminolysis procedure gave better coupling efficiency and better-defined PVDF-MA macromonomers. Thus synthesized PVDF-MA macromoner was then copolymerized with methyl methacrylate (MMA), resulting in the complete conversion of the macromonomer and synthesis of novel block copolymers (Scheme 6.8) [61]. In addition, such PVDF-xanthates are precursors for PVDF-*b*-PVAC block copolymers (Scheme 6.9) [62].

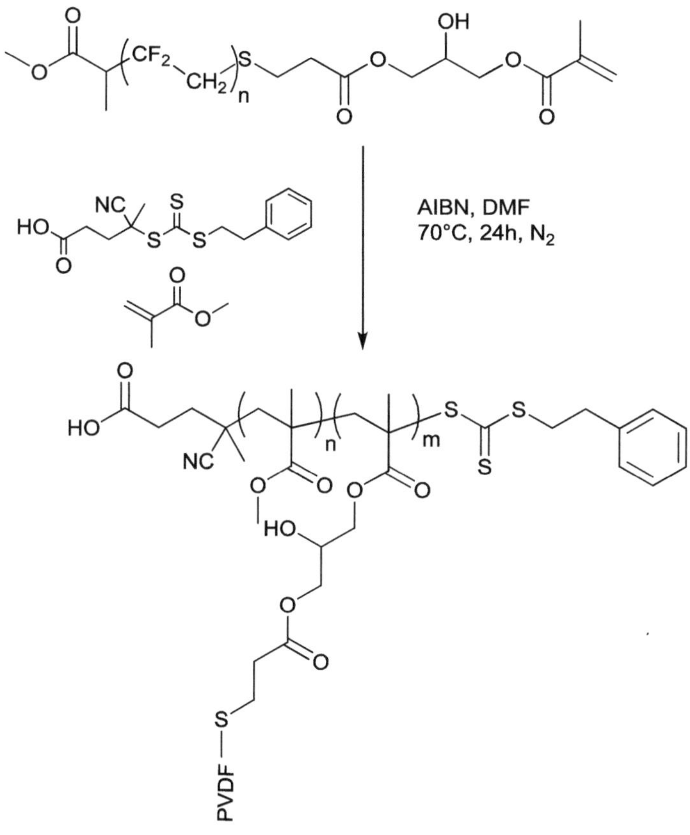

Scheme 6.8: RAFT copolymerization of PDVF-methacrylate macromonomer with MMA. Reproduced with permission from M. Guerre, B. Ameduri and V. Ladmiral, *Polymer Chemistry,* 2016, 7, 2, 441. ©2016, RSC [61].

6.3.2.2 RAFT/MADIX Copolymerization of Fluoromonomers

First Kostov *et al.* developed a fluorinated xanthate, $C_6F_{13}C_2H_4OCOCH(CH_3)SC(S)OEt$, to achieve efficient MADIX copolymerization of VDF with 3,3,3-trifluoropropene (TFP) [62]. The resulting fluorinated poly(VDF-*co*-TFP)-macroxanthate was subsequently chain extended by a further MADIX polymerization of VAc to produce novel poly(VDF-*co*-TFP)-*b*-oligo(VAc) block copolymers (Scheme 6.9).

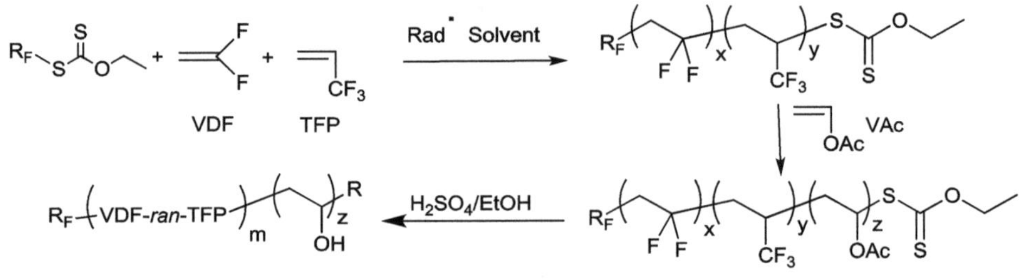

Scheme 6.9: Sequenced MADIX terpolymerization of VDF, TFP, and VAc controlled by a fluorinated xanthate for the synthesis of poly(VDF-*co*-TFP)-*b*-oligo(VAc) block copolymers further hydrolyzed into original surfactants.
(Source: G. Kostov, F. Boschet, J. Buller, L. Badache, B. Ameduri and S. Brandstadter, *Macromolecules,* 2011, 44, 7, 1841) [62].

Hydrolysis of the oligo(VAc) segment under acidic conditions produced amphiphilic poly(VDF-*co*-TFP)-*b*-oligo(vinyl alcohol) diblock copolymers. These block oligomers act as novel surfactants [63] endowed with low critical micellar concentrations and surface tensions (17 $mN.m^{-1}$ for 0.5% of copolymer in water), similar to perfluorooctanoic acid (PFOA). These block copolymers would be such alternatives to commercially available surfactants such as PFOA and perfluorooctane sulfonic acid (PFOS) (regarded as toxic, persistent and bioaccumlulable, and even suspected to be mutagenic) due to their small size and the high stability of the perfluorinated groups [63–65].

The xanthate mediated MADIX polymerization was subsequently extended successfully to develop polymeric surfactants for $scCO_2$ media. The developed surfactants displayed very interesting CO_2-philicity behavior. The synthesis involves the copolymerization of VDF and PMVE from hydrophilic poly(N,N-dimethylacrylamide) macro-RAFT agents (Scheme 6.10) [66].

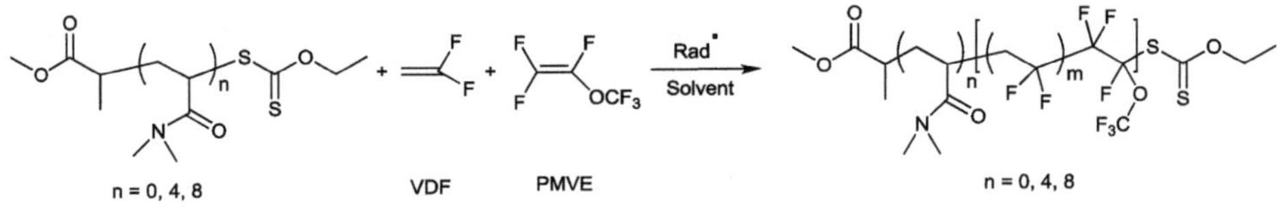

Scheme 6.10: Synthesis of PDMAc-*b*-poly(VDF-*co*-PMVE) block copolymers by RAFT/MADIX copolymerization of vinylidene fluorine (VDF) and perfluoromethyl vinyl ether (PMVE) from a macroxanthate based on oligo(N,N-dimethyl acrylamide) for the synthesis of block copolymers. (Source: E. Girard, J. D. Marty, B. Ameduri and M. Destarac, *ACS Macro Letters,* 2012, 1, 2, 270) [66].

A similar strategy was successfully applied for the CR copolymerization of VDF and TrFE for development of electroactive polymers (Scheme 6.11) [58]. Arkema claimed

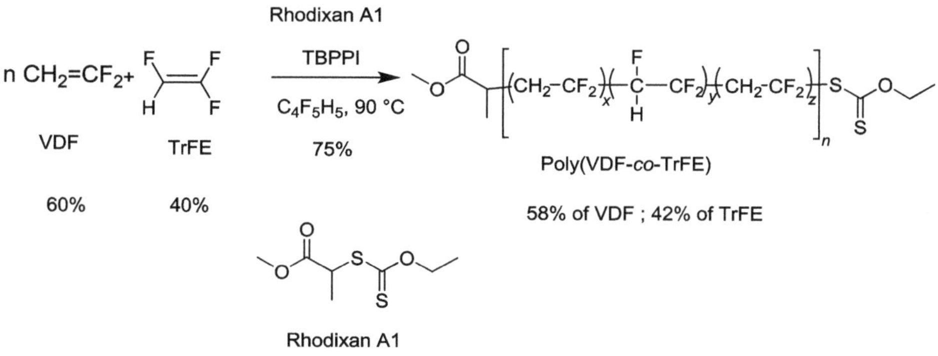

Scheme 6.11: Synthesis of block copolymers by MADIX polymerization VDF and TrFE. (Source: B. Ameduri and A. Allaaddine, US Patent 2015/0119523 2015 (deposited by Arkema and CNRS). [58].

the successful synthesis of block copolymers containing VDF, TrFE, and 1234yf units by MADIX polymerization [58]. In addition, another comonomer, *tert*-butyl 2-trifluoromethacrylate (MAF-TBE) (which can not be polymerized under radical conditions) was copolymerized with VDF, using bis (4-*tert*-butyl cyclohexyl) peroxydicarbonate as the initiator and *O*-ethyl-*S*-(1-methyloxycarbonyl) ethyl xanthate as the CTA via RAFT/MADIX polymerization (Scheme 6.12) producing copolymers with controlled molar masses ranged between 1,500 to 40,000 g.mol^{-1} and dispersities lower than 1.8 [67]. Chain extension of these poly(VDF-*co*-MAF-TBE) copolymers bearing a xanthate end-group was carried out either with VAc or VDF to produce poly(VDF-*co*-MAF-TBE)-*b*-poly(VAc) and poly(VDF-*co*-MAF-TBE)-*b*-PVDF block copolymers, respectively.

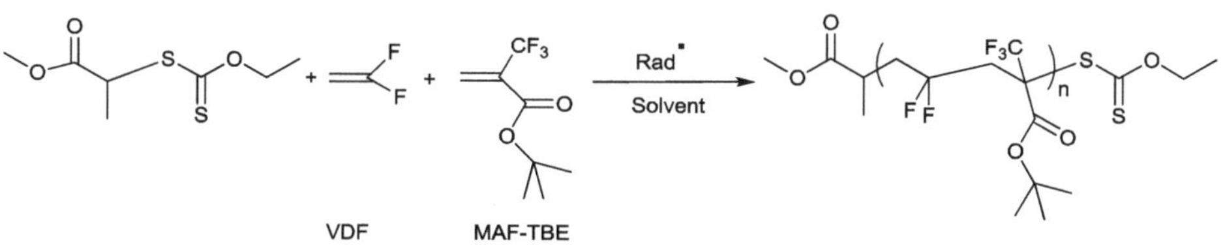

Scheme 6.12: Radical copolymerization of vinylidene fluoride (VDF) with *t*-butyl 2-trifluoromethacrylate (MAF-TBE) controlled by *O*-ethyl-*S*-(1-methyloxycarbonyl) ethyl xanthate. (Source: Y. Patil and B. Ameduri, Polymer Chemistry, 2013, 4, 9, 2783) [67].

Interestingly, such PVDF-based fluorinated copolymers have successfully been employed as compatibilizers for the preparation of poly(VDF-*co*-HFP) copolymer/functional silica nanocomposites as novel proton exchange membrane fuel cells (PEMFCs) by reactive extrusion [68]. The synthesis of these composites involved in situ generation of the inorganic phase (via sol-gel chemistry from mercaptopropylethoxysilane). For the best performance, the compatibilizer has to be incorporated between the organic and inorganic phases. The resulting composites bearing mercaptan functionalities were oxidized into SO_3H groups under various conditions. The resulting original membranes exhibited theoretical ion exchange capacity (IEC) of ca. 2 meq/g. (whereas the experimental ones ranged between 1.0 and 1.3 meq/g) and proton conductivities of these membranes reached 30–80 $mS.cm^{-1}$ at room temperature and 100% RH. Interestingly, their water swelling rates were unexpectedly lower than that of commercially available Nafion 212 [68]. Liu *et al.* [69, 70] reported the MADIX copolymerizations of CTFE (or HFP) with n-butyl vinyl ether (BVE) initiated under ^{60}Co γ-ray irradiation in the presence of S-benzyl O-ethyl dithiocarbonate (Scheme 6.13). The synthesized poly(CTFE-*alt*-BVE) copolymers were used as macroCTAs for the chain extension of VAc, leading to poly(CTFE-*alt*-BVE)-*b*-poly(VAc) diblock copolymers [69], whereas the poly(HFP-*alt*-BVE) copolymer end-capped with a sulfonic acid group was prepared by oxidation of the xanthate end group of the poly(HFP-*alt*-BVE) copolymer [70].

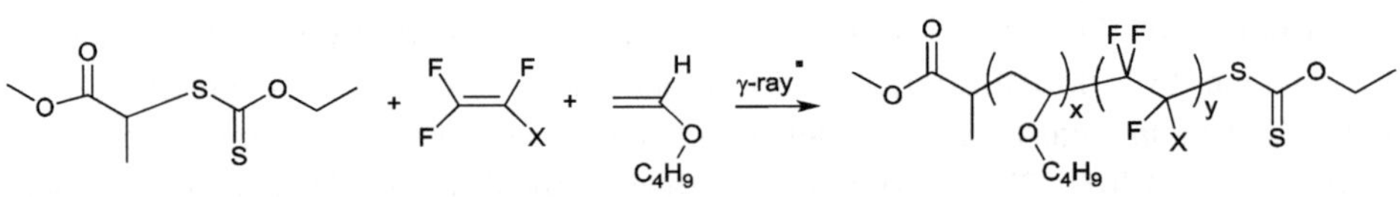

Scheme 6.13: Xanthate-mediated copolymerization of HFP (and/or CTFE) and *n*-butyl vinyl ether (BVE) under ^{60}Co γ-ray irradiation. (Source: L. Liu, D. Lu, H. Wang, Q. Dong, P. Wang and R. Bai, *Chemical Communications*, 2011, 47, 27, 7,839 and P. Wang, J. Dai, L. Liu, Q. Dong, B. Jin and R. Bai, *Polymer Chemistry*, 2013, 4, 6, 1,760) [69, 70].

6.3.2.3 RAFT (Co)polymerization of Fluoromonomers controlled by Trithiocarbonates

Wang's team [71] synthesized telechelic poly(VDF-*co*-HFP) copolymer by RDRP and then used it as a macro-CTA for a subsequent RAFT polymerization of a methacrylate imidazolium to prepare a series of triblock copolymers composed of poly(VDF-*co*-HFP) and an ionic liquid segment composed of imidazolium methacrylate monomer. The authors observed that due to the enhanced segmental mobility, the polymers bearing trifluorosulfonate counterions displayed higher ionic conductivities than their BF_4^- homologues. The temperature dependence of the ion mobilities of the triblock copolymers, as described using a Vogel-Tamman-Fulcher equation, indicates a strong correlation between ion conduction and polymer segmental dynamics.

6.3.3 Atom Transfer Radical Polymerization of Fluoroalkenes

Wang *et al.* [72] reported the unexpected controlled radical copolymerization of CTFE with chloromethylstyrene (CMS) using atom transfer radical polymerization (ATRP) between an electron rich CMS inimer (that acts as both monomer and initiator) and an electron-withdrawing monomer (CTFE) that yielded hyperbranched poly(CTFE-*co*-CMS) copolymers. The authors claimed that ATRP occurred on the chlorine atoms of CMS but surprisingly did not occur from CTFE. The attempts to homopolymerize CTFE by ATRP systematically failed, favoring the idea that the poly(CTFE-*co*-CMS) copolymers exhibit such an alternating structure, evidenced by elemental analyses. This hyperbranched polymer, soluble in common organic solvents, was amorphous with a glass transition temperature of 88 °C (intermediate between those of PCTFE and polystyrene (PSt), 57 and 100 °C, respectively).

In conclusion, among the RDRP methods employed for the synthesis of fluoropolymers from fluoroalkenes, ITP, RAFT/MADIX polymerization has emerged as the best possible technique. Controlled polymerization of fluoroalkenes has the potential to lead to many exciting new materials with unique properties.

6.4 Controlled Radical Polymerization of Fluorinated Acrylates/Methacrylates/Styrenes

6.4.1 Nitroxide Mediated Polymerization of Fluorinated Acrylates/Methacrylates

Nitroxide Mediated Polymerization (NMP) is based on the termination of propagating radical to TEMPO-based organic compounds. In this case, the polymer chains contain TEMPO-end groups. There are only few examples of TEMPO-mediated polymerization of fluorinated acrylates/methacrylates (FA/FMA). In 2012, Barth *et al.* [73] reported the TEMPO-mediated polymerization of 1H, 1H, 2H, 2H-tridecafluorooctyl methacrylate (TDFOMA) in bulk.

Martinelli *et al.* [74] prepared a series of amphiphilic triblock copolymers based on PSt modified with polyethylene glycol, polysiloxane, and perfluoroalkyl side chains. In this case, the triblock copolymers had been prepared by sequential NMP process. However, the surface composition did not follow the sequential position of different blocks as observed from elemental analysis. The migration of perfluoroalkyl side chains toward the surface was more prominent decreasing the surface energy of the block copolymer.

6.4.2 Atom Transfer Radical Polymerization of Fluorinated Acrylates/Methacrylates

ATRP process is based on the reversible electron transfer between propagating radical and transition metal complex. In this case, an alkyl halide is used as initiator and a transition metal complex act as the catalyst. Scheme 6.14 displays the chemical structures of some ligands used to prepare the transition metal complex for the successful polymerization of FA/FMA. There are reports on development of

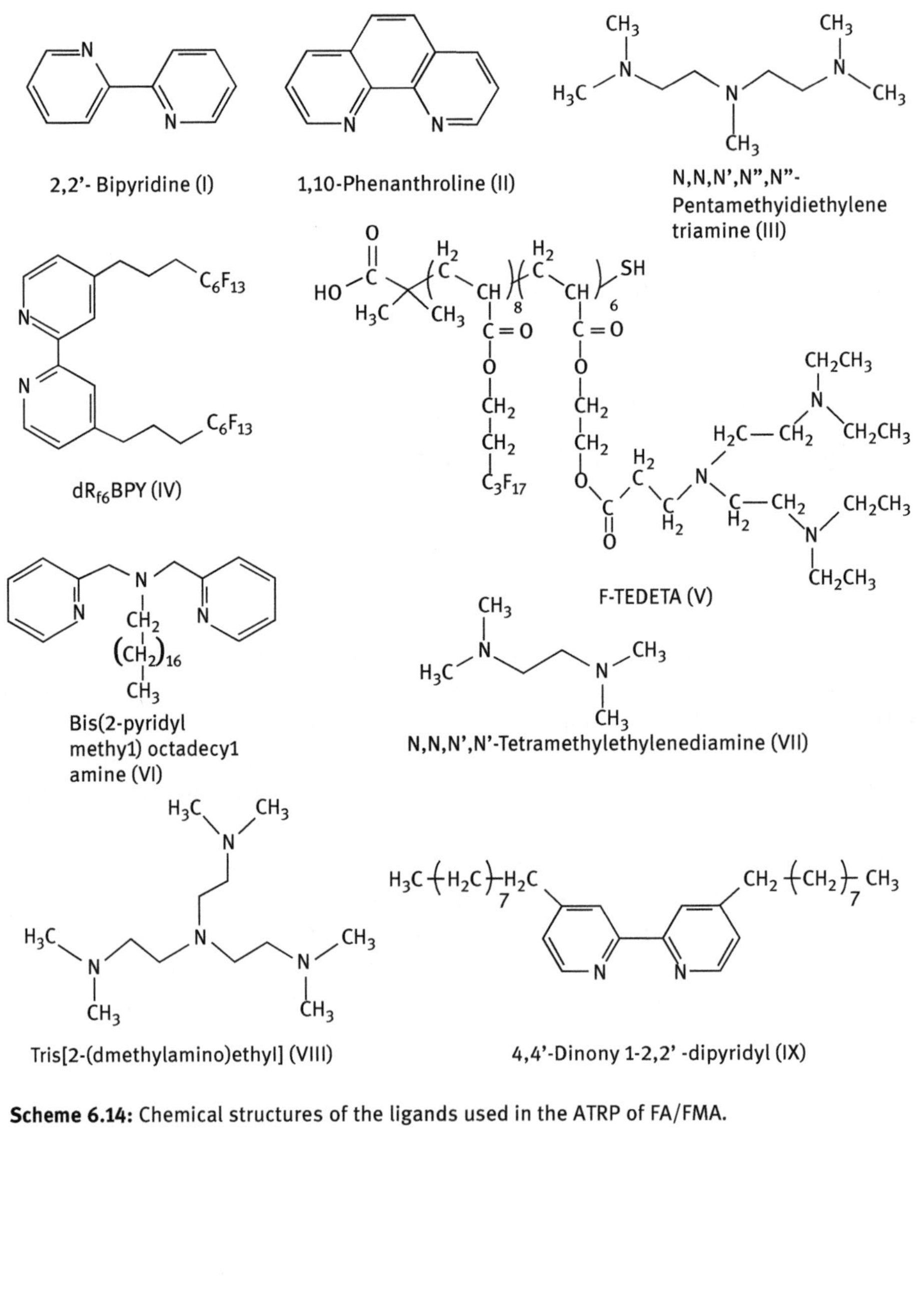

Scheme 6.14: Chemical structures of the ligands used in the ATRP of FA/FMA.

fluorinated homopolymers, random copolymers, and block copolymers by using the ATRP process. This process also offers the scope to prepare polymer brushes grown on a substrate. Researchers took this opportunity to modify verious surfaces and termed the polymerization process as surface-initiated ATRP (SI-ATRP). There is another category of ATRP process named activators generated by electron transfer ATRP (AGET-ATRP) where the catalyst is generated in reaction medium.

6.4.2.1 Fluorinated Homo & Random Copolymers via ATRP

In 2011, He *et al.* [75] reported the ATRP of 2,2,2-trifluoroethyl methacrylate (TFEMA). In this case, the authors studied the polymerization using different types of ligands like **I, III,** and **VIII.** Interestingly, the polymer displayed more controlled molecular weights and narrow dispersities with **I** as ligand, as compared to **III.** However, the polymerization using **VIII** as the ligand led to lower monomer conversion with gel formation. The inequality between activation and deactivation process with ligands of different denticity (the number of donor groups in a single ligand that bind to a central atom in a coordination complex) was supposed to be responsible for this phenomenon. ATRP of pentafluorophenyl methacrylate (PFPMA) was reported by Singha and coworkers [76]. The polymerization was carried out by using ligand **III** (Scheme 6.15). Interestingly, the polymer had an activated ester which can be easily replaced by a primary amine. Using azobisisobutyronitrile (AIBN) as a thermal radical source, the peptide could be grafted onto the polymer with a 50% grafting density, as demonstrated by SEC and NMR analysis.

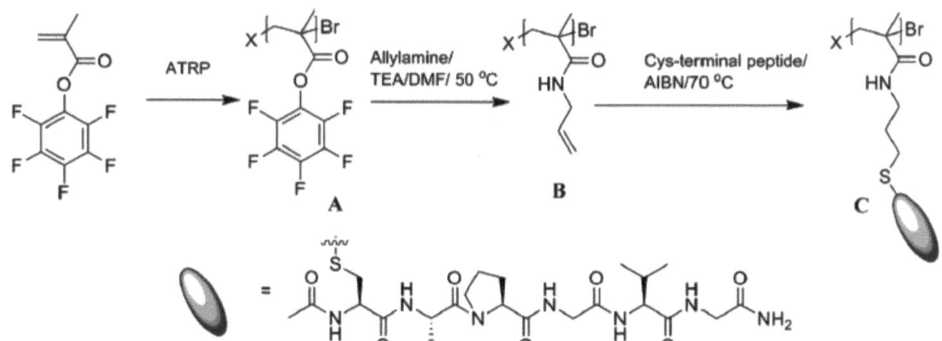

Scheme 6.15: Preparation of biofunctionalized polymethacrylate via tandem "ester-amide/thiol-ene" post-polymerization modification of poly(pentafluorophenyl methacrylate) obtained by ATRP. Reproduced with permission from N.K. Singha, M.I. Gibson, B.P. Koiry, M. Danial and H.-A. Klok, *Biomacromolecules*, 2011, 12, 8, 2,908. ©2011, ACS [76].

Durmaz and coworkers reported the preparation of fluorinated alternating copolymer by using ATRP and Suzuki coupling processes [77]. In this case, a fluorinated copolymer composed of perfluorooctylethyl acrylate (FOEA) and methyl methacrylate (MMA) and PSt was prepared separately via the ATRP process. These were used as macromonomers in Suzuki polycondensation reactions, producing the alternating copolymer as shown in Scheme 6.16.

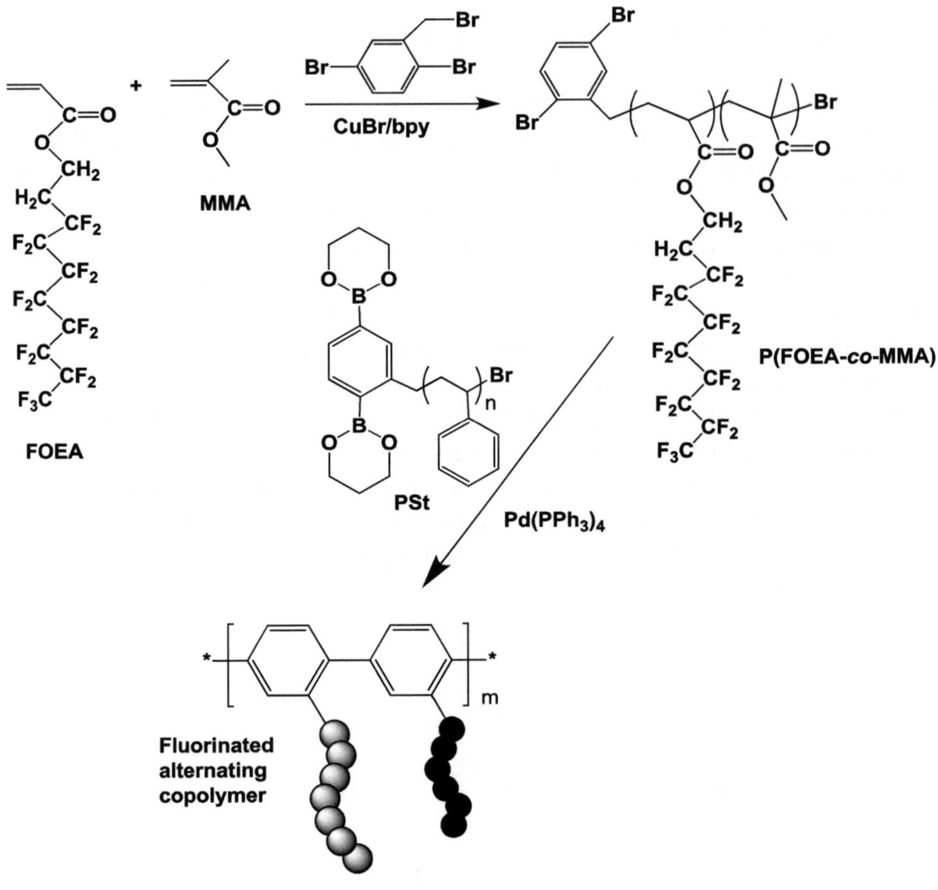

Scheme 6.16: Preparation of fluorinated alternating copolymer via consecutive ATRP and Suzuki coupling reaction. Source: Y.Y. Durmaz, E.L. Sahkulubey, Y. Yagci, E. Martinelli and G. Galli, *Journal of Polymer Science Part A: Polymer Chemistry,* 2012, 50, 23, 4,911. [77].

6.4.2.2 Fluorinated Block Copolymers via ATRP

Due to certain unique properties, fluorinated block copolymers have attracted research interest in recent years. In most cases, researchers took the advantage of the amphiphilic nature of diblock fluoropolymers (DBF), which let them produce a wide range of morphologies in water or organic solvents. The amphiphilicity in DBF arises due to the presence of a hydrophobic fluorinated block and any hydrophilic segment. In 2009, a general approach was reported by Lee *et al.* [78]. They prepared an amphiphilic DBF composed of a hydrophobic poly(1H, 1H-dihydroperfluorooctyl methacrylate (FOMA) and polyethylene oxide (PEO), a well-known hydrophilic polymer. In this case, the well-defined DBF had been prepared via ATRP using PEO-Br as macro-initiator. The DBF displayed different morphologies upon self-assembly in different organic solvents. The DBF adopted micellar morphology with hydrodynamic radii in the range of 60–300 nm in chloroform solvent. However, vesicles were formed in dimethyl formamide (DMF) and lamellar morphology was obtained in methanol. The preparation of PEO-based amphiphilic DBF was also reported by Jiang *et al.* [79]. They adopted ATRP to polymerize 2,2,3,4,4,4-hexafluorobutyl methacrylate (HFBMA) and used the resulting fluorinated polymer as a macro-initiator to polymerize hydrophilic oligo(ethylene glycol) methyl ether methacrylate (OEGMA). In this case, the authors considered the thermo-responsive property of poly(ethylene glycol) (PEG) segment and demonstrated the responsive nature of the micelles (30–40 nm at room temperature and 40–300 nm at 60 °C) formed by the DBF in its aqueous solution. Interestingly, a cotton fabric dipped in the miceller solution at room temperature showed water contact angle (WCA) of 150° after drying. However, when a dried cotton fabric was dipped in the hot micellar solution of DBF, it was quite hydrophilic. The same team [80] also used ATRP to prepare amphiphilic DBF composed of PHFBMA and poly(glycidyl methacrylate) (PGMA). This DBF, PHFBMA-*b*-PGMA was made amphiphilic by the ring opening reaction of epoxy groups in PGMA. In this case, the DBF self-assembled as hollow microspheres with a mean diameter of 50–80 nm in water. Peng *et al.* [81] prepared DBFs with a series of FA/FMA and *tert*-butyl acrylate (*t*BA). The hydrolysis reaction converted the tert-butyl groups to -COOH making the DBFs amphiphilic in nature. The spontaneous self-assembly of DBFs in water produced micellar morphology in the size range of 20–45 nm in diameter. Interestingly, the micelles with fluorinated core provided a strong signal in MRI examination. Thus, the DBF was successfully used as ^{19}F MRI imaging agent. In addition to the above examples of amphiphilic DBFs, a non-amphiphilic DBF based on PSt and PHFBMA was prepared by adopting ATRP [82]. In this case, the authors studied the self-aggregation phenomena of the prepared PSt-*b*-PHFBMA in THF/EtOAc mixture. As EtOAc is a non-solvent for fluorinated block, the DBF showed a transition in morphology from micelles (90–100 nm) to vesicles (150–200 nm) upon gradual increase of the EtOAc content.

Apart from the DBFs, ATRP has also been successfully used to prepare triblock fluoropolymer (TBF). In this regard, the approach by Guo and coworkers deals with a simple synthetic route to prepare amphiphilic ABC-type TBF [83]. In this case, the authors prepared a PEO-based macro-initiator to polymerize St and perfluorohexylethyl methacrylate (FHEMA) via consecutive ATRP. Importantly, all the prepared TBFs had controlled molecular weights and moderately narrow dispersities (1.31–1.47). The TBF showed very good protein resistance because of the presence of PEO-segment.

Rabnawaz and Liu [84] reported a novel TBF which is simultaneously photo-cleavable and photo-cross-linkable. The amphiphilic triblock terpolymer poly(ethylene oxide)-ortho-nitrobenzyl-poly [2-(perfluorooctyl) ethyl methacrylate]-*block*-poly (2-cinnamoyloxy ethyl methacrylate) (PEO-ONB-PFOEMA-*b*-PCEMA) was prepared via consecutive ATRP by using PEO-ONB-Br as macro-initiator. In this case, the ONB unit is photo-cleavable and PCEMA is photo-cross-linkable. The fluorinated segment PFOEMA had been incorporated to achieve superhydrophobicity and oleophobicity. Upon photo-cross-linking, the H_2O and CH_2I_2 contact angles increased to 154° and 136° from 52° and 32°, respectively, in the micellar films.

Polymers bearing polyhedral oligomeric silsesquioxanes (POSS) moieties are important, mainly because of their excellent surface properties. Thus, an improvement in surface properties is expected in a POSS-containing fluorinated polymer. Qiang *et al.* [85] adopted ATRP to prepare POSS-containing star-shaped PMMA-*b*-PTFEMA. In this case, POSS-Cl had been used as ATRP initiator to prepare the star shaped DBF, which displayed a honeycomb morphology resulting in superior hydrophobicity (WCA 95°-110°).

6.4.2.3 Hybrid Fluoropolymer and Brushes via ATRP

SI-ATRP is one of the important aspects of ATRP [26a]. It offers the scope to prepare well-defined hybrid polymer and polymer brushes with improved properties. There are two types of approaches known as "*grafting to*" and "*grafting from*" under SI-ATRP. The former approach deals with the attachment of a preformed polymer to a substrate. In this case, the polymer should have a functional end-group to bind a substrate. The latter approach enables one to control the growth of polymer chain from an initiator-functionalized surface (Scheme 6.17) [86]. Both approaches have attracted considerable attention to develop new macromolecular architectures.

Huang and He [87] adopted the "*grafting from*" approach to develop a hybrid DBF via SI-ATRP from nano-silica (nSiO$_2$) surface. In this case, nSiO$_2$ was chemically modified to be used as an ATRP initiator. The consecutive polymerization of MMA and dodecafluoroheptyl methacrylate (F_{12}HMA) was carried out to prepare the hybrid fluoropolymer SiO_2-*g*-PMMA-*b*-PF_{12}HMA (Scheme 6.18).

The hybrid DBF exhibited higher surface roughness and lower surface energy (10.97 mN/m) due to the accumulation of fluorinated segment over the surface. In a similar fashion, Huang *et al.* [88] recently developed an amphiphilic hybrid fluoropolymer as SiO_2-*g*-P(PEGMA-*b*-F_{12}HMA). The DBF produced spherical nanoparticles (~150 nm in water and ~170 nm in THF solution) due to self-assembly. Functionalized nSiO$_2$ was also used as ATRP initiator by Yu and coworker [89]. In

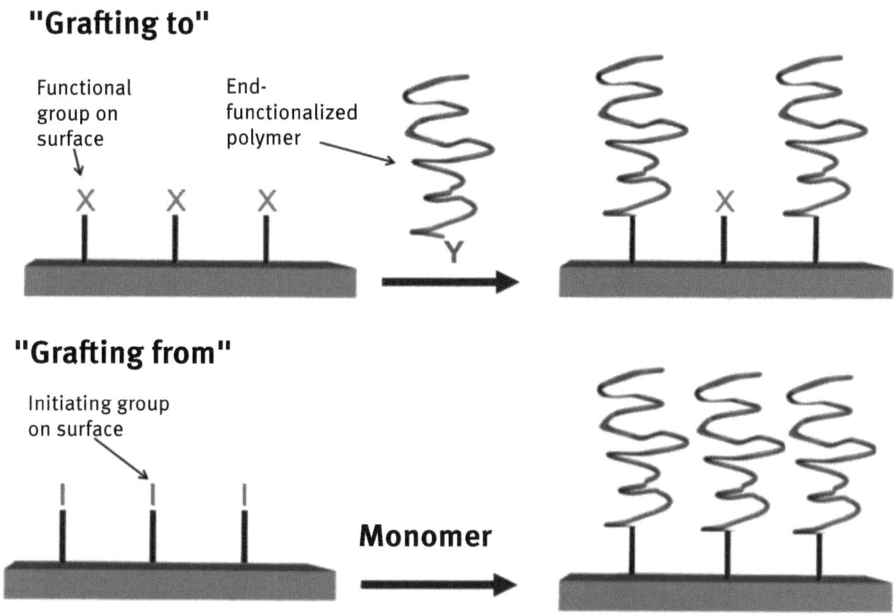

Scheme 6.17: The preparation of polymer brushes via "grafting to" and "grafting from" approaches. Reproduced with permission from P.M. Mendes, *Chemical Society Reviews*, 2008, 37, 11, 2,512. ©2008, RSC [86].

this case, the hybrid DBF was prepared by the consecutive ATRP of 3-methacryloxypropyltrimethoxysilane (MPTS) and HFBMA using SiO_2-Br as the initiator. The hybrid DBF created a nano-phase roughness as observed from the SEM and AFM analyses. As a result, it displayed superhydrophobicity with water contact angles (WCAs) in the range of 135° to 161°.

Apart from the hybrid polymers, the preparation of fluoropolymer brushes is of great significance as they exhibit interesting properties. PTFEMA brushes on poly(ethylene terephthalate) (PET) fabrics were synthesized via SI-ATRP in a "grafting from" approach [90]. Interestingly, the resulting PET fabrics displayed antifouling and superhydrophobic properties (Scheme 6.19).

Shinohara *et al.* [91] reported the preparation of fluoropolymer brushes on nano-imprinted surfaces. Nano-imprinting (NI) is a technology used to fabricate nanotopography on a polymer surface. In this case, a regular pillar pattern was created on a poly[MMA-*co*-2-(2-bromoisobutyryloxy) ethyl methacrylate (BIEM)] thin film via a NI process. The SI-ATRP of FOEA from that imprinted thin film produced a nano-structured surface covered with fluoropolymer. The polymeric film displayed superhydrophobicity due to the synergistic effect of surface geometry from NI and low surface energy provided by the fluorinated segments.

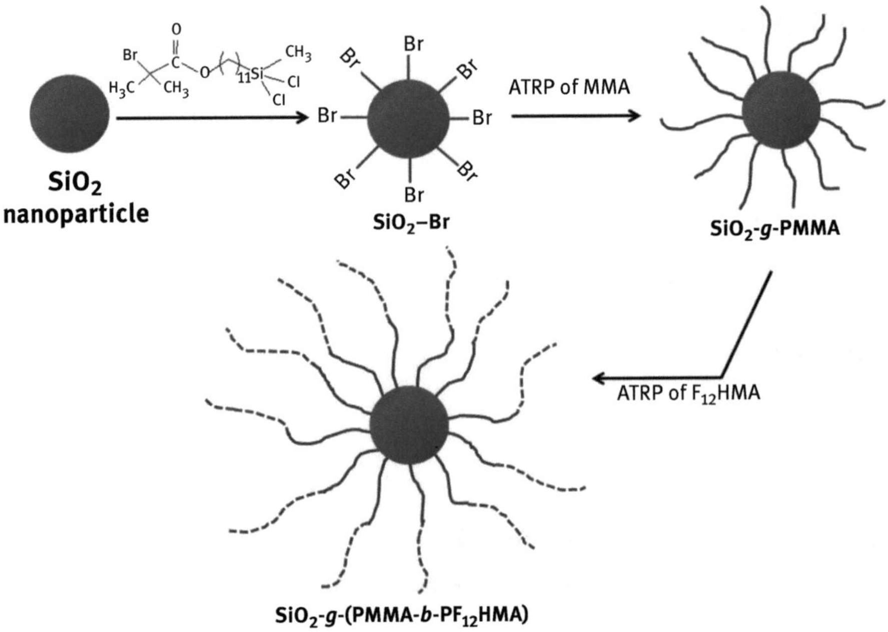

Scheme 6.18: Preparation of SiO_2-grafted fluorinated block copolymer via SI-ATRP process. Reproduced with the permission from H. Huang and L. He, *RSC Advances*, 2014, 4, 25, 13,108. ©2014, RSC [87].

Silicon wafers have also been used as initiating surfaces for the polymerization of fluoromonomers. The aggregation behavior of polymer chains grown on flat surfaces like silicon wafers has attracted considerable research interest. The surface properties such as adhesion, friction, and wettability may be tuned by the orientation of fluoropolymer chains. According to Yamaguchi *et al.* [92], the molecular aggregation in a polymer brush was dependent on the molecular weight distribution (MWD). In case of polymer with broad MWD, the pendant fluorinated groups (R_f) were perpendicular to the surface (Scheme 6.20). On the other hand, the R_f groups were parallel to the surface when the polymer exhibited narrow dispersity. This parallel orientation of R_f groups resulted in lower water resistance as compared to the polymer brush with perpendicular R_f groups.

The study by Bhairamadgi *et al.* [93] revealed that surface properties like friction and adhesion may be regulated by the number of fluorine atom in the monomer. In this case, fluoropolymer brushes on a silicon wafer were prepared by the combination of a thiol-yne click reaction and ATRP (Scheme 6.21). The copolymerization of ethyl methacrylate (EMA), TFEMA, HFBMA, and 2-perfluorooctylethyl methacrylate (FOEMA) produced polymer brushes with increasing numbers of F atoms per monomer (0, 3, 7, and 17). As expected, the fluoropolymer brush displayed the lowest adhesion forces when the monomer contained the highest number of F atoms.

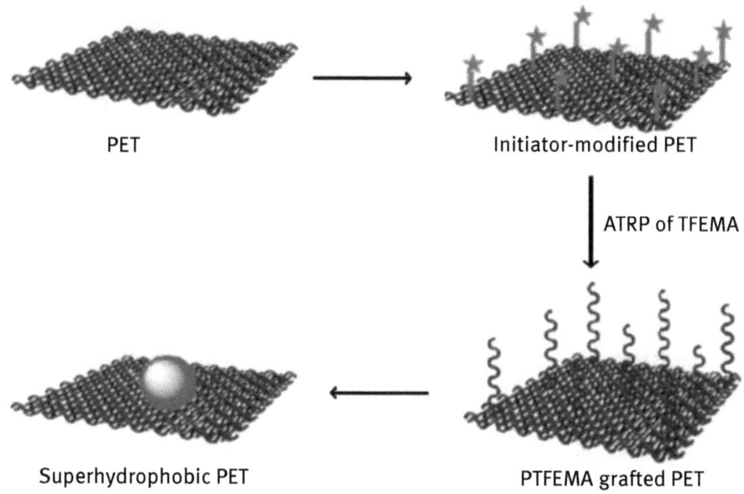

Scheme 6.19: Development of superhydrophobic surface via SI-ATRP of TFEMA on PET fabrics. Reproduced with permission from C.-H. Xue, X.-J. Guo, J.-Z. Ma and S.-T. Jia, *ACS Applied Materials and Interfaces*, 2015, 7, 15, 8,251. ©2015, ACS [90].

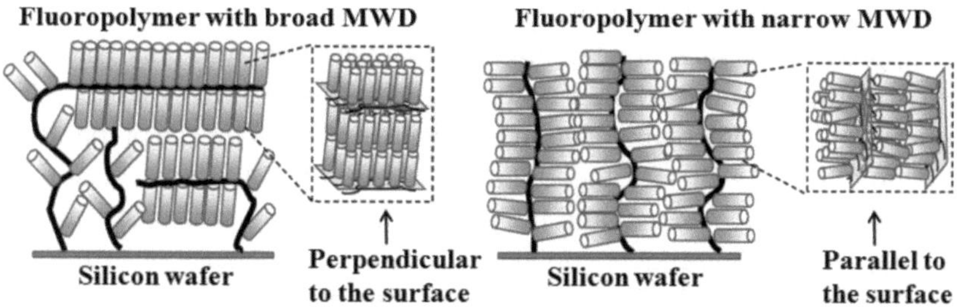

Scheme 6.20: Molecular orientation of fluorinated segments in a fluoropolymer brush prepared via SI-ATRP on silicon wafer. Reproduced with permission from H.Yamaguchi, M. Kikuchi, M. Kobayashi, H. Ogawa, H. Masunaga, O. Sakata and A. Takahara, *Macromolecules*, 2012, 45, 3, 1,509. ©2012, ACS [92].

6.4.3 AGET ATRP of Fluoromonomers

According to the previous discussions, it is obvious that ATRP has achieved considerable success in the polymerization of fluoromonomers. However, the process suffers some limitations in case of industrial scale-up. The air-sensitivity of the catalyst used in ATRP was found to be a big issue in commercial prospects. To overcome this issue, Matyjaszewski and his coworkers [94] developed the AGET ATRP process where catalyst Cu(I) complex is generated in situ by reduction of stable Cu (II) complex by the reducing agents like Cu(0), tin(II)2-ethylhexanoate ($Sn(EH)_2$),

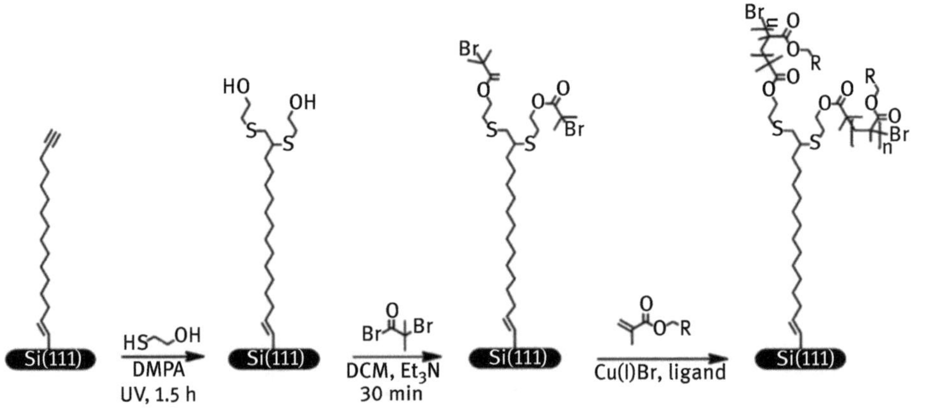

Scheme 6.21: Surface functionalization of silicon wafer with polyfluoromethacrylates via thiol-yne click reaction and ATRP. Reproduced with permission from N.S. Bhairamadgi, S.P. Pujari, C.J.M. Rijn and H. Zuilhof, *Langmuir*, 2014, 30, 42, 12,532. ©2014, ACS [93].

ascorbic acid, and so on. Soon after this development, this process was established as a simple and versatile method in RDRP. Researchers took advantage of this to simply prepare tailor-made fluoropolymers.

An early approach was made by Sun and Liu [95] who prepared poly(HFBMA)-*b*-poly(isobutyl methacrylate) (PHFBMA-*b*-PIBMA) via sequential AGET ATRP using Cu(0) as reducing agent. The polymerization was well-controlled for the preparation of macro-initiator PHFBMA-Br and the DBF, PHFBMA-*b*-PIBMA. In this case, the DBF displayed lower surface energy and higher fluorine content at the polymer-air interface as compared to the random copolymer. Due to the presence of hydrophobic fluorinated segment, the DBF self-assembled into micelles (100–150 nm) in water, showing core-shell morphology in TEM analysis.

Zhan and coworkers [96] developed a novel approach to prepare tailor-made fluorinated hybrid materials by the combination of SI-ATRP and AGET-ATRP. In this case, a fluorinated copolymer, poly(butyl acrylate-*co*-2-(N-ethyl perfluorooctane sulphamido) acrylate) was synthesized via SI-AGET ATRP using SiO_2-NH-Br as initiator and $Sn(EH)_2$ as the reducing agent. Interestingly, hybrid fluoropolymer displayed superhydrophobicity with WCA of 170.3° and anti-icing property due to the formation of a nano-structured morphology by the covalently bonded SiO_2 nanoparticles.

AGET-ATRP was also successfully applied for the emulsion polymerization of fluoromonomers. Shu and coworkers [97] reported the emulsion polymerization of TFEMA via AGET-ATRP using disodium 4-(10-(2-bromo-2-methyl-propanoyloxy)decyloxy)-4-oxo-2-sulfonatobutanoate as both initiator and surfactant (inisurf). In this case, ascorbic acid was used as the reducing agent.

To reduce the amount of transition metal catalyst to be used in ATRP, Jakubowski *et al.* [98] developed another process with continuous regeneration of the catalyst in the reaction medium. This process was termed as activators regenerated by electron transfer (ARGET) ATRP. ARGET-ATRP process needs very small amounts of Cu catalyst and higher amount of reducing agent as compared to AGET-ATRP process. ARGET-ATRP was also found to be successful in preparing fluorinated homopolymers and block copolymers. Schreiber *et al.* [99] employed ARGET-ATRP to prepare DBF containing 1H,1H,2H,2H-perfluorodecyl acrylate (FDA) and *n*-butyl acrylate (*n*BA) or *t*-butyl acrylate (*t*BA). In this study, a fluorinated ligand F-TEDETA ("V" in Scheme 6.14), derived from N,N,N′,N′-tetraethyldiethylenetriamine (TEDETA), was successfully used to achieve a faster rate of polymerization and narrow dispersity. According to Zheng and coworkers [100], ARGET-ATRP was found to be successful for emulsion (co)polymerization of 2,2,3,4,4,4-hexafluorobutyl acrylate (HxFBA). In this study, fluorinated acrylic copolymer consisting of MMA, HxFBA, and 2-hydroxy propyl acrylate (HPA) was prepared with narrow dispersity using O-phenanthroline as ligand.

6.4.4 RAFT Polymerization of Fluoromonomers

RAFT polymerization process is a powerful tool to prepare polymers with controlled molecular weight and well-defined architecture [28]. As mentioned above, this process requires a reversible chain transfer agent like di-thio or tri-thio carbonates/carbamates, which take part in the addition-fragmentation reaction with active and dormant chains. At the end of the polymerization, the macromolecules contain RAFT functionality, which can take part in a further chain extension reaction to prepare block copolymers. Some terminated polymer chains or dead chains derived from initiator are also formed. Thus, these systems require a higher ratio of RAFT agent to initiator in order to obtain a polymer with narrow dispersity. This was found to be challenging in preparing a fluoropolymer with well-defined architecture via the RAFT polymerization technique. In this case, the selection of the proper RAFT agent is crucial. In addition to the xanthates reported in Section 6.3.2.1, Scheme 6.22 displays some important RAFT agents used for the polymerization of FA/FMA.

The polymerization of 1H,1H,5H-octafluoropentyl acrylate (OFPA) was carried out in bulk using **1** as RAFT agent [101]. Although the polymer had narrow *Đ*, the conversion was only 53% at the polymerization temperature of 70 °C. A similar observation was reported for the solution polymerization of 2,2,3,3,4,4,4-heptafluorobutyl acrylate (HFBA) in 1,4-dioxane using RAFT agent **4** [102]. However, an increase in polymerization temperature from 70 °C to 90 °C resulted in 90% conversion with faster rate of polymerization. In this case, the polymer had a broad dispersity. Gibson and coworkers [103] adopted the RAFT process to polymerize PFPMA and subsequently employed post-polymerization modification with primary amines. This approach opened the route to synthesize libraries of water-soluble functional polymers.

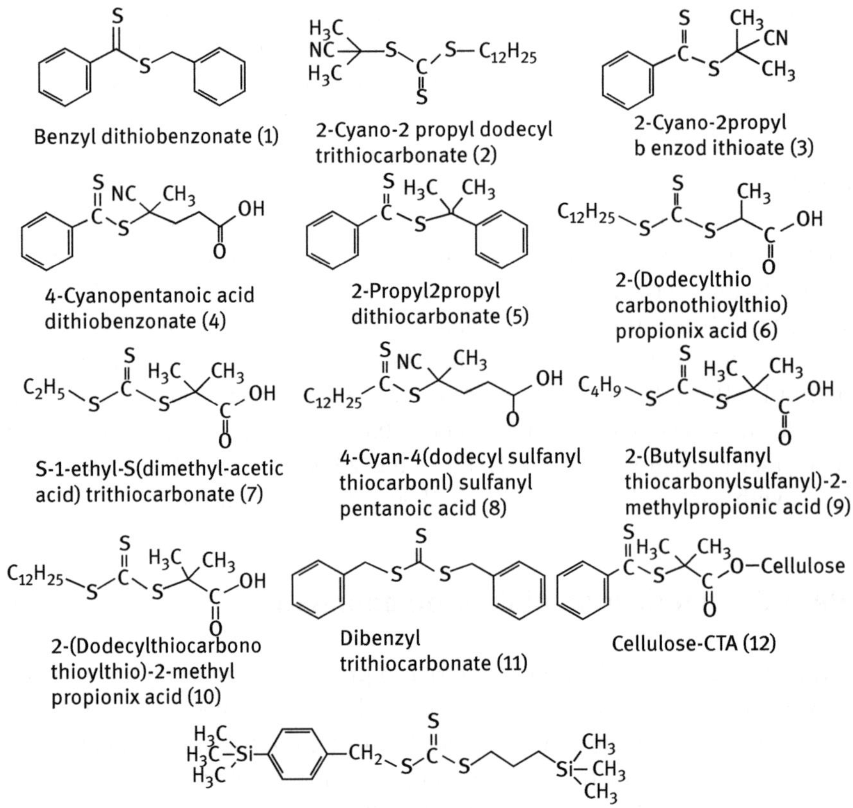

Scheme 6.22: Chemical structures of the RAFT agents used to polymerize fluorinated (meth) acrylates.

Some authors employed protonated monomers to prepare fluorinated random copolymers via RAFT polymerization techniques. Thus, they determined the reactivity ratios between the fluorinated and hydrocarbon monomers. In 2009, Zhang *et al.* [104] reported the RAFT copolymerization of 2-(perfluorohexyl)ethyl methacrylate (FHEMA) and butyl methacrylate (BMA) in miniemulsion using **3** as RAFT agent. According to their study, FHEMA had higher reactivity than BMA. In FHEMA, the propagating radical was more stabilized due to the electron-withdrawing inductive effect from the fluoroalkyl pendant group. However, a contradiction to this observation was reported for the RAFT mediated copolymerization of BA with fluorinated monomer HFBA using **11** as the RAFT agent [105]. The RAFT mediated copolymerization of BA with 2,2,2-trifluoroethyl acrylate (TFEA) and 2,2,3,3,3-pentafluoropropyl acrylate (PFPA) displayed a higher reactivity of the fluorinated monomers. However,

HFBA displayed a lower reactivity than BA. A similar observation was also reported by Koiry *et al.* [106]. In this case, the copolymerization of HFBA and BA was carried out in 1,4-dioxane using **4** as RAFT agent. All the aforementioned studies on the RAFT copolymerization of FA/FMA displayed an ideal copolymerization system producing perfectly random copolymers. Zaitsev *et al.* [107] reported the preparation of fluorinated alternating copolymer from electron rich N-vinyl pyrrolidone (NVP) and electron deficient 1,1,1,3,3,3-hexafluoroisopropyl-α-fluoroacrylate (FPFA) carried out in THF in the presence of RAFT agent **1**.

Most articles have shown that a polymer prepared by RAFT polymerization usually contains the RAFT functionality as end-group, which can further be used for chain extension reactions to produce block copolymers. Several researchers took this opportunity into account to prepare fluorinated block copolymers with desired architectures. Most of the authors took advantage of the hydrophobic and oleophobic nature of the fluorinated segment to synthesize amphiphilic block copolymers. In this regard, a simple approach was reported by Guo *et al.* [108] who prepared amphiphilic poly(methacrylic acid)-*b*-poly(2,2,2-trifluoroethyl methacrylate) (PMAA-*b*-PTFEMA) via RAFT polymerization in one pot. In this case, the block copolymer was formed via a polymerization-induced self-assembly process which originated from the water-soluble PMAA Macro-RAFT agent. Another amphiphilic DBF composed of poly(heptadecafluorodecylacrylate-*co*-acrylic acid) [P(HFDA-*co*-AA)] and polyacrylonitrile (PAN) was prepared by Grignard and coworkers. Electrospinning of the DBF solution in DMF produced a superhydrophobic surface with WCA upto 155.7° [109]. The RAFT polymerization process has also been used by Li *et al.* [110] to prepare PMMA-*b*-TFEMA, which displayed significant water and oil-resistance properties (θ_{water} = 104.3° and θ_{oil} = 80.0°). Due to self-assembly, the DBF displayed micellar aggregates having diameter 400–600 nm in aqueous solution. A reactive DBF, PTFEA-*b*-PPEGMA prepared by Yi *et al.* [111] was found to be a promising material for preparing fluorinated thermosets. The amphiphilic DBF was incorporated into epoxy to produce a nanostructured and fluorinated thermoset resin. The material showed a WCA of 102° and surface energy of 16.4 mN/m. Koiry *et al.* [112] used PEG-containing amphiphilic DBF as a surf-RAFT agent (macro-RAFT agent acting as surfactant also) in the miniemulsion polymerization of styrene. In this case, they prepared an amphiphilic PPEGMA-*b*-PHFBA using RAFT agent **3**. The DBF produced spherical micelles in aqueous solution. Those micelles were further being used as polymerization site for the miniemulsion polymerization of styrene producing core-shell particles where polystyrene occupied the core and DBF was in the shell. Some authors used RAFT polymerization technique to prepare TBF with hydrophilic-lipophilic-fluorophilic nature [113, 114]. In this case, the authors used sequential RAFT polymerization of different monomers along with the fluorinated monomer to demonstrate the triphilic nature of the TBF. Interestingly, the TBFs displayed multiple morphologies like core-shell-corona micelles and multicompartment micelles due to their self-aggregation.

6.4.4.1 RAFT polymerization in emulsion

Due to its environmental friendliness, emulsion polymerization is an extremely important tool for industrial production of different polymers for paint and coatings purposes, as well as for other applications. Since fluorinated polymers have low surface energies, they can be used in paints to improve the water and oil-resistance property of the paints or coatings. Researchers have thus incorporated fluorinated segments into polymers synthesized by emulsion polymerization techniques. Thus, the RAFT polymerization technique has been applied with success in emulsion polymerizations. However, the extremely hydrophobic nature of fluorinated monomer has raised issues of colloidal instability. To address this issue, some authors have adopted miniemulsion polymerization techniques in the RAFT polymerization process.

The RAFT miniemulsion polymerization of HFBA and subsequent chain-extension with BA was reported by Chakrabarty and Singha [115]. In this case, PHFBA-*b*-PBA displayed core-shell morphology in its miniemulsion where PBA formed the core and PHFBA formed the shell as established by the elemental analysis. This nanophase separation in the DBF, achieved by the RAFT mediated miniemulsion polymerization, has resulted in improved hydrophobicity with a WCA of 112.5°. An interfacial RAFT miniemulsion polymerization technique was adopted to prepare hollow fluorinated polymer particles [116]. An emulsifier-free method had been used by Chen *et al.* [117] to polymerize TFEMA via the RAFT technique. An amphiphilic RAFT agent, produced by the polymerization of acrylic acid (AA), was used as stabilizer and macro-RAFT agent to produce a fluorinated gradient copolymer. The copolymer underwent self-assembly in different organic solvents. A similar approach had also been adopted by Wang and coworkers [118], who prepared gradient copolymer of PSt and PHFBA.

6.4.4.2 Fluoropolymer nanocomposites via RAFT polymerization

Due to the exceptional nature of fluoropolymers, nanocomposites based on fluoro (meth)acrylates were found to be interesting materials. Due to the high cost of fluoropolymers, incorporation of nanomaterials in the fluoropolymer matrix enables reduction of cost, in addition to improved properties such as resistance to flammability, water, and oil. The earlier discussion on SI-ATRP described the preparation of fluoropolymer nanocomposites where the materials showed superhydrophobicity. The RAFT polymerization technique also offers the scope for surface-initiated polymerization (SI-RAFT) and modification of surfaces by grafting polymers. Liu *et al.* [119] described a novel approach to modify ramie fibers with PTFEMA via RAFT polymerization. This study revealed the preparation of RAFT functionality on cellulose fibers and subsequent polymerization of TFEMA in $scCO_2$. The fluoropolymer-modified fibers displayed excellent hydrophobicity with WCA up to 149°.

Yang *et al.* [120] used the SI-RAFT process to polymerize 1H,1H,2H,2H-heptadecafluorodecyl acrylate (HFDA) and TFEA from the surface of $BaTiO_3$. This fluoropolymer/$BaTiO_3$ nanocomposite was found to be a promising ferroelectric material for energy storage applications. Table 6.2 summarizes the literature on the polymerization of FA/FMA using different RAFT agents.

Table 6.2: Examples of fluoromonomers polymerized by using different RAFT agents shown in Scheme 6.22 [101–121].

RAFT agent used	FA/FMA used	References
1	OFPA, FPFA	[101,107]
2	HFBA	[115,121]
3	FHEMA	[104,112]
4	HFBA, TFEMA, PFPMA	[102,103,106,108]
5	TFEMA, TFEA	[110,111]
6	TFEMA, HFBA	[117,118]
7	HFDA, TFEA	[120]
8	DFHA	[116]
9	FDA	[113]
10	FDA	[109]
11	TFEA, PFPA, HFBA	[105]
12	TFEMA	[119]
13	PFBA	[114]

Chakrabarty and Singha reported the morphology and hydrophobicity of PHFBA/clay nanocoposite [121]. In this case, RAFT miniemulsion polymerization technique was adopted for the in situ preparation of nanocomposite. Addition of certain comonomers favored the interactions between PHFBA and nanoclay. The nanocomposite showed higher surface roughness and a WCA of 128.1° that is significantly higher than that of the pristine PHFBA.

6.4.5 Reverse Iodine Transfer Polymerization (RITP) of FA/FMA

The RITP process is an emerging RDRP technique, as other RDRP methods suffer from certain limitations like expensive and colored catalysts (for RAFT and NMP) and removal of catalyst (for ATRP), and so on. This process can also be applied to polymerize FA/FMA with good results.

Bouilhac and coworkers [122] reported the RITP of FDA in $scCO_2$. The molecular weights were well-controlled for the polymerizations with targeted molecular weights in the range of 10–100 kg/mol. However, broad dispersities ($Đ$ >1.5) were observed. The same team reported [123] RITP of FDA initiated by AIBN in trifluorotoluene. The polymerization took place with very high conversion (> 90%).

6.4.6 Single Electron Transfer-Living Radical Polymerization (SET-LRP) of FA/FMA

SET-LRP is a novel method to produce polymers with functional chain ends and controlled molecular weights. The polymerization is catalyzed by Cu(0) in the form of Cu-wires, powder, nanoparticles, and so on. It also requires the presence of nitrogen containing ligands and polar solvents to stabilize the Cu(II) complex. In this regard, SET-LRP of hydrophilic and hydrophobic acrylates in fluorinated solvents like 2,2,2-trifluoroethanol (TFE), 2,2,3,3-tetrafluoropropanol (TFP) were found to be successful, as the solvents contain both hydrophilic and hydrophobic segments [124, 125]. The hydrophilic segment is responsible for the disproportionation of Cu (I), and the hydrophobic segment can solubilize the hydrophobic monomer.

Samanta and coworkers reported the SET-LRP of 1H,1H,2H,2H-perfluorooctyl acrylate (PFOA), HFBA, 1H,1H,5H-octafluoropentyl methacrylate (OFPMA) and 1H,1H,5H-octafluoropentyl acrylate (OFPA) using TFE as solvent [126]. The polymerization of fluorinated acrylates was carried out at 25 °C whereas the methacrylates were polymerized at 50 °C. In this case, hydrazine activated Cu(0) wire was used as catalyst and Me_6-TREN (VIII) was used as ligand. SEC analysis showed controlled molecular weights and narrow dispersities in all the fluorinated polymers. A "thiol-bromo" click reaction was carried out in the chain end-functionality which was identified by ^{1}H NMR and MALDI-TOF MS analysis.

The same group reported the SET-LRP of activated fluoromonomers like 1,1,1,3,3,3-hexafluoroisopropyl acrylate (HFIA) and 1,1,1,3,3,3-hexafluoropropyl methacrylate (HFIMA) using Cu(0) wire as catalyst, **VIII** as ligand and TFE as solvent [127]. In this case, the authors successfully carried out the transesterification reaction with the fluoropolymers containing activated esters using two mild bases 1,8-diazabicycloundec-7-ene (DBU) and 1,5,7-triazabicyclo[4.4.0]dec-5-ene (TBD). This strategy was found to be an important tool to fabricate complex macromolecular architectures.

6.4.7 RDRP of Fluorinated Styrenes

Fluorinated monomers with long perfluoroalkyl side chains (>C_7F_{15}) are toxic, bioaccumulable, and resistant to degradation. [63–65, 128] Considering them as pollutants, The United States Environmental Protection Agency launched its PFOA Stewardship Program to eliminate the production and use of those fluorinated chemicals by 2015. In this regard, the monomers having fluorinated phenyl rings had been found to be suitable alternative. The polymers derived from those monomers, are capable of showing improved surface properties because of their tightly packed structures.

RDRP of these monomers was successfully carried out to prepare well-defined architectures for manifold applications in drug delivery, in thin-film lithography, and in antibiofouling surfaces, and so on. Scheme 6.23 displays the chemical structures of some vinyl monomers containing fluorinated aromatic side groups. Most of these researchers adopted ATRP to polymerize those monomers [16]. In this case, monomer A, the fluorinated analogue of styrene was found to be widely used, and it shows a much higher reactivity than styrene in radical polymerization.

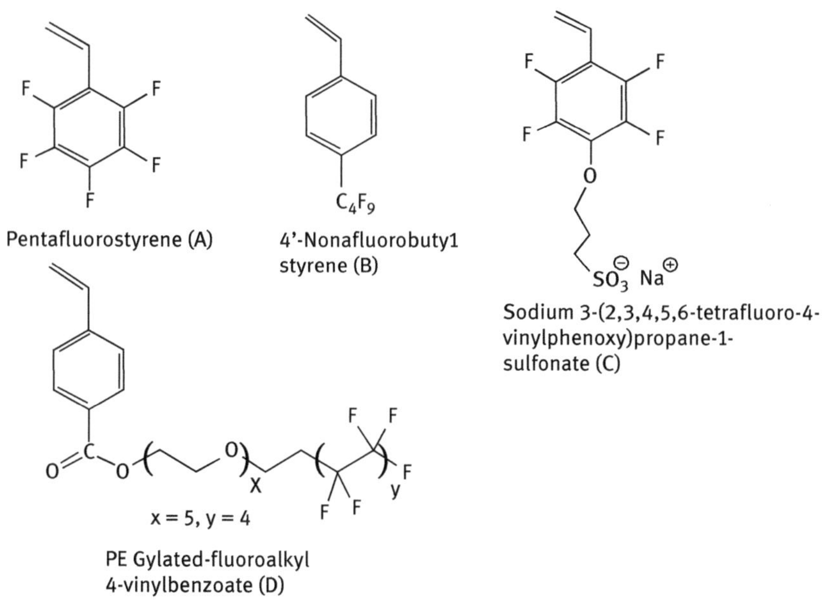

Scheme 6.23: Chemical structures of styrenic monomers bearing fluorinated aromatic rings polymerized by RDRP technique.

The rate of polymerization of monomer **A** was found to be dependent on the nature of solvent. ATRP in aromatic solvents like benzene or toluene produced higher conversion due to the *Π-Π* stacking between solvent and monomer [129]. This type of interaction increased the stability of the propagating radical. Tan and coworkers [130] prepared an amphiphilic DBF based on monomer A via ATRP. The DBF displayed micellar structure (20–50 nm) in its aqueous solution because of the hydrophobicity of the fluorinated aromatic segment.

Fluorinated aromatic rings are prone to nucleophilic substitution reactions because of electron deficiency provided by the electronegative fluorine atoms. An interesting approach to produce monomer **C** by the derivatization of monomer A was reported by Dimitrov *et al.* [131]. In this case, the authors carried out nucleophilic substitution of F-atom at para-position of **A** by hydroxyl group followed by sulfopropylation to produce **C** (Scheme 6.24). Since the monomer was soluble in water, the ATRP of C followed by chain extension was carried out in a water/methanol (3:1) mixture. The nucleophilic substitution reaction and ATRP process was also adopted by Pollack and coworkers [132] to prepare a well-defined hyperbranched fluoropolymer.

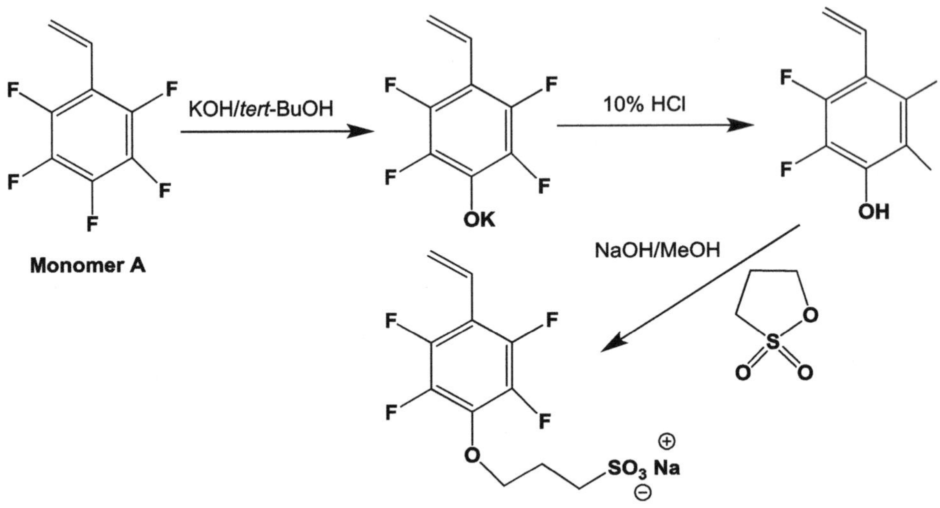

Scheme 6.24: An approach to derivatize pentafluorostyrene (monomer A).
Source: I. Dimitrov, K. Jankova and S. Hvilsted, *Journal of Fluorine Chemistry*, 2013, 149, 30. [131].

Styrenic monomers can also be derivatized by fluoroalkyl group to prepare fluorinated styrenes. In this regard, Martinelli and coworkers [133] adopted a simple approach by coupling 4-Vinylbenzoic acid and PEGylated-fluoroalkyl surfactant to produce an amphiphilic fluorinated monomer **D**. In this case, ATRP of monomer **D** using PSt-Br as macro-initiator produced an amphiphilic DBF which displayed considerably lower surface energy due to the formation of a nano-structured surface.

Ceretta and coworkers [134] used another novel approach to prepare monomer **B**, a fluoroalkyl subtitiuted styrene, via Ullmann coupling between 1-iodoperfluorobutane and 4-bromoacetophenone followed by a reduction and dehydration. In this case, the authors adopted the ITP technique to achieve narrow dispersity. Interestingly, the polymer prepared by ITP displayed higher contact angle hysteresis as compared to the same polymer synthesized by conventional radical polymerization.

6.5 Conclusions and Future Outlook

Due to the nature of F-monomers and their reactivities, specific reactions of RDRP are suitable only for specific monomers. For F-alkenes, ITP, MADIX, and reactions involving borane are the most suitable techniques. Industrial application of these products as thermoplastic elastomers have existed for several decades. As for F-(meth)acrylates and styrenes, ATRP, NMP, RAFT, and RITP have been the most suitable methods.

Acknowledgements: The authors thank post-doc researchers and PhD students (mentioned as coauthors in the list of references), industrial companies and colleagues are also granted for fruitful discussions and for building valuable collaborations and/or for sponsoring various studies and/or supplying free samples as well as the French National Agency (ANR, PREMHYS project) and Council of Scientific and Industrial Research, New Delhi, India for financial support.

References

[1] S. Ebnesajjad in Fluoroplastics, Volume 2: Melt Processible Fluoropolymers; The Definitive User's Guide and Databook, William Andrew Publishing, Norwich, New-York, 2003.

[2] J. Scheirs in Modern Fluoropolymers, Wiley, New-York, 1997.

[3] G. Hougham, P.E. Cassidy, K. Johns and J. Davidson in Fluoropolymers: Synthesis and Applications, Vol. 1 and 2, Plenum Publish, New-York, 1999.

[4] B. Ameduri and B. Boutevin in Well Architectured Fluoropolymers: Synthesis, Properties and Applications, Elsevier, Amsterdam, 2004.

[5] a) B. Ameduri, B. Boutevin and G. Kostov, Progress in Polymer Science, 2001, 26(1), 105; b) B. Ameduri and B. Boutevin, Journal of Fluorine Chemistry, 2005, 126, 2, 221; c) A.L. Moore in Fluoroelastomers Handbook; the Definitive User's Guide and Data Book, Plastic Design Library, William Andrew Publishing, Norwich, New York, 2006.

[6] P. Crouse, G. J. Puts, B. Ameduri, Chemical Reviews 2019, 118, 1763–1805.

[7] Z. Cui, E. Drioli and Y.M. Lee, Progress in Polymer Science, 2014, 39(1), 164.

[8] a) S. Desbief, B. Grignard, C. Detrembleur, R. Rioboo, A. Vaillant, D. Seveno, M. Voué, J. De Coninck, A. M. Jonas, C. Jérôme, P. Damman and R. Lazzaroni, Langmuir, 2010, 26(3), 2057; b) M. Licchelli, M. Malagodia, M.L. Weththimunia and C. Zanchia, Progress in Organic Coatings, 2013, 76, 2–3, 495.

[9] F. Mikes, H. Teng, G. Kostov, B. Ameduri, Y. Koike and Y. Okamoto, Journal of Polymer Science Part A: Polymer Chemistry, 2009, 47(23), 6571.

[10] a) R.J. Plunkett in High Performance Polymers: Their Origin and Development, ACS Meeting in New York, April 1986; b) R.B. Seymour and G.S. Kirshenbaum in History of High Performance Polymers, Elsevier, New York, 1987.

[11] A. Taguet, B. Ameduri and B. Boutevin, Advances in Polymer Science, 2005, 184, 127.

[12] J.S. Humphrey and R. Amin-Sanayei in Vinylidene Fluoride Polymers, Encyclopedia of Polymer Science and Technology, 3rd ed.; Mark, H. F., Ed.; Wiley; New York, 2004, 4, 510–533.

[13] B. Ameduri, Chemical Reviews, 2009, 109(12), 6632.

[14] B. Ameduri, Macromolecules, 2010, 43(24), 10163.

[15] a) K. Matyjaszewski, Y. Gnanou and L. Liebler in Macromolecular Engineering, Wiley-VCH, Weiheim, Germany, 2007; b) K. Satoh and M. Kamigiato, Chemical Reviews, 2009, 109, 11, 5120; c) B.M. Rosen and V. Percec, Chemical Reviews, 2009, 109, 11, 5069; d) K. Matyjaszewski, Science, 2011, 333, 6046, 1104; e) K. Matyjaszewski, Macromolecules, 2012, 45, 10, 4015; f) M.J. Monteiro and M.F. Cunningham, Macromolecules, 2012, 45, 12, 4939; g) K. Matyjaszewski and M. Moller in Polymer Science, A comprehensive Reference, Elsevier, Amsterdam, 2012; h) N.V. Tsarevsky and B.S. Sumerlin in Fundamentals of Controlled/Radical Polymerization, Royal Society of Chemistry, Oxford, UK, 2013.

[16] N.M.L. Hansen, K. Jankova and S. Hvilsted, European Polymer Journal, 2007, 43(2), 255.

[17] G.D. Fu, E.T. Kang, K.G. Neoh, C.C. Lin and D.J. Liaw, Macromolecules, 2005, 38(18), 7593.

[18] a) M. Eberhardt and P. Theato, Macromolecular Rapid Communications, 2005, 26(18), 1488; b) Y. Inoue, J. Watanabe, M. Takai, S. Yusa and K. Ishihara, Journal of Polymer Science Part A: Polymer Chemistry, 2005, 43, 23, 6073.

[19] a) R.R. Matheson Jr., The commercialization of controlled polymer synthesis, The Knowledge Foundation, Cambridge, MA, USA, December 4–5, 2000; b) K. Matyjaszewski and J. Spanswick, Materials Today, March 2005, 26–32.

[20] a) M. Tatemoto and T. Nakagawa, German Patent 1978/2,729,671 (assigned to Daikin Kogyo Co Ltd Jap) (Chem. Abst. 88 137374m); b) M. Tatemoto and T. Nakagawa, US Patent 1979/4,158,678 (assigned to Daikin Kogyo Co Ltd Jap); c) M. Tatemoto, Eur. Pat. 1990/399,543 (assigned to Daikin Ind. Ltd); d) I. Wlassics, G. Rapallo, M. Apostolo, N. Bellinzago and M. Albano, Eur. Pat. Appl. EP 1999/0,979,832 A1 (assigned to Ausimont S.p.A.).

[21] M. Destarac, Macromolecular Reaction Engineering, 2010, 4(3–4), 165.

[22] B. Ameduri, Journal of the Taiwan Institute of Chemical Engineers, 2014, 45(1), 3124.

[23] a) M. Tatemoto, The first Regular Meeting of Soviet-Japanese Fluorine Chemists, Tokyo, February 15–16, 1979; b) M. Tatemoto, US 1979/4158678, (assigned to Daikin Kogyo Ltd); c) M. Tatemoto and T. Shimizu in Modern Fluoropolymers, Wiley, New York, 1997.

[24] a) T. Otsu and M. Yoshida, Macromolecular Rapid Communications, 1982, 3(2), 127; b) A. Sebenik, Progress in Polymer Science, 1998, 23, 5, 875; d) T. Otsu and A. Matsumoto, Advances in Polymer Science, 1998, 136, 75.

[25] a) M.K. Georges, R.P.N. Veregin, P.M. Kazmaier and G.K. Hamer, Macromolecules, 1993, 26 (11), 2987; b) V. Sciannamea, C. Jérome and C. Detrembleur, Chemical Reviews, 2008, 108, 3, 1104; c) J. Nicolas, Y. Guillaneuf, C. Lefay, D. Bertin, G. Gigmes and B. Charleux, Progress in Polymer Science, 2013, 38, 1, 63.

[26] a) K. Matyjaszewski, Macromolecules, 2012, 45(10), 4015; b) M. Ouchi, T. Terashima]] and M. Sawamoto, Chemical Reviews, 2009, 109, 11, 4963.

[27] a) G. David, C. Boyer, J. Tonnar, B. Ameduri, P. Lacroix Desmazes and B. Boutevin, Chemical Reviews, 2006, 106(9), 3936; b) J. Tonnar and P. Lacroix Desmazes, Angewandte Chemie International Edition, 2008, 47, 7, 1294.

[28] a) S. Perrier and P. Takolpuckdee, Journal of Polymer Science Part A: Polymer Chemistry, 2005, 43(22), 5347; b) C. Barner-Kowollik, Handbook of RAFT Polymerization, Wiley-VCH, Weinheim, 2008; c) C. Boyer, V. Bulmus, P. Davis Thomas, V. Ladmiral, J. Liu and S. Perrier, Chemical Reviews, 2009, 109, 11, 5402; d) G. Moad, E. Rizzardo and S.H. Thang, Australian Journal of Chemistry, 2009, 62, 11, 1402; e) A. Gregory and M.H. Stenzel, Progress in Polymer Science, 2012, 37, 1, 38; e) Y. Chen, W. Luo, Y. Wang, S. Sun, M. Han and C. Zhang, Journal of Colloid and Interface Science, 2012, 369, 1, 46; f) D.J. Keddie, G. Moad, E. Rizzardo and S. H. Thang, Macromolecules, 2012, 45, 13, 5321; g) M.R. Hill, R.N. Carmean and B.S. Sumerlin, Macromolecules, 2015, 48, 16, 5459.

[29] a) D. Charmot, P. Corpart, H. Adam, S.Z. Zard, T. Biadatti and G. Bouhadir, Macromolecular Symposia, 2000, 150(1), 23; b) D. Taton, M. Destarac and S.Z. Zard, "Macromolecular Design by Interchange of Xanthates: background, design, scope and applications" in Handbook of RAFT Polymerization, C. Barner-Kowollik,; Ed., Wiley-VCH, Weinheim, 2008, Chap. 10, 373.

[30] S. Yamago, B. Ray, K. Iida, J.I. Yoshida, T. Tada, K. Yoshizawa, Y. Kwak, A. Goto and T. Fukuda, Journal of the American Chemical Society, 2004, 126(43), 13908; b) S. Yamago, Chemical Reviews, 2009, 109, 11, 5051; c) S. Yamago and Y. Nakamura, Polymer, 2013, 54, 3, 981.

[31] a) A. Debuigne, R. Poli, R. Jérôme, C. Jérôme and C. Detrembleur, Progress in Polymer Science, 2009, 34(3), 211; b) L.E.N. Allan, M.R. Perry and M.P. Shaver, Progress in Polymer Science, 2012, 37, 1, 127; c) M. Hurtgen, C. Detrembleur, C. Jérôme and A. Debuigne, Polymer Reviews, 2011, 51, 2, 188; d) A. Kermagoret, C. A. Fustin, M. Bourguignon, C. Detrembleur, C. Jérôme and A. Debuigne, Polymer Chemistry, 2013, 4, 8, 2575; e) A.N. Morin, C. Detrembleur, C. Jérôme, P. De Tullio, R. Poli and A. Debuigne, Macromolecules, 2013, 46, 11, 4303.

[32] a) T.C. Chung and A. Petchsuk, US Patent 2002/6,355,749 (assigned to Dai-Act); b) Z.-C. Zhang, Z. Wang and T.C. Chung, Macromolecules, 2007, 40, 15, 5235.

[33] a) A. Goto, T. Suzuki, H. Ohfuji, M. Tanishima, T. Fukuda, Y. Tsujii and H. Kaji, Macromolecules, 2011, 44(22), 8709; b) A. Goto, Y. Tsujii and H. Kaji, ACS Symposium Series, 2012, 1100, Chap. 20, 305; c) A. Ohtsuki, A. Goto and H. Kaji, Macromolecules, 2013, 46, 1, 96; d) A. Goto, Y. Tsujii, H. Kaji, N.V. Tsarevsky and B.S. Sumerlin in Fundamentals of Controlled/Radical Polymerization, Royal Society of Chemistry, Oxford, UK, 2013, Chap. 7, 250–286; e) A. Goto, A. Ohtsuki, H. Ohfuji, M. Tanishima and H. Kaji, Journal of the American Chemical Society, 2013, 135, 30, 11131.

[34] R. Timmerman, Journal of Applied Polymer Science, 1962, 6(22), 456.

[35] P. Lacroix-Desmazes, B. Ameduri and B. Boutevin, Collection of Czechoslovak Chemical Communications, 2002, 67(10), 1383.

[36] B. Boutevin, Journal of Polymer Science Part A: Polymer Chemistry, 2000, 38(18), 3235.

[37] M. Apostolo, V. Arcella, G. Storti and M. Morbidelli, Macromolecules, 2002, 35(16), 6154.

[38] C. Boyer, D. Valade, L. Sauguet, B. Ameduri and B. Boutevin, Macromolecules, 2005, 38, 25, 10353.

[39] C. Boyer, D. Valade, B. Ameduri, P. Lacroix-Desmazes and B. Boutevin, Journal of Polymer Science Part A: Polymer Chemistry, 2006, 44(19), 5763.

[40] a) S. Beuermann and M. Imran-ul-Haq, Macromolecular Symposia, 2007, 259(1), 210; b) M. Imran-ul-Haq, N. Förster, R. Vukicevic, K. Herrmann, R. Siegmann and S. Beuermann in Controlled/Living Radical Polymerization: Progress in RAFT, DT, NMP & OMRP, K. Matyjaszewski, ACS Symposium Series, American Chemical Society, Washington, DC, Chap. 15, 2009, 11, 233–243.

[41] J.M. DeSimone, E.E. Maury, Y.Z. Menceloglu, J.B. McClain, T.J. Romack and J.R. Combes, Science, 1994, 265(5170), 356.

[42] a) M. Tatemoto and S. Morita, US Patent 4361678, 1982 (assigned to Daikin); b) F. Boschet and B. Ameduri, Chemical Reviews, 2014, 114, 2, 927.

[43] T. Yagi, N. Tsuda, T. Noguchi, K. Sakaguchi, Y. Tanaka and M. Tatemoto, Eur. Patent-1990/ Appl. 0,422,644 (assigned to Daikin Industries, Ltd.).

[44] G. Mladenov, B. Ameduri, G. Kostov and R. Mateva, Journal of Polymer Science Part A: Polymer Chemistry, 2006, 44(4), 1470.

[45] N. Durand, B. Ameduri and B. Boutevin, Journal of Polymer Science Part A: Polymer Chemistry, 2011, 49(1), 82.

[46] A.D. Asandei, O.I. Adebolu and C.P. Simpson, Journal of the American Chemical Society, 2012, 134(14), 6080.

[47] P. Cernoch, S. Petrova, Z. Cernochova, J.-S. Kim, C.P. Simpson and A.D. Asandei, European Polymer Journal, 2015, 68, 460.
[48] A.D. Asandei, O.I. Adebolu, C.P. Simpson and J.-S. Kim, Angewandte Chemie International Edition, 2013, 52(38), 10027.
[49] Y. Patil and B. Ameduri, Progress in Polymer Science, 2013, 38(5), 703.
[50] A.D. Asandei, US Patent 0057419 A1, 2015 (assigned to the University of Connecticut).
[51] C.P. Simpson, O.I. Adebolu, J.-S. Kim, V. Vasu and A.D. Asandei, Macromolecules, 2015, 48 (18), 6404.
[52] a) M. Tatemoto and T. Amano, Eur. Patent 77,998, 1982 (assigned to Daikin); b) M. Yamabe, G. Kojima, H. Wachi and S. Kodama, US Patent 4,418,186, 1983 (assigned to Asahi Glass Co.); c) B. Otazaghine, L. Sauguet, M. Boucher and B Ameduri, European Polymer Journal, 2005, 41, 8, 1747; d) B. Ameduri, B. Boutevin, M. Armand and M. Boucher, Eur. Pat. 2004/ 1,242,485 and US 2006/014889 (assigned to Hydro-Quebec).
[53] R. Dams, Presented at the "17th European Symposium in Fluorine Chemistry" Conference, Paris, France, July, 21–25, 2013.
[54] C. Boyer, B. Ameduri and M.H. Hung, Macromolecules, 2010, 43(8), 3652–3663.
[55] a) G. Kostov, M.H. Hung and B. Ameduri, US Patent US2011/0,015,359 (assigned to DuPont Performance Elastomers and CNRS); b) G. Kostov, M. Holan, M.H. Hung and B. Ameduri, Macromolecules, 2012, 45, 18, 7375.
[56] A. Soulès, B. Ameduri, B. Boutevin and G. Calleja, Macromolecules, 2010, 43(10), 4489.
[57] M. H. Hung, and B. Ameduri, US Patent 2012/8,138,274, 2012 (assigned to DuPont Performance Elastomers and CNRS)
[58] B. Ameduri and A. Allaaddine, US Patent 2015/0119523, 2015 (deposted by Arkema and CNRS).
[59] G. Lopez, A. Thenappan and B. Ameduri, ACS Macro Letters, 2015, 4(1), 16.
[60] M. Guerre, B. Campagne, O. Gimello, K. Parra, B. Ameduri and V. Ladmiral, Macromolecules, 2015, 48(21), 7810.
[61] M. Guerre, B. Ameduri and V. Ladmiral, Polymer Chemistry, 2016, 7(2), 441.
[62] a) G. Kostov, B. Ameduri and S. Brandstadter, PCT/US Appl. 2007/017425 (assigned to Great Lakes-Chemtura, Chem. Co.; b) G. Kostov, F. Boschet, J. Buller, L. Badache, B. Ameduri and S. Brandstadter, Macromolecules, 2011, 44, 7, 1841.
[63] J. Kovarova and Z. Svobodova, Neuroendocrinology Letters, 2008, 29(5), 599.
[64] M. Lindstrom, Environmental Science & Technology, 2011, 45(19), 7954.
[65] A. Zaggia and B. Ameduri, Current Opinion in Colloid & Interface Science, 2012, 17(4), 188.
[66] E. Girard, J.D. Marty, B. Ameduri and M. Destarac, ACS Macro Letters, 2012, 1(2), 270.
[67] a) B. Ameduri and Y. Patil WO2013/160621 et US 9,184,461 B2 (assigned to Arkema France, CNRS et l'Ecole Nationale Supérieure de Chimie de Montpellier); Arkema patent b) Y. Patil and B. Ameduri, Polymer Chemistry, 2013, 4, 9, 2783.
[68] S. Seck, V. Bounor-Legare, J.F. Gerard, B. Ameduri, J. Bigarre and P. Buvat, WO2014/173888, 2013 (assigned to CEA, INSA and CNRS).
[69] L. Liu, D. Lu, H. Wang, Q. Dong, P. Wang and R. Bai, Chemical Communications, 2011, 47(27), 7839.
[70] P. Wang, J. Dai, L. Liu, Q. Dong, B. Jin and R. Bai, Polymer Chemistry, 2013, 4(6), 1760.
[71] C. Chanthad, K.A. Masser, K. Xu, J. Runt and Q. Wang, Journal of Materials Chemistry, 2012, 22(2), 341.
[72] W. Wang, D. Yan, D. Bratton, S.M. Howdle, Q. Wang and P. Lecomte, Advanced Materials, 2003, 15(16), 1348.
[73] J. Barth, R. Siegmann, S. Beuermann, G.T. Russell and M. Buback, Macromolecular Chemistry and Physics, 2012, 213(1), 19.

[74] E. Martinelli, G. Galli and A. Glisenti, European Polymer Journal, 2014, 60, 69.

[75] G. He, G. Zhang, J. Hua, J. Sun, S. Hua, Y. Li, F. Liu, D. Xiao, H. Zou and G. Liu, Journal of Fluorine Chemistry, 2011, 132(9), 562.

[76] N.K. Singha, M.I. Gibson, B.P. Koiry, M. Danial and H.-A. Klok, Biomacromolecules, 2011, 12 (8), 2908.

[77] Y.Y. Durmaz, E.L. Sahkulubey, Y. Yagci, E. Martinelli and G. Galli, Journal of Polymer Science Part A: Polymer Chemistry, 2012, 50(23), 4911.

[78] M.Y. Lee, Y.T. Jeong, K.T. Lim, B.-C. Choi, H.G. Kim and S.Y. Gal, Molecular Crystals and Liquid Crystals, 2009, 508(1), 173.

[79] B. Jiang, L. Zhang, B. Liao and H. Pang, Polymer, 2014, 55(21), 5350.

[80] B. Jiang, L. Zhang, J. Shi, S. Zhou, B. Liao, H. Liu, J. Zhen and H. Pang, Journal of Fluorine Chemistry, 2013, 153, 74.

[81] H. Peng, I. Blakey, B. Dargaville, F. Rasoul, S. Rose and A.K. Whittaker, Biomacromolecules, 2009, 10(2), 374.

[82] Y.-N. Zhou, H. Cheng and Z.-H. Luo, Journal of Polymer Science Part A: Polymer Chemistry, 2011, 49(16), 3647.

[83] W. Guo, X. Tang, J. Xu, X. Wang, Y. Chen, F. Yu and M. Pei, Journal of Polymer Science Part A: Polymer Chemistry, 2011, 49(7), 1528.

[84] M. Rabnawaz and G. Liu, Macromolecules, 2012, 45(13), 5586.

[85] X. Qiang, X. Ma, Z. Li and X. Hou, Colloid and Polymer Science, 2014, 292(7), 1531.

[86] P. M. Mendes, Chemical Society Reviews, 2008, 37(11), 2512.

[87] H. Huang and L. He, RSC Advances, 2014, 4(25), 13108.

[88] H. Huang, J. Qu and L. He, Journal of Polymer Science Part A: Polymer Chemistry, 2016, 54(3), 381.

[89] H.-J. Yu, and Z.-H. Luo, Journal of Polymer Science Part A: Polymer Chemistry, 2010, 48(23), 5570.

[90] C.-H. Xue, X.-J. Guo, J.-Z. Ma and S.-T. Jia, ACS Applied Materials and Interfaces, 2015, 7(15), 8251.

[91] T. Shinohara, Y. Higaki, S. Nojima, H. Masunaga, H. Ogawa, Y. Okamoto, T. Aoki and A. Takahara, Polymer, 2015, 69, 10.

[92] H. Yamaguchi, M. Kikuchi, M. Kobayashi, H. Ogawa, H. Masunaga, O. Sakata and A. Takahara, Macromolecules, 2012, 45(3), 1509.

[93] N.S. Bhairamadgi, S.P. Pujari, C.J.M. Rijn and H. Zuilhof, Langmuir, 2014, 30(42), 12532.

[94] K. Min, H. Gao and K. Matyjaszewski, Journal of the American Chemical Society, 2005, 127 (11), 3825.

[95] Y. Sun and W. Liu, Journal of Fluorine Chemistry, 2011, 132(1), 9.

[96] X. Zhan, Y. Yan, Q. Zhang and F. Chen, Journal of Materials Chemistry A, 2014, 2(24), 9390.

[97] J. Shu, C. Cheng, Y. Zheng, L. Shen, Y. Qiao and C. Fu, Polymer Bulletin, 2011, 67(7), 1185.

[98] W. Jakubowski, K. Min and K. Matyjaszewski, Macromolecules, 2006, 39(1), 39.

[99] U. Schreiber, B. Hosemann and S. Beuermann, Macromolecular Chemistry and Physics, 2011, 212(2), 168.

[100] S. Zheng, B. Xiong and Y. Xiong, Pigment & Resin Technology, 2012, 41(2), 95.

[101] A.V. Markin, S.D. Zaitsev, O.S. Zotova and N.N. Smirnova, Journal of Chemical & Engineering Data, 2013, 58(11), 3201.

[102] B.P. Koiry, M. Moukwa and N.K. Singha, Journal of Fluorine Chemistry, 2013, 153, 137.

[103] M.I. Gibson, E. Fröhlich and H.-A Klok, Journal of Polymer Science Part A: Polymer Chemistry, 2009, 47(17), 4332.

[104] Q. Zhang, Q. Wang, Z. Luo, X. Zhan and F. Chen, Polymer Engineering & Science, 2009, 49(9), 1818.

[105] S. Chen and W.H. Binder, Polymer Chemistry, 2015, 6(3), 448.
[106] B.P. Koiry, H.-A. Klok and N.K. Singha, Journal of Fluorine Chemistry, 2014, 165, 109.
[107] S.D. Zaitsev, Y.D. Semchikov and E.V. Chernikova, Polymer Science Series B, 2009, 51(3), 84.
[108] L. Guo, Y. Jiang, T. Qiu, Y. Meng and X. Li, Polymer, 2014, 55(18), 4601.
[109] B. Grignard, A. Vaillant, J. Coninck, M. Piens, A.M. Jonas, C. Detrembleur and C. Jerome, Langmuir, 2011, 27(1), 335.
[110] G. Li, A. Xu, B. Geng, S. Yang, G. Wu and S. Zhang, Journal of Fluorine Chemistry, 2014, 165, 132.
[111] F. Yi, R. Yu, S. Zheng and X. Li, Polymer, 2011, 52(24), 5669.
[112] B.P. Koiry, A. Chakrabarty and N.K. Singha, RSC Advances, 2015, 5(20), 15461.
[113] K. Skrabania, H. Berlepsch, C. Böttcher and A. Laschewsky, Macromolecules, 2010, 43(1), 271.
[114] J.-N. Marsat, M. Heydenreich, E. Kleinpeter, H. Berlepsch, C. Böttcher and A. Laschewsky, Macromolecules, 2011, 44(7), 2092.
[115] A. Chakrabarty and N.K. Singha, Journal of Colloid and Interface Science, 2013, 408, 66.
[116] H. Chen and Y. Luo, Macromolecular Chemistry and Physics, 2011, 212(7), 737.
[117] Y. Chen, W. Luo, Y. Wang, C. Sun, M. Han and C. Zhang, Journal of Colloid and Interface Science, 2012, 369(1), 46.
[118] H. Wang, H. Zhou, Y. Chen and C. Zhang, Colloid and Polymer Science, 2014, 292(11), 2803.
[119] X. Liu, J. Chen, P. Sun, Z.-W Liu and Z.-T. Liu, Reactive and Functional Polymers, 2010, 70(12), 972.
[120] K. Yang, X. Huang, Y. Huang, L. Xie and P. Jiang, Chemistry of Materials, 2013, 25(11), 2327.
[121] A. Chakrabarty and N.K. Singha, Macromolecular Chemistry and Physics, 2015, 216(6), 650.
[122] C. Bouilhac, M. Chirat, C. Joly-Duhamel and P. Lacroix-Desmazes, Macromolecular Chemistry and Physics, 2013, 214(20), 2259.
[123] S. Clerc, J. Tonnar and P. Lacroix-Desmazes, European Polymer Journal, 2013, 49(3), 682.
[124] S.R. Samanta, H.-J. Sun, A. Anastasaki, D.M. Haddleton and V. Percec, Polymer Chemistry, 2014, 5(1), 89.
[125] S.R. Samanta, M.E. Levere and V. Percec, Polymer Chemistry, 2013, 4(11), 3212.
[126] S.R. Samanta, R. Cai and V. Percec, Polymer Chemistry, 2014, 5(18), 5479.
[127] S.R. Samanta, R. Cai and V. Percec, Polymer Chemistry, 2015, 6(17), 3259.
[128] H. Hori, E. Hayakaiva, H. Einaga, S. Kutsuna, K. Koike, T. Ibusuki, H. Kiatagawa and R. Arakawa, Environmental Science & Technology, 2004, 38(22), 6118.
[129] P.D. Pickett, S.C. Radzinski and E.S. Tillman, Journal of Polymer Science Part A: Polymer Chemistry, 2012, 50(1), 156.
[130] B.H. Tan, H. Hussain, Y. Liu, C.B. He and T.P. Davis, Langmuir, 2010, 26(4), 2361.
[131] I. Dimitrov, K. Jankova and S. Hvilsted, Journal of Fluorine Chemistry, 2013, 149, 30.
[132] K.A. Pollack, P.M. Imbesi, J.E. Raymond and K.L. Wooley, ACS Applied Materials & Interfaces, 2014, 6(21), 19265.
[133] E. Martinelli, S. Menghetti, G. Galli, A. Glisenti, S. Krishnan, M.Y. Paik, C.K. Ober, D. Smilgies and D.A. Fischer, Journal of Polymer Science Part A: Polymer Chemistry, 2009, 47(1), 267.
[134] F. Ceretta, A. Zaggia, L. Conte and B. Ameduri, Journal of Polymer Science Part A: Polymer Chemistry, 2013, 51(15), 3202.

Saswati Ghosh Roy and Priyadarsi De

7 Polymers prepared via reversible-deactivation radical polymerization (RDRP) for biomedical applications

7.1 Introduction

Synthetic polymers have now become very important and are an essential part of our everyday lives. Initially, polymers were exploited for the replacement of natural materials and used as coatings, adhesives, packaging materials, rubber-based items, and so on. In more recent years, synthetic polymers have facilitated the technological advancement with their applications in higher-value-added materials in optical, electronics, biomedical, and biotechnological fields. During the past few decades, synthetic functional polymers have gained enormous impact for their biomedical and pharmaceutical applications in drug delivery systems, as scaffolds for tissue-engineering and repair, bio-imaging, protein delivery, gene therapy, bioassays, or bio-separations [1, 2]. These applications have led to the increasing demand for the synthesis of tailor-made polymers with desired functionality and properties. In this context, reversible-deactivation radical polymerization (RDRP) provides an excellent tool for material design with controlled molecular weight, narrow molar mass distribution (MMD), composition, chain architectures, and site-specific functionalities. This technique allows polymer scientists to design new polymeric materials with targeted properties, and improve the properties of materials available in the market. In this chapter we will discuss about the design of various functional polymers via the RDRP technique for applications in diverse biomedical fields.

7.2 Types of Reversible-Deactivation Radical Polymerizations (RDRP)

More than half of all commercial synthetic polymers are synthesized using conventional free radical polymerization (FRP) and this process has numerous advantages over other polymerization techniques. Conventional FRP can be exploited for the (co) polymerization of a broad range of vinyl monomers [3]. However, this technique results in polymers with uncontrolled molecular weights and molecular architectures,

Saswati Ghosh Roy and Priyadarsi De, Polymer Research Centre, Department of Chemical Sciences, Indian Institute of Science Education and Research Kolkata, Mohanpur, Nadia, West Bengal, India

https://doi.org/10.1515/9783110643695-007

which is associated with composition, chain length, molecular shape, and α,ω-functionality. To circumvent these drawbacks, several novel controlled radical polymerization techniques were developed since the 1990s. Various modifications and improvements have been made throughout the 2000s that help us to understand these polymerization techniques better and to utilize them for a wide range of polymeric material design. Listed based upon the chronological development of these polymerization techniques, the most important techniques are nitroxide-mediated polymerization (NMP), atom transfer radical polymerization (ATRP), reversible addition-fragmentation chain transfer (RAFT) polymerization, and organometallic-mediated radical polymerizations (OMRP) [4].

Though such polymerization techniques have been called collectively by a variety of names (controlled radical polymerization; living radical polymerization; controlled/living polymerization); according to the recent recommendation of IUPAC they should be called RDRPs. The IUPAC recommendation discourages the use of the term "living" associated with other common name; as RDRPs proceed through radical intermediates and some radical-radical bimolecular coupling reactions are unavoidable. Furthermore, during the chain polymerization other side reactions such as chain transfer to monomer or solvent may occur. But the word "controlled" may be used in this context if it matches at all [5].

Each of RDRP methods involves the same basic equilibrium process that helps the polymer chains to grow with predetermined and relatively uniform molecular weight (in the ideal case) and chain end functionality. Besides this equilibrium process, there are three important requirements for the synthesis of polymers with controlled molecular weight. These are: (1) fast initiation relative to chain propagation, (2) radical concentrations should be less than the polymer chain concentrations, (3) the equilibrium between dormant species and active chain should be fast [6, 7]. If all requirements are met, polymer chains have begun to grow at the early stage of polymerization and only few monomer additions occur per activation cycle, as a result the majority of polymer chains are present in their dormant state. Hence, the extent of chain termination and chain transfer reactions are minimized. The following section will discuss in detail individual RDRP techniques that have been employed for the synthesis of various functional polymer architectures for use in biomedical applications.

7.3 Nitroxide-Mediated Polymerization (NMP)

NMP is an attractive RDRP technique, which produces colorless and odorless polymers. Following the vital work of Solomon and coworker [8, 9], Georges *et al.* first reported the NMP technique for styrene (St) using a bicomponent initiating system comprised of a traditional free radical initiator (e.g., benzoyl peroxide (BPO) and a stable nitroxide (2,2,6,6-tetramethylpiperidinyl-1-oxy (TEMPO)) [10]. However, the limitations of TEMPO-mediated NMP are slow polymerization rates (~ 25–70 h), high polymerization temperatures (120–150 °C) and limited numbers of appropriate monomers (only styrene and styrenic derivatives) [11, 10]. Development of more active 2nd generation acyclic alkoxyamines/nitroxides such as 2,2,5-trimethyl-4-phenyl-3-azahexane-3-oxyl (TIPNO) [12, 13], *N-tert*-butyl-*N*-(1-diethylphosphono-(2,2-dimethylpropyl)-*N*-oxyl (DEPN or SG1) [14, 15], and *N-tert*-butyl-*N*-(1-*tert*-butyl-2-ethylsulfinyl)propyl nitroxide (BESN) (Figure 7.1) overcome these limitations and are able to polymerize alkyl acrylates [16, 17], acrylic acid [18], acrylamides [19, 20], and dienes in a controlled manner. A very limited number of 2nd generation functionalized alkoxyamines, e.g., BlocBuilder alkoxyamine based on the SG1 nitroxide [21, 22], NHS-BlocBuilder [23], and azlactone-functionalized SG1-based (AzSG1) alkoxyamines [24] (Figure 7.1) have also been developed for post polymerization modifications [25, 26] or for conjugation with bio-molecules like peptides/proteins [27, 28].

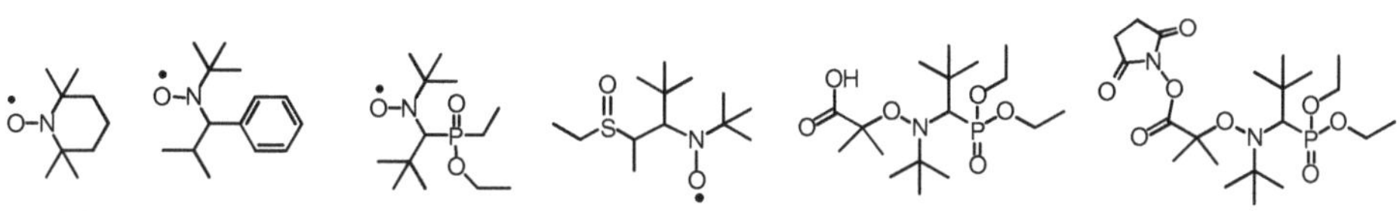

Figure 7.1: Structures of nitroxides used in NMP: TEMPO (a), TIPNO (b), SG1 or DEPN (c), BESN (d), BlocBuilder (e), and NHS-BlocBuilder (f).

7.3.1 Biomedical applications of NMP-based polymers

Due to the possibility of introduction of various functional groups into the polymer chain, the NMP technique has been employed for the construction and development of biomaterials, such as glycopolymers, protein-polymer conjugates, prodrugs, and so on for their applications in biomedical fields. Various example of biomaterials prepared thus far via NMP will be discussed in the following sections.

7.3.1.1 Glycopolymers

Carbohydrates are not only fundamental building blocks and energy storage vehicles for every living organism, but also play a key role in biological processes involving cell-cell interactions, such as inflammation, viral infection, signal transmission, and so on. Glycopolymers have pendant saccharide groups, and have been investigated extensively as carbohydrate mimics. Natural saccharides bind weakly to their receptors, but display higher affinities when they cluster or aggregate; this is known as the "cluster glycoside effect", which has potential application in biomedical fields.

However the synthesis and modification of saccharides are not straightforward processes. Therefore, it was hypothesized to synthesize glycopolymers as carbohydrate mimics. Synthetic glycopolymers are prepared from glycomonomers having polymerizable styrenic [29], acrylic [30], or methacrylic [31] moieties (Figure 7.2).

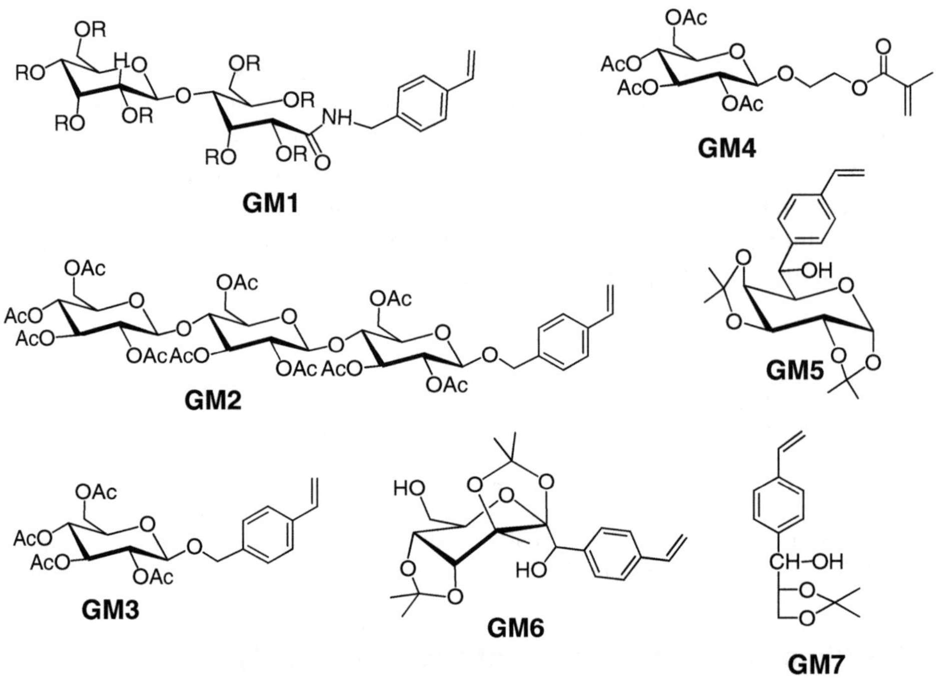

Figure 7.2: Structures of glycomonomers polymerized by NMP.

Fukuda and coworkers [32] first reported the synthesis of glycopolymers from *N*-(*p*-vinylbenzyl)-[O-*β*-D-galactopyranosyl-(1→4)]-D-gluconamide (VLA) (**GM1,** R = H) and its acetyl derivative (Ac-VLA) (**GM1,** R = Ac), using 2-(benzoyloxy)-1-(phenylethyl)-di-*tert*-butyl nitroxide (BS-DBNO) as alkoxyamine in the presence of dicumyl peroxide (DCP) in *N,N*-dimethylformamide (DMF) at 90 °C via the NMP technique. They also reported the synthesis of a well-defined novel glycolipid from Ac-VLA (**GM1,** R = Ac) using lipophilic DBNO-based alkoxyamine 1-(4-*N,N*-dioctadecylcarbamoylphenyl)ethyl-DBNO (DODA-PE-DBNO) as initiator followed by hydrolysis. These glycolipids were mixed with a phospholipid to form stable liposomes which have been used for lectin recognition [33]. Coupling reactions of polystyrenes (PSt)-TEMPO using divinylbenzene (DVB) as a cross-linking agent were performed in the presence of 4-vinylbenzyl glucoside peracetate (**GM3**) and 4-vinylbenzyl maltohexaoside peracetate (**GM2**) to obtain core-glyco conjugated star-shaped polymers. These star polymers produced amphiphilic star-shaped PSt with glucose and maltohexaose containing hydrophilic cores and PSt arms by deacetylation. Amphiphilic properties of star polymers have been altered by regulating the molar ratio of glycomonomer and DVB in the feed from 0.13 to 0.38. Amphiphilic stars showed encapsulation ability toward water-soluble molecules in chloroform with potential application in drug delivery [34, 35]. Miura and coworkers [36] reported the synthesis of glycopolymers with a narrow dispersity (*Đ*) by NMP using BPO-TEMPO from styrene carrying acetylated lactose monomer, followed by deprotection. This polymer showed a cylindrical structure and helical conformation, confirmed by circular dichroism (CD) spectroscopy. This polymer has shown a degree of polymerization (*DP*)-dependent affinity toward lectins, and the higher the *DP*, the stronger is the affinity.

The SG1 mediated NMP polymerization of methacrylate-based glycomonomer was reported to obtain poly(2-(2′,3′,4′,6′-tetra-*O*-acetyl-*β*-D-galactosyloxy)ethyl methacrylate-*co*-styrene) (P(AcGalEMA-*co*-St)) copolymer. A low percentage of St (10 mol%) decreases the average activation–deactivation equilibrium constant, thus allowing for a more controlled polymerization of glycomonomer. Amphiphilic block copolymers P (GalEMA-*co*-St)-*b*-PSt and PSt-*b*-P(AcGalEMA-*co*-St) were obtained from polymerization of glycomonomer AcGalEMA (**GM4**) by using either P(AcGalEMA-*co*-St)-SG1 or PSt-SG1 macroinitiators, followed by deacetylation of the AcGalEMA moiety. Self-assembled aggregation behavior of these amphiphilic polymers gave micellar structures and honeycomb-structured porous films with bioactivity [30]. Recently, synthesis of dextran-*b*-P(St-*co*-MMA) (MMA = methyl methacrylate) and dextran-*b*-PSt amphiphilic linear diblock copolymers has been achieved via NMP using dextran-SG1 as macro-initiator [37].

7.3.1.2 Peptide/protein-polymer conjugates

NMP has been successfully employed for the preparation of polymer protein/peptide bioconjugates. Wooley's group [38, 39] reported a novel strategy for the preparation of bioconjugate block copolymers supported on Wang's resin. Straightforward synthesis of polymer-peptide conjugates via NMP polymerization of St was reported using a SG1-functionalized peptide (SG1-GGGWIKVAV) initiator followed by subsequent cleavage of the solid support [40].

Well-defined and fluorescent *N*-succinimidyl ester-functionalized copolymers were easily synthesized under SG1-control without any purification beyond a simple precipitation [27]. Copolymers produced quantitative coupling with neuroprotective tripeptide, whereas, partial conjugation was obtained with lysozyme, used here as a model protein. Most recently, Nicolas and coworkers [24] reported one-step synthesis of azlactone-functionalized SG1-based alkoxyamine (AzSG1) for the NMP of St, *n*-butyl acrylate (*n*BA), and MMA in presence of a small amount of acrylonitrile (AN) as a co-monomer. All polymers were formed with controlled M_n and narrow dispersities ($Đ$ = 1.2–1.4). As a proof of concept, a well-defined Az-functional poly(ethylene glycol) (PEG)-based polymer was synthesized from oligo(ethylene glycol)methyl ether acrylate (OEGA) initiated by the AzSG1 alkoxyamine via NMP at 110 ºC in toluene. The resulting Az-POEGA was employed for the formation of a polymer-protein conjugate with lysozyme.

Molawi and Studer [41] have covalently attached the initiator moiety derived from TEMPO or a hindered TEMPO derivative with the side chain of L-serine via ether bond formation. The resulting serine-based alkoxyamine was successfully immobilized into various peptides using standard solution phase peptide synthesis. Similarly, TEMPO-based alkoxyamine-conjugated lysine and glycine residues were employed for the syntheses of a series of peptides by classical solution phase peptide coupling. The resulting peptides (having up to eight alkoxyamine moieties) were employed as initiators in NMP of St and *N*-isopropyl acrylamide (NIPAM) in highly controlled manner to obtain peptide–polymer conjugates with linear peptide backbones and a distinct number of polymeric side chains [42].

The combination of NMP and *N*-carboxyanhydride (NCA) polymerization has also been employed for the construction of peptide-polymer conjugates [43]. A dual TIPNO-based initiator with a primary amine group was employed for the ring opening polymerization (ROP) of *γ*-benzyl-L-glutamate-NCA (BLG-NCA), followed by NMP polymerization of St in one-pot without intermediate isolation. Well-defined (PBLG-*b*-PSt) bioconjugates were obtained with narrow $Đ$ values. This approach provides a flexible synthetic route to various combinations of homopolymer-peptide rod-coil block copolymers [44].

7.3.1.3 Drug delivery

Nanomedicines based on polymeric nanocarriers are enabling modern drug delivery application and bio-imaging [45]. RDRP techniques are the most efficient methods for the construction of well-defined polymer nanoparticles for drug delivery devices or polymeric prodrugs. Whereas, NMP was widely exploited for the preparation of glycopolymers and polymer-protein bioconjugates; synthesis of polymeric prodrugs via NMP is less explored. Harrisson *et al.* employed NMP for designing a new class of efficient drug-polymer conjugate prodrug nanoparticles with high drug loading capacity [46]. Gemcitabine-functionalized SG1-based macroalkoxyamine initiator was employed for the controlled growth of polyisoprene (PI) by NMP. The resulting amphiphilic polymer underwent self-assembly into stable, narrowly dispersed nanoparticles (130–160 nm) with remarkable colloidal stability. These nanostructures displayed efficient anticancer activity both in vitro (on various cancer cell lines) and in vivo (on human pancreatic carcinoma-bearing mice), and also suppressed the inherent toxicity of the employed chemotherapeutics. In another work they have reported a convenient way to synthesize degradable PEG-based copolymers for biomedical applications [47]. A series of functionalized amide-containing alkoxyamines (-COOH- or NHS-containing groups) based on the SG1 nitroxide was reported for the synthesis of functionalized materials from St, *n*BA, and MMA with a small amount of AN via NMP [48].

7.4 Atom Transfer Radical Polymerization (ATRP)

ATRP is a versatile and powerful technique that can be employed for the polymerization of a wide range of monomers including styrenics, (meth)acrylates, acrylonitrile, (meth)acrylamides, as well as water-soluble monomers, under mild reaction conditions and in various reaction media [49, 50]. Overall, remarkable growth and exploration of ATRP in the polymer field has been observed over the past 20 years, and is logical to continue its development for biomedical applications.

7.4.1 Synthesis of polymer bioconjugates

Coupling of synthetic polymers with bio-molecules (peptide, protein, oligonucleotides, carbohydrates, antibody-based drug molecules, and so on) is of great interest in biomedical and therapeutic applications, as it imparts improved pharmacokinetics, enhanced physical, and proteolytic stability. Polymeric therapeutics include polymeric drugs [51], polymer-drug conjugates [52], polymer–protein conjugates [53], polymeric micelles bound with drug molecules [54], and multi-component polyplexes prepared as non-viral vectors [55]. ATRP has been extensively used for preparing polymer-bioconjugate materials involving the processes of grafting polymer chains from, to, or by PEGylation, which can impart new properties to the materials.

7.4.1.1 "Grafting from" approach

In a "*grafting from*" (g-f) process polymer chains grow directly from an initiating site on a peptide, protein, or a surface [56]. Although modification of the peptide/protein with initiating moieties is essential, this method offers high yields and easy purification of the resulting conjugates [57]. Here, the ATRP initiating site is usually attached to the peptide/protein either covalently or through strong complexation. Short peptide sequences [58], biotin [59], and proteins such as streptavidin [60], chymotrypsin [61], and bovine serum albumin (BSA)-based [62] ATRP initiators have been successfully synthesized and used to prepare bioconjugates. Peptide-functionalized polymers prepared by this method were used for cell attachment and for in vivo targeting in drug delivery application (Figure 7.3) [63].

A solid-phase supported oligopeptide (SPPS) macro-initiator (Figure 7.3(a)) was employed for the solution phase ATRP of *n*BA to prepare SPPS-P*n*BA with controlled M_n and $Đ$ (1.19) [64]. Maynard's group [58] reported the strategy for the preparation of peptide-polymer conjugates with precise sites of attachment. Amino acid modified ATRP initiator (Figure 7.3(b)) was used for the site-specific SPPS synthesis, which was further employed for the polymerization of St, 2-hydroxyethyl methacrylate (HEMA) and HEMA modified *N*-acetyl-D-glucosamine (GlcNAc) as ATRP macro-initiator. Trzcinska *et al.* [65, 66] reported enzymatically degradable PMEO$_2$MA-Gly-Arg-Lys-Phe-Gly-dansyl (PMEO$_2$MA-GRKFG-Dns) and NIPAM-GRKFG-Dns bio-conjugates via ATRP polymerization of 2-(2-methoxyethoxy)ethyl methacrylate (MEO$_2$MA) or NIPAM using resin loaded Gly-Arg-Lys-Phe-Gly-dansyl-macro-initiator (Figure 7.3(c)). Dansyl-labeled pentapeptides were synthesized using SPPS from Fmoc-protected amino acids. Above the lower critical solution temperature (LCST) of MEO$_2$MA or PNIPAM, these bio-conjugates forms mesoglobules with peptide at the corona, which on tryptic hydrolysis cleaved to peptide fragments.

Protein-functionalized polymer bioconjugates by "*grafting from*" ATRP (Figure 7.4) were prepared for various biomedical applications. The first example of ATRP polymerization initiated from specific domains on proteins was carried out using a streptavidin-coupled biotinylated initiator (Figure 7.4(a)) for polymerization of NIPAM and polyethylene glycol methacrylate (PEGMA) at room temperature [67]. Gao *et al.* reported the synthesis of protein-polymer conjugates, where polymers were carefully grown from the *N*-terminus [68] and *C*-terminus [69] of proteins, which showed prolonged circulation time and improved drug accumulation in tumors. They prepared a myoglobin-POEGMA site-specific (*N*-terminal) and stoichiometric (1:1) conjugate via a g-f approach using myoglobin-ATRP initiator (Figure 7.4(b)). This conjugate showed a 41-fold increased blood exposure compared to the free protein, whereas GFP-C-POEGMA conjugate showed 15-fold increased blood exposure and 50-fold increase in tumor accumulation compared to the unmodified protein. They also reported the synthesis of site-specific in situ growth of a cyclized protein-polymer conjugate (c-GFP-POEGMA, GFP = green fluorescent protein) using *N*-terminus or *C*-terminus modified cyclized GFP (c-GFP-Br) macro-initiator. The c-GFP-POEGMA conjugate showed 9- and 310-fold improved thermal stability and considerably improved tumor retention as compared to the l (linear)-GFP-POEGMA [70]. In another example, genetically encoded initiator was employed for site-specific polymer growth from proteins. Amino acid-based initiator p-bromoisobutyryloxymethyl-L-phenylalanine was first generically encoded with GFP, and further employed as ATRP initiator to produce a polymer-GFP bioconjugate where the polymer is associated at the selected site on GFP [71]. Recently, preparation of protein–polymer hybrids (Figure 7.4(c)) via a g-f approach using genetically encoded non-canonical amino acid (nCAA) containing ester linked cleavable ATRP initiator has also been reported. The base labile ester bond allows the cleavage

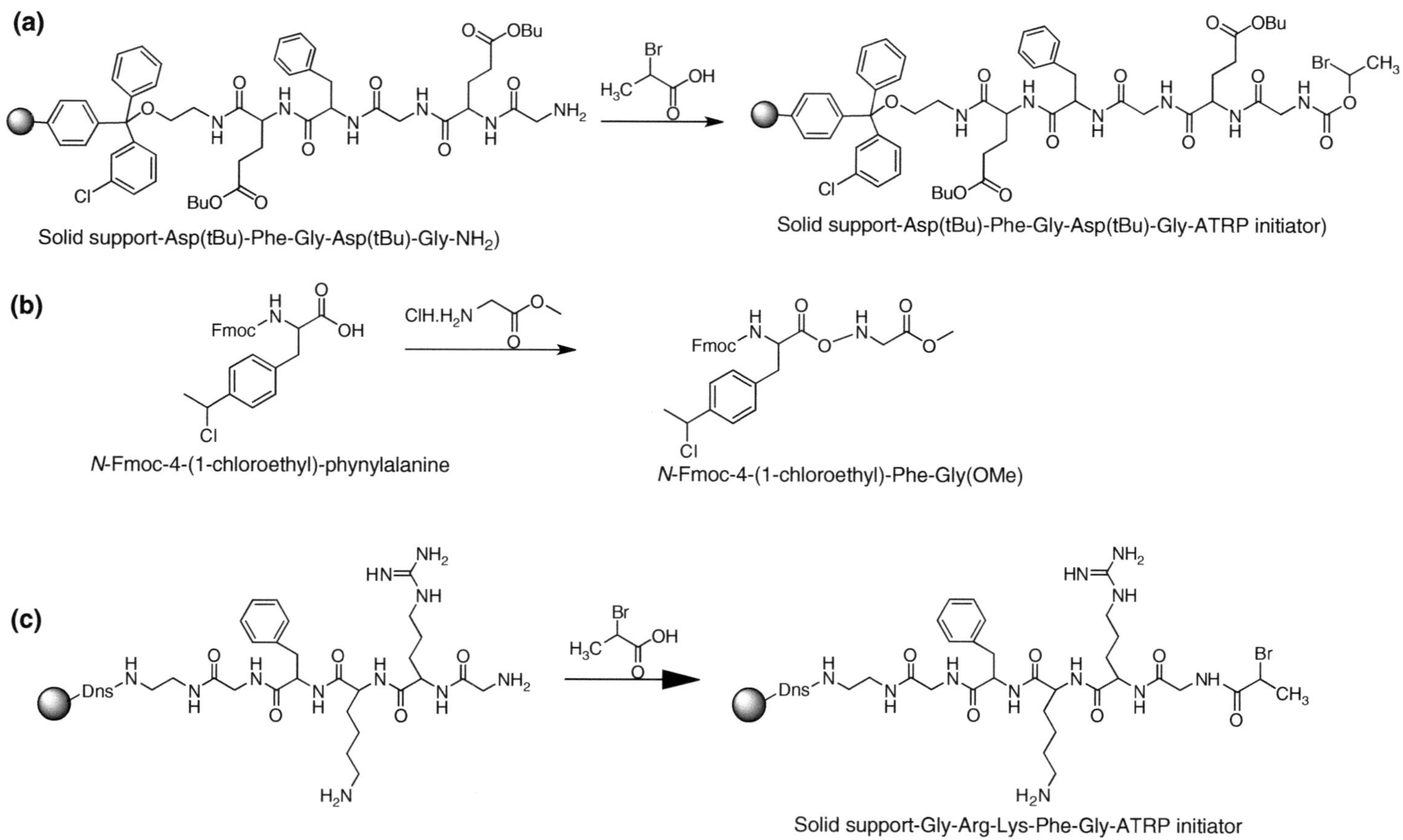

Figure 7.3: Peptide-macro initiator used in *grafting-from* approach in ATRP to prepare polymer-peptide bioconjugates. Figure 1.3(a) was adapted from Reference [64], Figure 1.3(b) was adapted from Reference [58] and Figure 1.3(c) was adapted from Reference [65].

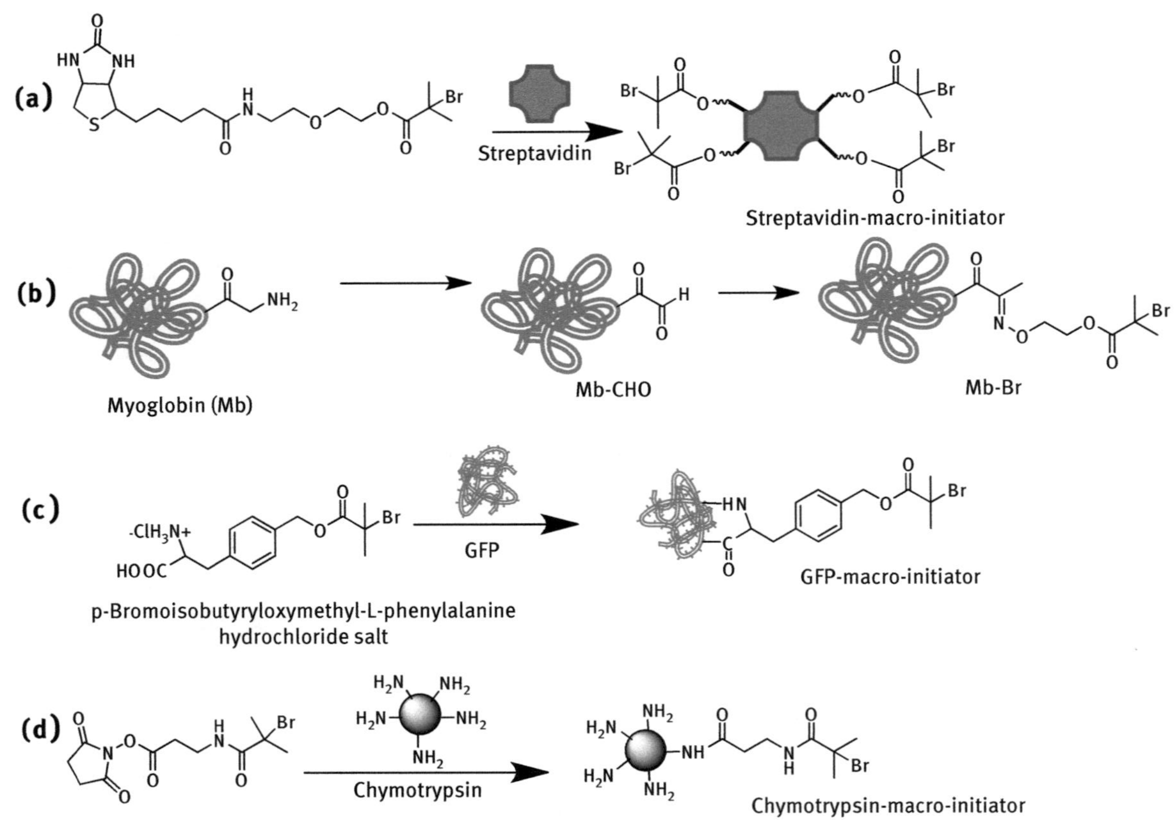

Figure 7.4: Protein-macro initiators used in *grafting-from* approach in ATRP for the synthesis of polymer-peptide bioconjugates. Figure 7.4(a), (b), (c), and (d) are adapted from Reference [67, 68], [72], and [61], respectively.

of grafted polymer from the protein [72]. Chymotrypsin-polymer poly(quaternary ammonium) (CT-PQA) conjugate (Figure 7.4(d)) was synthesized using a protein reactive, water-soluble chymotrypsin-ATRP-macro-initiator via a g-f approach. This conjugate showed stability at extreme temperatures and pH and increased bioactivity at low pH as compared to the native enzyme. This conjugate also showed increased enzymatic affinity towards peptide over a wide range of pH [61].

7.4.1.2 "Grafting to" approach

Since ATRP allows the synthesis of polymers with well-defined initiator end groups, it is an attractive route for the covalent attachment of polymer to the biomolecules to form bioconjugates via the "*grafting to*" (g-t) approach (Figure 7.5). For example, hydroxy-functionalized-POEOMA (HO-POEOMA) (Figure 7.5(a)) was prepared and employed for further functionalization of biotin, pyrene, and GRGDS peptide via a g-t approach. In addition, telechelic di-biotin polymer was also obtained via a combination of ATRP and click chemistry [73]. HEMA-functionalized elastin-like peptide tropoelastin Val–Pro–Gly–Val–Gly (VPGVG) (Figure 7.5(b)) was homopolymerized by ATRP.

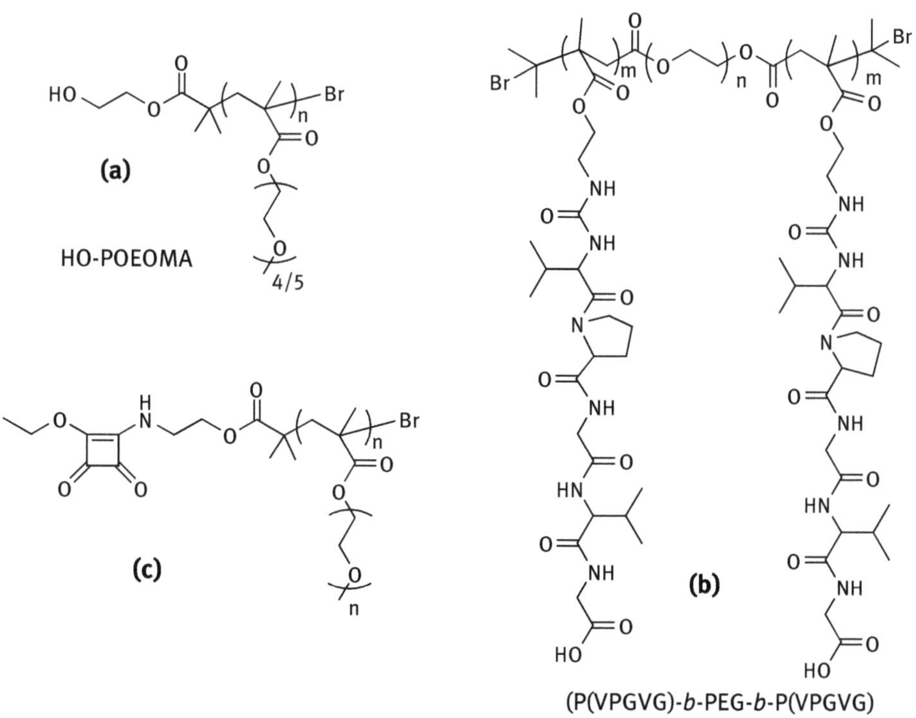

Figure 7.5: Functional polymers synthesized via ATRP for bioconjugation via "*grafting to*" approach. Structures (a), (b) and (c) in Figure 7.5 are adapted from references [62], [73] and [74].

It was also polymerized with a α,ω-di-functionalized PEG-macro-initiator to form an ABA triblock copolymer. This block copolymer displayed a pH dependent LCST due to the transition of peptide moieties from random to β-spiral at 40–70 ° C [74]. A new squaric acid-based ATRP initiator inspired to construct a library of linear, mid-functional, and 3-arm squaric acid functionalized poly(polyethylene glycol methacrylate) (PPEGMA) derivatives (Figure 7.5(c)). These polymers were further employed for the preparation of PPEGMA-BSA conjugates via g-t approach. [62].

7.4.1.3 PEGylation

The covalent binding of a protein/peptide to a PEG is known as "PEGylation." This approach offers to shape a new hybrid macromolecule (bioconjugate) with several potential beneficial effects including increased bioavailability and plasma half-lives, decreased immunogenicity, reduced proteolysis, and better solubility and stability. Abuchowski et al. [75] reported the first example of "PEGylation" in 1977, where methoxy-terminated PEGs (MPEGs) were covalently attached to bovine liver catalase. This PEG-catalase showed enhanced circulating time in the blood without any proof of an immune response. This result stimulated huge attention for the preparation of polymer-bioconjugates in the direction of drug and biomolecule delivery.

In this context, ATRP offers a wide range of polymer possibilities (Figure 7.6) due to its tolerance to most functional groups and can polymerize various sets of monomers. Comb-like polymers based on PEGMA and poly(ethylene glycol) acrylate (PEGA) were favorably employed as an alternative to traditional PEGylation. Due to their "umbrella-like" shape, branched PEG polymers exhibit higher flexibility of both the PEG chains and the polymer backbone and introduced a better resistance against proteolysis and antibodies action. Haddleton and coworkers employed ATRP to construct well-defined NHS-functional polymer PPEGMA (Figure 7.6 (a)) for the bioconjugation to lysine residues of lysozyme protein [76]. A recombinant human growth hormone (rh-GH)-PPEGMA hybrid was prepared from ATRP polymerization of PEGMA

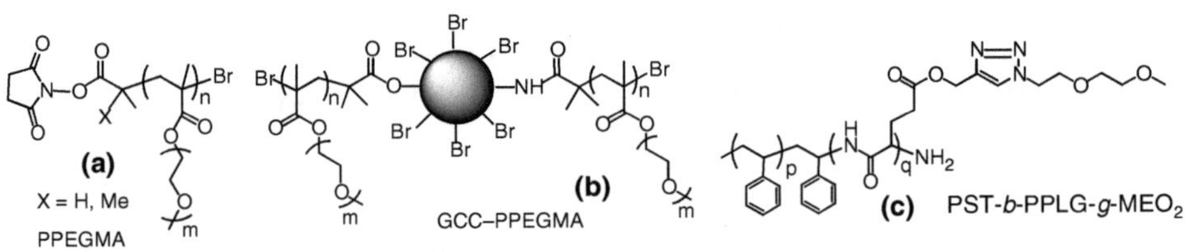

Figure 7.6: PEGylated polymers synthesized via ATRP for bioconjugation. Structures (a), (b), and (c) are adapted from reference [76, 77] and [79], respectively.

using rh-GH-ATRP initiator. This hybrid showed similar hormonal activity as that of native rh-GH in vivo when a daily dose was of 40 μg, even better bioavailability was observed with higher dose (120 μg) [56]. Red fluorescent nanoparticles (GCC NPs) functionalized ATRP initiator was employed for the PEGMA to obtain GCC–PPEGMA conjugate (Figure 7.6 (b)), which emit a bio-favorable deep red luminescence and exhibit excellent biocompatibility. The abundant hydroxyl groups on the surface of the nanoparticle (NP) conjugates can be further functionalized with diverse large biomolecules or other groups [77]. Another example of synthesis of PEGylated fluorescent NP from one-pot ATRP and "click chemistry" is reported, where PEGylated fluorescent NPs are biocompatible and further used as a labeling agent for KB cells [78]. PEGylated polypeptide block copolymers, PSt-*b*-poly(*γ*-propargyl-L-glutamate-*g*-ethylene oxide) PSt-*b*-(PPLG-*g*-MEO_2) Figure 7.6 (c), were synthesized using combinations of ATRP, ROP, and click chemistry [79].

7.4.2 Drug delivery devices

A variety of effective polymer-based controlled drug delivery systems (CDDS) has been projected and precise control over the hydrodynamic volume, morphology, chemical composition, and structure of polymers is obvious for generation of nanomedicines. These include polymer–drug conjugates, drug-encapsulated micelles, and vesicles prepared from amphiphilic or stimuli responsive doubly hydrophilic block copolymers, stars, NPs, cross-linked microgels/nanogels, and so on. Also, a variety of polymer-based gene delivery systems have been reported. This section describes ATRP-made polymer-based drug/gene delivery devices.

7.4.2.1 Polymeric nanostructures

Amphiphilic block copolymers consist of both hydrophobic and hydrophilic blocks that are covalently attached to one another. Due to the different solubilities of both the blocks in selective solvents, amphiphilic block copolymers can undergo self-assembly to form various nanostructures (micelles, vesicles, nanorods, and so on) in aqueous media. These nanostructures, with hydrophobic inner cores, can efficiently encapsulate hydrophobic therapeutics via hydrophobic interactions, electrostatic attractions, hydrogen, and/or covalent bonds and hydrophilic shell/corona ensures the water solubility and biocompatibility. Nanostructures can be programmed for triggered release of therapeutics in the presence of specific stimuli, such as temperature, pH, and so on or by passive diffusion.

Stimuli-responsive block copolymers have been designed and developed for drug delivery applications. This approach involves the design of double-hydrophilic block copolymers having a stimuli-responsive block that can form core of the micelle in response to external stimuli, such as pH, temperature, light, redox, and so on. An example of a pH-responsive block copolymer is poly(glycidol)-*b*-poly(4-vinylpyridine) (PGI-*b*-P4VP). The PGI block is water-soluble, whereas the P4VP segment is soluble at acidic pH, but aggregates at basic pH. Therefore, this block copolymer forms micelle having a P4VP core surrounded with PGI corona in basic aqueous media [80]. Another example of a pH responsive block copolymer is poly[2-(methacryloyloxy) ethylphosphorylcholine]-*b*-poly[2-(dimethylamino)ethyl methacrylate]-*b*-poly[2-(diisopropylamino)ethyl methacrylate] (PMPC-*b*-PDMAEMA-*b*-PDPAEMA) triblock copolymer, as well as PMPC-*b*-P(DMAEMA-*stat*-DPAEMA) block-statistical copolymer. These polymers form tuneable pH-responsive vesicles in aqueous solution due to different block lengths of PDPAEMA and introduction of PDMAEMA at different positions in the polymer blocks. These polymer NPs can efficiently encapsulate the anticancer drug (doxorubicin, DOX) and showed retarded release profile at physiological pH [81].

Another way of block copolymer nanostructure preparation is from block-copolymers with thermo-responsive blocks. Poly(sodium 2-acrylamido-2-methyl-propanesulfonate-*b*-NIPAM) (PAMPS-*b*-PNIPAM) with variable block length is an example of a thermo-sensitive anionic block copolymer. These polymers show temperature (LCST of PNIPAM ≥ 32 °C) – induced self-assembled aggregation behavior in aqueous solution at different NaCl ionic strengths. Spherical core-shell type micelles are formed with a higher PNIPAM/PAMPS ratio and are independent of ionic strength, whereas core-shell type micelles are formed with lower PNIPAM/PAMPS ratio at high ionic strength. ABA and BAB are two triblock thermos-responsive copolymers (where A=PNIPAM, B=PHEMA) synthesized by ATRP. In these triblock copolymers PHEMA is the water-soluble block and PNIPAM is the water-soluble thermo-responsive block that can aggregate above LCST (≥ 32 °C). The ABA and BAB copolymers separately formed branch and flower micelles, respectively, in water, and the aggregate formed is mainly influenced by the course of the phase transition [82].

Another approach to nanostructure formation is from the double hydrophilic block copolymers with dual pH and thermo-responsive blocks. A novel double hydrophilic block copolymer (PNIPAM-*b*-P4VP) is the example of dual pH and thermoresponsive block copolymer. This block copolymer showed temperature dependent aggregation behavior at 36 °C, and critical pH dependent aggregation at pH 4.7. In aqueous solution, the copolymer presents as unimer at pH 2.8 at 25 °C but forms spherical core-shell micelle when the temperature was raised to 50 °C at pH 2.8, where the PNIPAM block formed the core and P4VP segment is in the shell. On the other hand, reverse spherical core-shell micelle was formed when pH was increased from 2.8 to 6.5 at 25 °C, with the core formed by the P4VP block and the shell formed by the PNIPAM segment [83]. Dual pH and thermo-responsive block copolymer poly(ethylethylene phosphate)-*block*-poly[2-(dimethylamino)ethyl methacrylate] (PEEP-*b*-PDMAEMA) was synthesized via combination of ROP and ATRP. This diblock copolymer formed different self-assembled aggregates with different particle size and morphologies depending upon pH and temperature. These nanoparticles have ability to condense with DNA and have potential application in gene delivery [84]. Several other dual pH and thermo-responsive triblock copolymers [85], penta-block copolymers [86], dense copolymer brushes [87], stars [88], and so on are reported, which can form pH- and temperature-induced self-assembled micellization in aqueous media. "Schizophrenic" hemo-compatible copolymer micelles from poly(*N*-isopropyl acrylamide)-*b*-poly(sulfobetaine methacrylate) (PNIPAM-*b*-PSBMA) via switchable thermo-responsive transition of nonionic/zwitterionic block showed extremely high anticoagulant activity and anti-hemolytic activity in human blood over a wide range of temperatures (4 to 40 °C) [89].

Amphiphilic stimuli-responsive block copolymers comprised of hydrophobic blocks with cleavable hydrophobic pendants (e.g., *o*-nitrobenzyl [90], coumarin [91], spiropyran [92] BODIPY [93], and so on) have also been investigated. These block copolymers undergo micellization in water with hydrophobic pendent cleavable block at the core of the micelles, which can be disrupted in the presence of external stimuli (light, pH, and so on) due to the cleavage of pendant hydrophobic groups in the hydrophobic segment. Such types of degradation ensure the controlled release of hydrophobic drug molecules. Coumarin-containing hydrophobic block copolymer micelles can further undergo core cross-linking due to photo-dimerization of coumarin in the presence of visible light. This photo dimerization is reversible and is cleavable when micelles are illuminated in UV light [91]. Stimuli-responsive degradable amphiphilic block copolymers can also be designed, having cleavable linkages within the polymer backbone or in the side chain pendant, using ATRP. For example, well-defined reductively degradable amphiphilic block copolymers consisting of disulfide linkages (-S-S-) positioned repetitively on hydrophobic chain pendants were synthesized via ATRP [94, 95]. Synthesis and self-assembly of dual redox and thermo-responsive double hydrophilic block copolymers having pendant -S-S- linkage have also been reported [96]. Redox-responsive polymer–drug conjugates based on hydrophilic diblock copolymer have been synthesized for intercellular drug delivery, where paclitaxel drug was covalently attached via a -S-S- linkage [97].

7.4.2.2 Glycopolymer-based nanostructures

Usually hydroxyl groups of polysaccharide chains are modified with an ATRP initiator (e.g., 2-bromoisobutyryl bromide) to form ATRP macro-initiators, which are further employed for the synthesis of amphiphilic block copolymers. A mannose-containing glycopolymer library, prepared via a combination of ATRP and "*azide-alkyne*" cycloaddition, has been used to give novel insights into the mechanisms of

HIV infection [98]. Maltoheptaose-*b*-PNIPAM (Mal7-*b*-PNIPAM) block copolymers were prepared via *azide-alkyne* cycloaddition, between an alkynyl-functionalized Mal7 and N_3-PNIPAM synthesized by ATRP. The self-assembly of such systems showed well-defined vesicular morphologies [99]. However, diblock copolymer (Mal7-*b*-PMMA) synthesized by similar method underwent self-assembly aggregation in water into large compound micelles (LCMs) and reverse micelle-type nanoparticles that efficiently encapsulated and release hydrophobic guest molecules [100]. Thermo-responsive glycopolymers prepared from copolymerization of NIPAM and *N*-allylacrylamide via ATRP, followed by subsequent thiol–ene reaction of 1-thiosugars, showed lectin-binding ability [101]. Amphiphilic glycopolymers were synthesized by chemical modification of hydroxyethyl acrylate (HEA) and *n*BA units by d-(+)-glucosamine or *N*-(4-aminobutyl)-d-gluconamide, present in di- and triblock copolymers PBA-*b*-PHEA and PHEA-*b*-PBA-*b*-PHEA, and these materials may find potential applications in lectins binding [102]. Synthesis of various glycopolymers via ATRP (Figure 7.7) and carbohydrate drug and neoglycopolymer via combination of ATRP and *azide-alkyne* cycloaddition has been summarized recently [103].

ATRP of acrylate and methacrylate analogues of sugar moieties, followed by block copolymerization for the synthesis of glycopolymer-based amphiphilic block copolymers, is an alternative approach for the preparation of nanostructures for drug delivery applications. Methacrylate analogues of glucofuranoside and galactopyranoside were employed for the preparation of fluorescent ABA triblock copolymers, which are biocompatible and very attractive materials for the synthesis of biohybrid macromolecules [104]. Triple stimuli (temperature/pH/photo)-responsive amphiphilic glycopolymer, poly[DMAEMA-*co*-6-*O*-methacryloyl-D-galactopyranose)-*b*-poly(4-(4-methoxyphenylazo) phenoxy methacrylate) (P(DMAEMA-*co*-MAGP)-*b*-PMAZO) was synthesized by ATRP, followed by hydrolysis of the isopropylidene protecting groups. This triple responsive glycopolymer nanoparticle showed interesting properties in both in vivo and in vitro environments [105]. Poly[methacryloyl-d-glucopyranoside] (GP) and PEO-*b*-GP glycopolymers were synthesized via ATRP followed by deacetylation of acetylated polymers. Improved biocompatibility with osteoblast cells was observed in the presence of PEO-GP with polymer concentration from 0.1 to 1,000 μmol/L [106]. Star-shaped porphyrin-cored PCL-*b*-poly(gluconamidoethyl methacrylate) block copolymers (SPPCL-PGAMA) prepared via combination of ROP and ATRP have potential in drug delivery, photodynamic therapy, and very specific recognition with Concanavalin A (ConA) for targeted drug delivery application [107]. Well-defined linear and/or comb-like PSt-*co*-AcGEMA polymers have lectin-recognition capabilities, which specifically recognize ConA. These bioactive polymers have potential applications as templates, for bioanalysis and as cell culture materials [108]. A multivalent glycopolymer with pendant saccharides were synthesized by surface-initiated (SI)-ATRP of 2-O-(*N*-acetyl-*β*-d-glucosamine)ethyl methacrylate (GlcNAcEMA), which showed very selective, specific, and strong

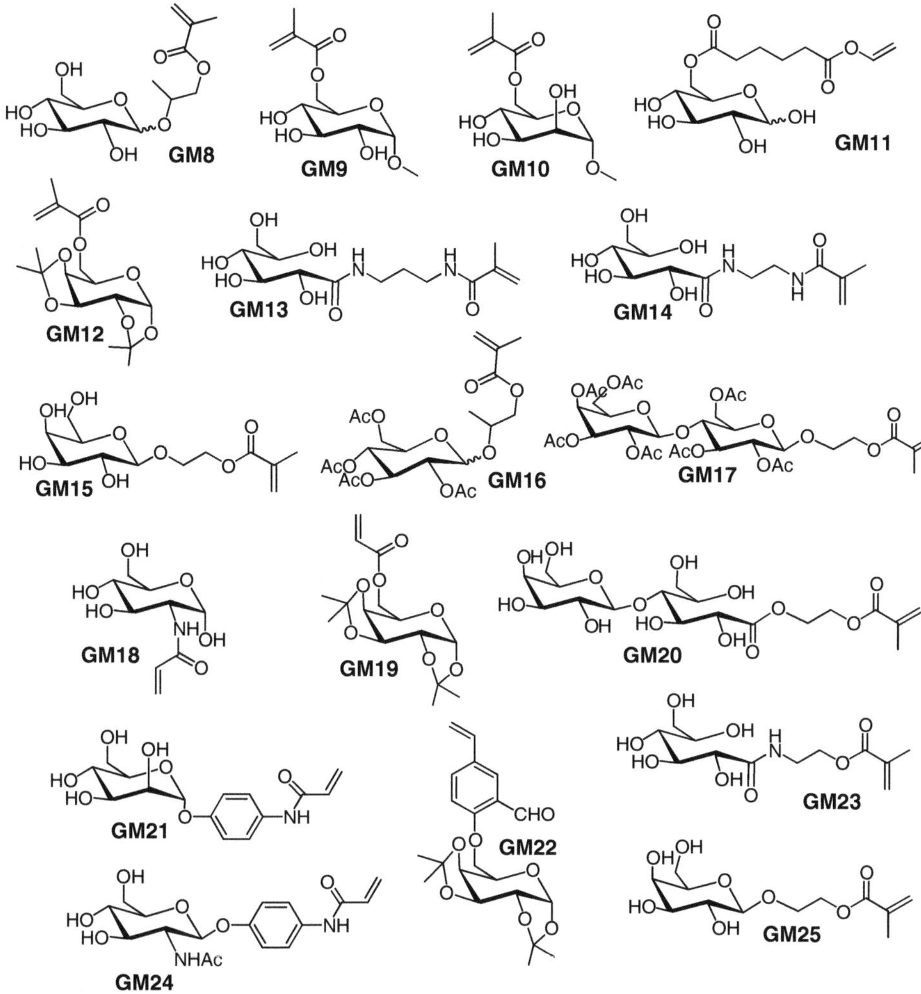

Figure 7.7: Glycomonomers polymerized via ATRP and RAFT.

interactions with lectins [109]. Recently, water-soluble BODIPY-conjugated glycopolymers have been reported as fluorescent probes for live cell imaging [110].

7.4.2.3 Polyester-based nanostructures

ATRP was also employed for the preparation of well-defined polyester-based amphiphilic block copolymers from hydrophilic methacrylates such as poly(2-

methacryloyloxyethyl phosphorylcholine) [111], PDMAEMA [112] and PHEMA [113]. These polymers show good cytocompatibility and can efficiently incorporate hydrophobic drug molecules.

7.4.2.4 Micro/nano-gels

Micro/nano-gels are cross-linked hydrogel particles with micro- or nanoscopic dimensions having three-dimensional networks, formed by chemical and/or physical cross-linking of polymer chains. They exhibit biocompatibility, high water content, desirable mechanical properties, a large surface area for multivalent bioconjugation, and an interior network for the incorporation of biomolecules. The size of nanogels can be tuned from optimal diameter for increased blood circulation time to nanoscopic diameter (< 200 nm), which enables better cellular uptake. Finally, the interior network allows for encapsulation of bioactive molecules (such as drugs, proteins, carbohydrates, and nucleic acids). Therefore, nanogels (NG) of well-defined polymers are synthesized in the presence of the cross-linker via RDRP polymerization techniques. For example, pH-sensitive poly[(*N*-2-hydroxyethyl acrylamide)-*co*-acrylic acid (AAc)] (P(HEAA-*co*-AAc)) NGs with controlled size and morphology were synthesized via inverse-microemulsion ATRP. Due to ultrastability, super low fouling ability, and environment-responsive controlled release properties, these NGs have great potential for drug delivery applications [114]. Site-specific functional quaternized NGs (QNG) were synthesized using dithiopropionyl poly(ethylene glycol) dimethacrylate (DMA) as a cross-linking agent. These QNG are biodegradable in nature due the presence of -S-S- linkages, and they form stable complexes with *p*DNA and siRNA at relatively low weight ratios of QNG to *p*DNA (5:1) and QNG to siRNA (15:1) [115]. Novel dual responsive (pH and reduction) mPEG-*b*-PDEA NGs were prepared via quaternization of DEA with bromine from *N,N*'-bis(bromoacetyl) cystamin. Drug release properties of DOX loaded NGs can be regulated by the pH and the concentration of glutathione (GSH) [116]. Recently, cationic nanostructured polymeric NGs prepared by ATRP for in vitro and in vivo siRNA delivery in mammalian models have been reported. NG reduces enzymatic degradation of siRNA within polyplexes, and nuclease resistance, without compromising gene knockdown potency, may be enhanced by methylation of siRNA [117]. The detailed description of polymer NGs for drug/nucleic acid delivery and NG engineering for biomedical applications has been reported elsewhere [118].

Functional NGs were prepared for bioconjugation using ATRP in inverse miniemulsion polymerization in the presence of functional ATRP initiators such as 2-hydroxyethyl 2-bromoisobutyrate (HOEtBriB) or copolymerization with functional monomers. For example, NGs prepared from HEA using HOEtBriB could conjugate with biomolecules. For example, POEOMA was synthesized via Activator generated by electron transfer (AGET)-ATRP inverse miniemulsion of OEOMA initiated by HOEtBriB was further functionalized with biotin, pyrene and GRGDS peptide [119]. A well-defined protein–nanogel hybrid (GPF-PEOMA nanogel) was synthesized, where GFP is site specifically attached into the nanogel and preserved its native tertiary structure [120]. Future directions of NG research should be toward improvement in design of micro/nanogels with specific targeting residues that can enable highly selective uptake into specific cells, particularly cancer cells.

7.4.3 Bioactive scaffold and bioactive surfaces for tissue engineering

Bioactive scaffolds are three-dimensional porous solid biomaterials which play a distinctive role in tissue regeneration and repair. During the past two decades, several works have been carried out to develop potentially applicable scaffold materials for tissue engineering. Polymeric scaffolds are attractive for this purpose due to their controllable microstructure, strength, rate of degradation, porosity, and their shapes and sizes. RDRP methods enable the preparation of well-defined polymers with complex architectures having incorporated reactive functional groups. An important discovery in this area of scaffold formation is the presence of adhesion domains in the functional polymers or other extracellular glycoproteins, which contain cell-binding peptide motif Arg-Gly-Asp (RGD). For example, a scaffold with arranged functional groups was prepared from end-functionalized poly(ε-caprolactum) (PCL) with either a polymerization-initiating group or a peptide sequence, followed by sequential electro-spinning. Surface-initiated (SI) PCL-POEGMA bottlebrushes have resulted in high performance antifouling and cell resistant coatings. These scaffolds with tailorable properties are potential materials for many applications in tissue engineering and regenerative medicine [121]. Functional PMA brushes grafted from the chitosan microsphere surfaces via SI-ATRP and subsequent conjugation of RGD peptides on the pendent carboxyl groups of PMAA give RGD peptide-functionalized chitosan microsphere scaffolds for fast cell expansion of human umbilical vein endothelial cells (HUVECs) [122]. Polystyrene-based macro-initiator electro-spun fibers were grafted with polymer brushes poly(trimethylsilyl-propargyl-hexaethylene glycol methacrylate-*co*-OEGMA) via SI-ATRP and attached with RDG peptide to prepare scaffolds for tissue engineering applications [123]. The scaffold was further modified via grafting of PEGMA onto the fiber surface by SI-ATRP. These scaffolds show potential application in vascular tissue engineering [124]. Modification of PCL nanofibers with biomimetic polydopamine (PDA) layer and preparation of four types of antifouling polymer brushes was carried out by SI-ATRP for the preparation of cell adhesion with mouse embryonic fibroblasts (MEFs) [125]. Scaffold prepared from polymeric nanofibrils

(NFs) obtained from electro-spun nanofibers and subsequent surface-decoration with poly(methacrylate) derivatives via ATRP showed higher degrees of protein sequestration and enhanced cell proliferation [126]. The details of polymer-protein conjugation scaffolds for drug delivery and tissue engineering are discussed elsewhere [127].

A physical gel scaffold prepared form ABA triblock copolymer consisting of PNIPAM (A block) and *O*-phosphoethanolamine (PEA) grafted PAA (PAA-*g*-PEA) (B block) has great potential for bone tissue engineering [128]. Biodegradable P (HEMA-*co*-PCL) has been developed for tissue engineering applications, in cardiac and other areas [129]. A thermo-responsive gelatin–PNIPAM injectable hydrogel scaffold is another approach that has been developed for in vivo bone defect regeneration [130]. PNIPAM hydrogel has also been used for the preparation of tissue engineering scaffolds [131]. Several approaches have been made for the preparation of hydrogel-, microgel-, and nanogel-based tissue engineering scaffolds for controlled delivery of proteins and drug release, tissue engineering, and regenerative medicine [132, 133].

The interaction of cells with material surfaces is an important subject for cell adhesion, growth, signaling, and segregation. In this respect, ATRP provides numerous approaches for the preparation of bioactive surface. For example, bioactive surfaces based on four polymer brushes prepared via SI-ATRP of OEGMA, HEMA, carboxybetaine acrylamide (CBAA), and *N*-(2-hydroxypropyl) methacrylamide (HPMA) from gold-coated silicon wafers with self-assembled monolayers (SAMs). These surfaces showed superior antifouling ability towards four human body fluids (blood plasma, cerebrospinal fluid, urine, and saliva) and four animal fluids (foetal bovine and calf sera, egg, and milk) [134]. Bioactive surfaces were prepared using a straightforward and facile strategy. SI-ATRP is employed to grow antifouling polymers preserving the end groups, which were converted to a reactive cyclopentadienyl (Cp). This Cp formed Diels–Alder adduct with maleimide-BSA, which possesses excellent antifouling properties [135]. POEGMA modified gold surface via SI-ATRP was decorated with peptide used to immobilize tissue plasminogen activator (tPA) surfaces, which displayed the defensive properties against the plasminogen activator inhibitor-1 [136].

Materials surface modified with polymers containing quaternary ammonium (QA) groups have shown great antibacterial activity [137]. The quaternized PVDF-*g*-PDMAEMA and PVDF-*g*-PDMAEMA-*b*-PHEMA membranes prepared via SI-ATRP showed excellent antibacterial activity against *Staphylococcus aureus* (*S. aureus*) [138]. Quaternized DMAEMA-*g*-agarose copolymers in aqueous media and their hydrogels on substrate surfaces showed excellent antibacterial activity against *S. aureus and Escherichia coli (E. coli)* [139]. Similar types of quaternized polymer PCL-*g*-poly(dimethylaminopropyl methacrylamide)-Arg–Glu–Asp–Val showed improved antibacterial activity, anti-thrombogenicity, and endothelial cell proliferation [140].

7.4.4 Polymers for bioimaging and photodynamic therapy

Colloidal inorganic NPs have drawn a great deal of interest as bioimaging materials due to their unique electronic and optical properties. One approach of binding NPs is the coating of NP with polymers having multi-dented ligating sites. For this purpose several di/multi-block copolymers were synthesized with pendant anchoring groups that enable the binding of nanocrystals. For example, (HOOC-PEGMA-*b*-PDMAEMA-*b*-(PDMAEMA-*co*-POMA)) triblock copolymer was synthesized via ATRP, which can simultaneously solubilize inorganic nanoparticle, condense siRNA, destabilize endosome, provide bioconjugation sites, and render outstanding colloidal stability in serum. Specifically, the hydrocarbons of the hydrophobic PDMAEMA-*co*-POMA block bind with the (quantum dots) QD through hydrophobic–hydrophobic interactions, whereas the hydrophilic PEGMA and PDMAEMA blocks render water solubility to the entire nanoparticle. The PDMAEMA block at pH = 7.4 binds with siRNA molecules and also capable of disrupting endosome membrane to facilitate drug cytosolic release [141]. There are several block copolymers that have been reported for the stabilization of NPs via ligand exchange process [142]. Detail discussion on polymers for imaging applications has been discussed in detailed in some reviews [143].

Recently, photodynamic therapy (PDT) is becoming an emerging clinical modality in treatment of several malignant tumors. Usually PDT consists of photosensitizing drug, light, and oxygen that produce singlet oxygen at appropriate wavelength and causes tissue damage through apoptotic and/or necrotic mechanism. This is a non-invasive technique as compared to conventional therapies, as it selectively destroys the cancer cells and builds up immunological responses against them. Several well-defined polymer-photosensitizer conjugates have been developed using ATRP technique. For example, a combined chemotherapy/PDT delivery strategy in one system was prepared from star-porphyrin-cored SPPCL-*b*-PGAMA copolymers via a combination of ROP and ATRP. This star polymer formed micelle that can efficiently encapsulate paclitaxel drug and also acts as a nanosized photosensitizing agent due to the presence of the porphyrin core [107]. A mixed micelle system with pH-responsive and photoactive properties based on star polymer mPEG-*b*-PDPAEMA with poly(ethylene glycol)-conjugated zinc tetraphenylporphyrin core can encapsulate DOX and simultaneously acts as a photosensitizing agent to facilitate combined chemo-photodynamic therapy in one delivery system [144]. A PDT agent prepared from zinc phthalocyanine-encapsulated haemoglobin-conjugated PEG-*b*-PAA-*b*-PSt micelles showed excellent photocytotoxicity to Hela cells in vitro [145].

7.4.5 Biodegradable polymers

Synthetic biodegradable polymers find diverse biomedical applications due to their tailorable designs or modifications. ATRP is an attractive tool for the synthesis of tailor-made biodegradable polymers from vinyl monomers in combination with other techniques, such as ROP or degradable cross-liker, initiators, "*click*" chemistry. As polymer formed in ATRP via a radical chain growth mechanism, it is possible to introduce degradability into ATRP-produced polymers via combination of ATRP with ROP. For example, bottlebrushes with poly-L-lysine (PLL) backbone with PSt and POEGMA as side-chains are biodegradable polymers prepared via combination of ROP and ATRP [146]. Three types of block copolymers PCL-poly(glycidyl methacrylate) (PCL–PGMA), PCL–PMAA, and PCL–PHEMA, with biodegradable PCL segment were synthesized [147]. Biodegradable comb-shaped polymer with a poly(*N*-3-hydroxypropyl)aspartamide (PHPA) backbone and PDMAEMA side chains-based gene carrier has also been reported [148]. Biodegradable polymers were also prepared using polysaccharide backbones in comb-shaped and bottle brush-shaped polymers [149].

Introduction of disulfide bonds into ATRP-produced polymers is another attractive option for controlled degradation that can be incorporated in a variety of ways. Functional ATRP initiators can be prepared in such a way that polymer chains can grow in two directions, resulting in the existence of -S-S- bonds in the middle of each polymer chain, which can be cleaved into two thiol-terminated chains upon exposure to a variety of reducing agents. Examples include the polymerization of linear polymethacrylates [150], polymer-antibody conjugate [151], comb [152], branched [153], and star polymers [154]. The coupling of polymer chains via "*click coupling*" is another way to introduce degradable bonds into polymers prepared by ATRP [155].

The above discussion confirmed several biological applications of ATRP-derived polymers. But, note that there are still issues for the use of ATRP-derived polymers for biological systems [156], because of the difficulty to remove traces of toxic copper from the polymer matrix. Very recently, bioinspired catalyst [157], bacterial redox systems [158], and so on, have been developed to prepare polymers in a manner analogous to ATRP technique.

7.5 Bio-Application of RAFT

Among the various RDRP techniques, RAFT is the most chosen route in terms of tolerance to monomer and initiator, as well as versatility toward experimental conditions such as solvent and temperature. In this section, we will discuss recent developments in the area of RAFT polymerization for the design and the synthesis of polymers for bio-applications, such as glycopolymers, gene/drug delivery systems, biomaterials, bioconjugates, gels, and so on.

7.5.1 Bioconjugates

Attachment of polymers to biomolecules can facilitate their stability, solubility, and biocompatibility. Bioconjugation to synthetic polymers can also introduce additional functionality to the bioconjugates, thereby inducing novel self-assembly, designing morphology and phase behavior. Bioconjugates that can be prepared via the RAFT method include peptide/protein-polymer conjugates, DNA/RNA functionalized polymers, glycopolymers, and so on.

7.5.1.1 Polymer-bioconjugates via a "grafting from" approach

The in situ or "*grafting from*" approach to bioconjugate formation has many advantages, as discussed earlier in ATRP "*grafting from*" approach. In this case the *R* group of RAFT agents contain –COOH group or activated ester group for the direct attachment of biomolecules. Therefore, biomolecule-attached RAFT agent (or biohybrid RAFT agent) was then directly employed for the polymerization of various vinyl monomers to form bioconjugates (Figure 7.8). This approach was exploited to attach different biomolecules such as sugar (galactose), phospholipid, biotin, peptide, protein [159], lysozyme [160] (Figure 7.8(a)), BSA, and so on. For example, a biotin-modified RAFT agent synthesized by coupling of biotinylated -COOH and –OH of RAFT in the presence of dicyclohexylcarbodiimide (DCC) and dimethylaminopyridine (DMAP) (Figure 7.8(b)), and was employed for controlled synthesis of cleavable biotin containing polymers [161]. Four different peptide-macro RAFT agents were used in controlled polymerization for a number of monomers (NIPAM, DMA, *n*BA, and MMA) [162]. BSA-macro RAFT agent using the selective reaction of thiol-pyridyl disulfide (PDS) (Figure 7.8(c)) between *S,S*-bis (α,α'-dimethyl-α''-acetic acid) trithiocarbonate and α,ω-PDS was employed for the preparation of heterobiofunctional protein–polymer conjugates [163]. BSA-macro RAFT agent was synthesized by reaction of excess pentafluorophenyl (PFP) functionalized trithiocarbonate RAFT agent with BSA (Figure 7.8(d)) was employed to generate BSA-poly [2,2-dimethyl-1,3-dioxolane)methyl) acrylamide] conjugates [164]. BSA-RAFT agent was also synthesized using thiol-maleimide coupling chemistry, (Figure 7.8(e)) followed by the in situ polymerization of NIPAM to form thermoresponsive polymer-protein conjugates [165].

7.5.1.2 Bioconjugation via "grafting to" approach

Post-polymerization bioconjugations to the RAFT-made polymers have been generally achieved via modification of RAFT end groups as a linker. In many cases, aminolysis or reduction of the thiocarbonylthio moiety has been employed to yield

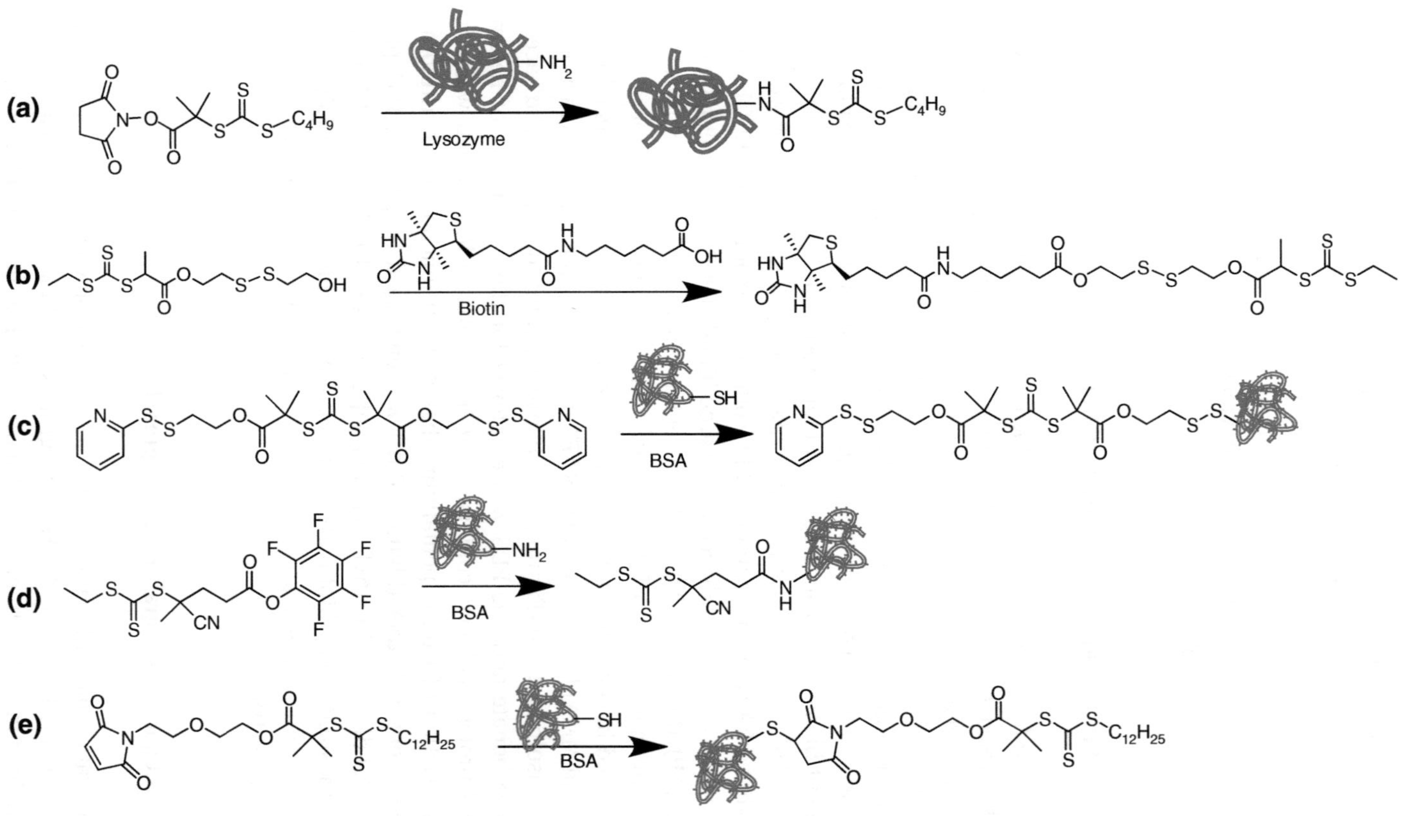

Figure 7.8: Synthesis of protein-based RAFT agent for the synthesis of polymer-protein bioconjugates. Figure 7.8(a), (b), (c), (d), and (e) are adapted from Reference [160–165], respectively.

thiol-ended polymers [166, 167]. Introduction of pyridyl disulfide (PDS) functionality is an alternative approach, which has been applied for bioconjugation in recent years [168]. The PDS group can be incorporated in the polymer chain end via either *R* or *Z* groups of RAFT agents. Using functional RAFT agent, PDS functionalized PHEA, PEGMA, and PNIPAM were prepared [169–171] (Figure 7.9(a) & (b)). The pyridyl ethyl disulfide bond does not disturb the RAFT mechanism during the polymerization, and the resulting polymer having PDS group at the chain-end is used to conjugate with thiol functionalities in biomolecules (Figure 7.9(a)) [172, 173]. The disulfide bond in the conjugate can readily be reduced for the release of the biomolecules from the bioconjugates.

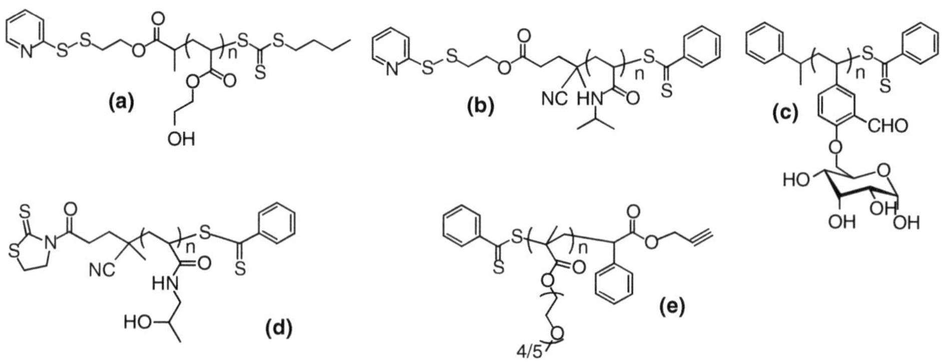

Figure 7.9: Functionalized polymers synthesized via RAFT for the preparation of bioconjugates. Structures (a), (b), (c), (d) and (e) are adapted from reference [169], [171], [177], [178] and [180], respectively.

Bioconjugation via thiol-ene reactions has also been exploited by modification of RAFT end group-terminated polymers to generate thiols, which subsequently react with biomolecules containing a double bond [174]. For example, PNIPAM prepared via RAFT polymerization was aminolyzed to yield thiol-terminated chains, which subsequently reacted with excess 1,8-bis-maleimidodiethyleneglycol to give maleimide-terminated polymer. This polymer was further reacted with BSA and ovalbumin to yield polymer–protein conjugates by a "*grafting-to*" approach [175].

Pound and coworkers introduced a new method of bioconjugation via the reaction between protein amine groups with aldehyde functionalized RAFT-agent end groups [176]. The hydrolysis of RAFT end groups of poly(vinylpyrrolidone) (PVP) at pH 4.5, yielding hydroxyl end groups which were further transformed into aldehyde groups via thermolysis. This aldehyde functionality was then reacted with the amine groups on proteins, peptides, or oligonucleotide biomolecules. Similarly, proteins were successfully attached on a micellar surface with aldehyde groups via oxime coupling (Figure 7.9(c)) [177]. Synthesis of lysozyme-polymer conjugates using thiazolidine-2-thione coupling chemistry and investigation of bioactivity of the conjugates have been reported (Figure 7.9(d)) [178]. Several authors [179, 180] have reported the synthesis of azide-functionalized RAFT agents for application in "*azide-alkyne-click*" type conjugation with biomolecules having alkyne functional groups (Figure 7.9(e)). A series of heterotelechelic polymers with an azido or alkyne group at the chain end have been synthesized. Then, the azido or alkyne group was employed in a click reaction with desired biomolecules.

7.5.1.3 DNA/RNA conjugation

Gene therapy has huge potential for the treatment of a range of acquired and genetic disorders such as cancer, haemophilia, and cystic fibrosis. A variety of viral and non-viral gene delivery agents were developed and verified for their gene delivery efficacies [181]. Among the most commonly used cationic polymers used for gene delivery applications, poly(ethyleneimine) (PEI)-based systems have been studied extensively for gene therapy due to their high proton sponge effect, ability to avoid lysosomal trafficking, various mechanisms of uptake, efficient DNA complexation, and facile DNA release. Other commonly used cationic polymers prepared via RAFT with controlled M_n and narrow $Đ$ are PDMAEMA, 3-(dimethylamino)propyl methacrylamide (DMAPMA), 2-dimethylaminoethyl acrylate (DMAEA), 3-aminopropyl methacrylamide (APMA), 2-aminoethyl methacrylamide (AEMA), 2-methacryloxyethyl phosphorylcholine (MPC), and so on (Figure 7.10). However, due to their high toxicity, non-specific interactions with cells and proteins in vitro and in vivo, and the ability to alter the immune response of these cationic polymers, the researchers were encouraged to prepare biocompatible cationic polymers [182]. Recently, Sun and Gao employed a general and facile approach via ATRP to prepare biocompatible, cationic multiamino polymers derived from methacrylate monomers of naturally occurring Boc-protected amino acids (AA), that could tightly condense DNA [183]. Most recently, De and coworkers reported RAFT polymerization of a series of naturally

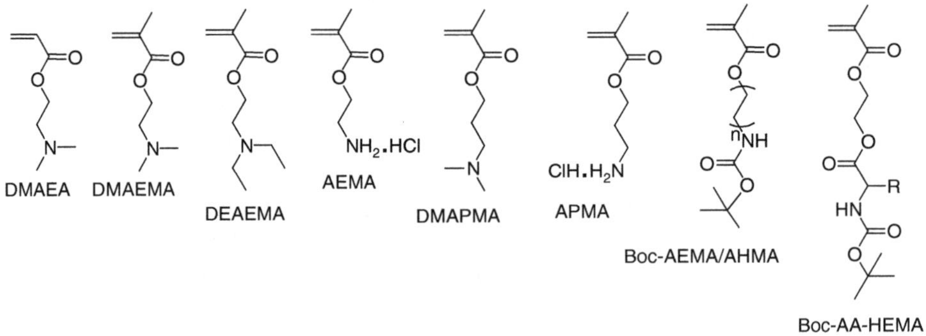

Figure 7.10: Cationic monomers used in RAFT polymerization for gene delivery applications.

occurring amino acids or short peptide-based acrylate and methacrylate monomers (Boc-amino acid/peptide methacryloyloxyethyl ester, Boc-AA/Pep-HEMA) to prepare their respective boc protected cationic polymers P(Boc-AA/Pep-HEMA) with controlled M_n and $Đ$ [184, 185] followed by boc deprotection. Side chain alanine- and phenylalanine-based cationic homopolymers P(H_3N^+-Ala/Phe-HEMA) [186] and their amphiphilic block copolymers mPEG$_{2k/5k}$-*b*-P(H_3N-Ala/Phe-HEMA) were synthesized with controlled M_n and $Đ$ (1.12–1.34) via RAFT. These block copolymers are biocompatible and has plasmid DNA (*p*DNA)-binding ability as studied by using agarose gel retardation assay and AFM (Figure 7.11) [187]. A series of pH responsive cationic linear and branched polymers from Boc-leucine/isolucineacryloyloxyethyl/methacryloyloxyethyl ester (Boc-Leu/Ilu-HEA/HEMA) [188, 189], Boc-tryptophanmethacryloyloxyethyl ester (Boc-Trp-HEMA) [190, 191], Boc-valineacryloyloxyethyl ester (Boc-Val-HEA) [192] were synthesized via RAFT followed by trifluoroacetic acid (TFA) deprotection. Some of these materials have been shown to be biocompatible, and therefore it is anticipated that these polymers will be potential candidates for DNA binding and as nonviral vectors for gene delivery.

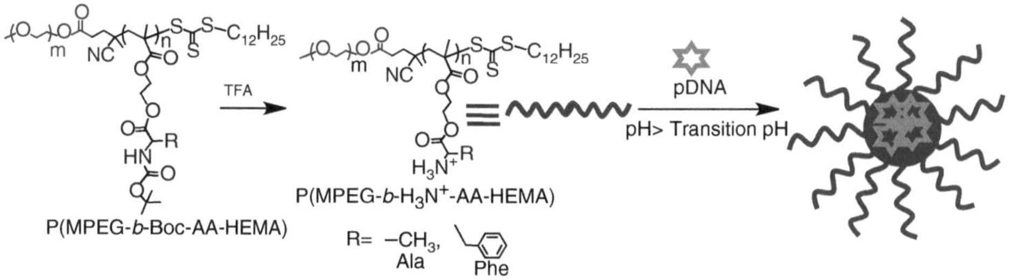

Figure 7.11: Complexation of amino acid-based cationic, pH responsive block copolymers prepared via RAFT polymerization with *p*DNA. (Source: S. Kumar, R. Acharya, U. Chatterji and P. De, *Langmuir*, 2013, 29, 15,375) [187].

7.5.2 Glycopolymers

The first report of glycopolymer synthesis via RAFT is the polymerization of 2-methacryloxyethyl glucoside (2-MAOEGlc, G1) in water at 70 °C in the presence of (4-cyanopentanoic acid)-4-dithiobenzoate (CTP) as chain transfer agent (CTA) [193]. The same initiator/CTA system was also employed in the polymerization of methyl 6-*O*-methacryloyl-D-glucoside (6-*O*-MMAGlc, G2) in water [194]. Homopolymers containing a gluconolactone derivative GAEMA (G6) and GAPMA (G7) were prepared via RAFT in a controlled fashion in water/DMF mixtures using CTP; was further employed as macro-CTAs to synthesize polycationic block copolymers of AEMA, APMA, or MPC. The P(APMA-*b*-GAPMA) showed nanoparticles (~ 100 nm) formation via complexation with *p*DNA at physiological and slightly acidic pH. The glycopolymers were found to be nontoxic and PGAPMA even showed enhanced cell proliferation. PAPMA was found to be highly toxic over a range of concentration, whereas P (APMA-*b*-GAPMA) was found to be biocompatible [195]. RAFT polymerization of (2-(2,3,4,6-tetra-Oacetyl-*β*-D-glucosyloxy)ethyl methacrylate) (PAcGlcEMA) G11, followed by acetyl deprotection and end group modification, produced PAcGlcEMA-PDS. This polymer formed a PGlcEMA-GSH bioconjugate, which showed enhanced binding affinity with the protein ConA. The antioxidant activity due to disulfide linkages present in this bioconjugate has potential for biodetection, biomimetics, and targeted antioxidant delivery [196].

The amphiphilic diblock glycopolymer of PLA and 1,2:3,4-di-*O*-isopropylidene-6-*O*-acryloyl-*R*-D-galactopyranose (G15) forms micelles in water, which were further stabilized by cross-linking of the shell using a diacrylate in a chain extension step [197]. Light responsive block copolymer poly(AzoMA)-*b*-poly(*β*-GalEtMA) prepared via RAFT, followed by acetyl deprotection, formed well-defined micelles in aqueous media due to cis-trans isomerization. This micellar nanostructure can efficiently encapsulate the model hydrophobic compound (Nile Red) and also exerted low cell cytotoxicity [198]. Micelles were obtained from RAFT-made block copolymer having Boc-Val-HEMA and *β*-D-glucose pentaacetatemethacryloyloxyethyl esters (Ac-G-HEMA), after either Boc group or acetyl group deprotection. This block copolymer showed enhanced lectin-binding behavior. This enhanced lectin-binding ability is due to simultaneous binding with both amino acid/sugar entities on the shell, as compared to the only glycopolymer nanoparticle [199]. Block and statistical copolymers of 2-(*α*-D-mannopyranosyloxy)ethyl methacrylate (ManEMA) and DMAEMA showed binding ability with ConA. Block and statistical copolymers showed similar *p*DNA condensing abilities [200].

The cluster glycoside effect is the phenomenon through which carbohydrates on a cell surface and lectins interact. This can be achieved via self-assembly of amphiphilic polymers, or by attaching glycopolymers onto inorganic particles. Surface-modified honeycomb-structured porous film prepared using poly(styrene-*co*-maleic anhydride) (PSMA) followed by cross-linking with 1,8-diaminooctane was further anchored with RAFT agent. Then this surface-modified RAFT agent was employed for the graft copolymerization of NIPAM and *N*-acryloyl glucosamine (AGA). This graft copolymer, PNIPAM-*g*-PAGA, showed selective binding ability with lectin ConA, which is regulated with the LCST of PNIPAM [201]. As gold nanoparticles (AuNPs) have been used as signal transducers, saccharide-modified AuNPs have been used to monitor biological phenomena. For example, thiol-terminated *p*-acrylamidophenyl *R*-mannoside G19 and acrylamidophenyl-*N*-acetyl-*β*-glucosamine G20-based homopolymers and copolymers prepared via RAFT followed by end group reduction were then grafted to AuNPs that can bind with lectin [202]. Glycopolymer poly(acrylamidophenyl *α*-mannose-*co*-acrylamide)-modified AuNPs showed strong binding ability with lectins and have potential application as biomarkers in immunochromatographic assays [203]. α-Galactose (a-Gal) and α-mannose (a-Man)-based glycopolymers modified AuNPs showed specific interaction ability with sugar recognition proteins (lectins and Shiga toxins) [204].

7.5.3 Drug delivery

RAFT-made polymers have been progressively exploited in potential drug delivery applications. To date, various RAFT-made block copolymer micelles, vesicles, star polymers, nanoparticles, and capsules have been investigated as potential advanced drug carriers and also polymer-drug conjugates as prodrugs (Figure 7.12).

7.5.3.1 Amphiphilic nanostructures

Huge attention has been given to the synthesis of RAFT-made amphiphilic block copolymers for the construction of CDDS due to the ability to control the block lengths that affect the critical micelle concentration (thus stability), hydrodynamic size, and morphology. It is also possible in the case of RAFT-made polymers to introduce chemical functionalities in the micelle corona and core that stabilize the supramolecular structure via covalent bonds or conjugating with biologically active molecules, such as cell specific targeting moieties and therapeutics.

In general, the corona of micelles should contain a suitable polymer that can regulate the nano-assembly, minimize the immunological reactions in vivo, and have prolonged blood-residence time. Amphiphilic block copolymers having a corona composed of an inert polymer such as PEG, PHPMA, poly(*N*-acryloylmorpholine) (PAM), and many others, have been synthesized by the RAFT for potential drug delivery applications. In general, PEG-protected micelles have been synthesized by PEG-based macro-CTA [205]. PAM has nonimmunogenic and long blood-circulation properties similar to those of PEG. An example is well-defined polymeric core-corona micelles formed from the block copolymer, poly(*n*-butyl methacrylate)-*b*-PAM (PBMA-*b*-PAM), showed biocompatibility and stability under physiological serum condition as verified in BSA. This nanostructure has potential applications in drug delivery or nano-scale biosensors and bioreactors [206]. Another example is water-soluble star polymer-DOX conjugate carriers, which were prepared by the grafting of poly(amido amine) (PAMAM) dendrimers by hetero-telechelic HPMA copolymers by RAFT, followed by DOX conjugation via hydrazone bonds that further enable the pH-controlled drug release in the target tumor tissue [207].

Recently, amphiphilic side-chain fatty acid-derived polymer nanostructures are also considered to be potential candidates for construction of CDDS [208, 209]. Another approach is amino acid-based tadpole-shaped organic/inorganic hybrid

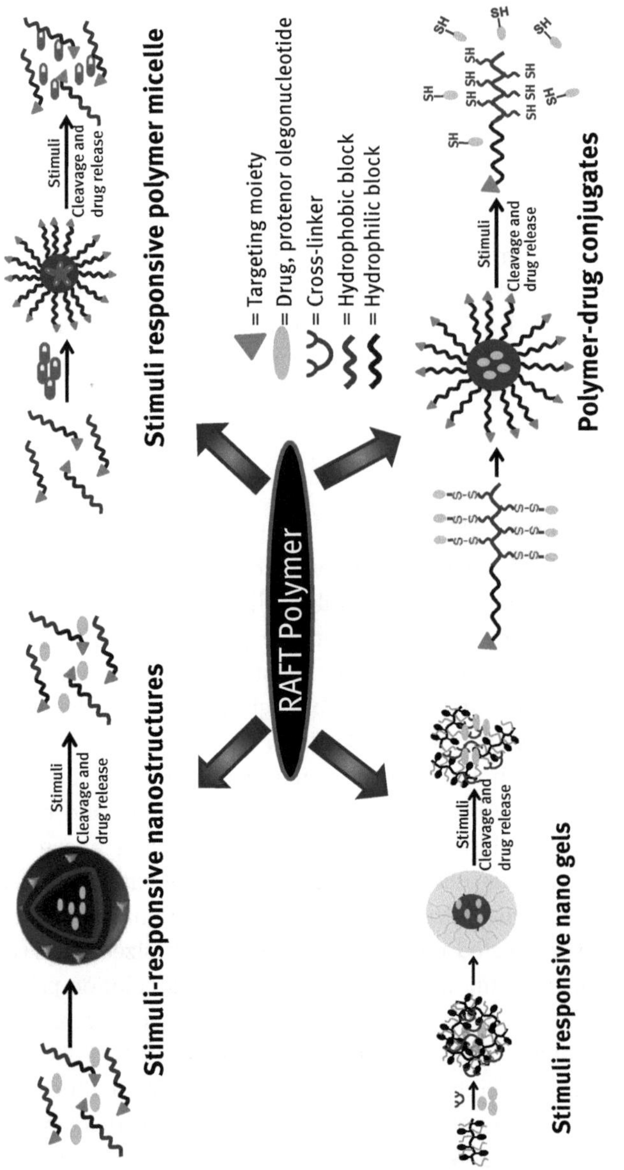

Figure 7.12: Drug delivery devices prepared from RAFT-made polymers.

amphiphilic polymer (H_3N^+-Leu-HEMA), which is end functionalized with hydrophobic polyhedral oligomericsilsesquioxane (POSS). This material showed aggregation behavior in aqueous medium, which can efficiently encapsulate Nile Red as a hydrophobic probe [210]. Amphiphilic block copolymers of isobutylene (IB) with Boc-Ala-HEMA, Boc-Leu-HEMA and POEGMA prepared via a combination of cationic polymerization and RAFT techniques showed self-assembly to form micellar aggregates in water that also have potential for CDDS formation [211, 212]. However, this kind of classical micellar system releases drug molecules when the system would be diluted below the critical micellar concentration (CMC), hence, they are less ideal for delivery applications.

7.5.3.2 Stimuli-responsive nanostructures

A wide variety of stimuli-responsive polymers (such as pH-, temperature-, light-, chemical, and biological molecule sensitive) with varied polymeric architectures (linear, branched, star, brush, comb, macrocyclic polymers, and so on) has been synthesized via RAFT and used to prepare nanostructures, e.g., micelles, vesicles, rod, capsule, and so on, for the construction of potential CDDS. Synthesis of pH-responsive double hydrophilic block copolymers from primary, tertiary amine, and carboxylic group containing monomers was achieved via RAFT in aqueous or organic media. It was possible to regulate the micellization behavior and/or the hydrodynamic dimensions of the micelles by changing the pH of the solutions and the composition of the polymers. These drug-encapsulated nanostructures are further used for pH-controlled drug release [213]. Recently, side chain amino acid-based [214] and short peptide-based [215, 216] pH responsive cationic polymers were found to be biocompatible and have the ability to bind with *p*DNA, make them promising candidates for biomedical applications. Side chain cholesterol-based pH and thermo-responsive copolymer formed well-defined self-assembled micelle in water, which can efficiently encapsulate a model hydrophobic dye [217].

A large variety of stimuli-responsive cross-linked micellar systems and star polymers have been designed using RAFT techniques [218]. Glutathione-sensitive cross-linked micelles [219] and hyperbranched polymers [220] have been reported, which have potential for release of drugs in cell cytoplasm. A series of pH responsive hyperbranched, star and brush polymers has also been described for the encapsulation of drug molecules [221, 222]. For example, water soluble star polymer nanocarriers from poly(amido amine) (PAMAM)-*g*-HPMA bound with DOX through hydraxone bonds show pH responsive drug release at targeted tumors [223]. Factors influencing the design of nanoparticle therapies, as well as the development of RDRP techniques, have been reviewed recently [224].

7.5.3.3 Functionalized nanostructures with targeting moiety

Functionalization of the outer surface of micelles/vesicles with biomolecules is of great interest for cell-specific targeting of CDDS. The RAFT technique allows the synthesis of α- and ω-functional CTAs and pendant-group functional block polymers suitable for conjugation with cell-specific targeting moieties (e.g., vitamins, peptides, antibodies, and so on) to form cell-targeted CDDS [225]. Bio-molecule functionalized amphiphilic block copolymers can be prepared via in situ polymerization, using biomolecule-modified RAFT agents or biomolecule functionalized monomers [226]. For example, De *et al* have synthesized α-azido terminal thermoresponsive block copolymers from an azido-modified RAFT agent [227], which efficiently coupled with propargyl-modified folate (folate receptors overexpressed by certain cancer cells) via orthogonal click addition.

7.5.3.4 Polymer-drug conjugates

The RAFT technique offers a brilliant route for the preparation of well-defined, narrow dispersity polymers with functional groups that can form covalent conjugation with drugs and/or other functional elements for drug delivery applications. PHPM-based core-cross-linked star polymers with methotrexate-conjugated cross-linked PHPMA core and PHPMA arms were synthesized via RAFT. This novel drug-polymer conjugate showed enhanced stability against premature degradation and enzyme responsive drug release properties [225]. Drug-conjugated copolymer made via the RAFT method showed drug release in human serum conditions due to the hydrolysis of ester linkers. This copolymer-conjugate is also biocompatible up to the polymer concentration of 20 mg/mL [228]. RAFT made HPMA-based copolymer-drug conjugates showed potent cytotoxicity, strong synergy, and are potential as chemotherapeutic agent for the treatment of acute myeloid leukemia [229]. Four different drugs, including cytarabine, daunorubicin, GDC-0980, and JS-K were studied in vitro for the two-drug combinations treatment of acute myeloid leukemia.

7.5.3.5 Micro/nanogel particles and hydrogels

Nanogels can be prepared via direct RAFT polymerization of vinyl monomers in the presence of a cross-linker. Example include the synthesis of nanogels of MEO_2MA or copolymers of MEO_2MA and PEGMA in the presence of poly(ethylene glycol) dimethacrylate (PEGDMA) cross-linker via RAFT-mediated aqueous dispersion. These nanogels showed enhanced stability in BSA and 100% fetal bovine serum (FBS) solutions [230]. Polymethacrylate/poly(*N*-vinyl pyrrolidone) hydrogel capsules prepared via RAFT showed specific interaction with cells through antibody/antigen recognition [231]. Injectable hydrogels prepared from poly(ethylene glycol) diacrylate (PEGDA) and thiolated hyaluronic acid via RAFT have potential for stem cell delivery and tissue regeneration [232].

Thermo-reversible physical hydrogels were prepared from RAFT-synthesized PNIPAM-*b*-PDMA-*b*-PNIPAM triblock copolymers [233]. This hydrogel showed similar mechanical properties as that of collagen, a biopolymer widely used in tissue-engineering applications. Most recently, De and coworkers have developed thermoresponsive cross-linked hydrogels from MEO_2MA [234, 235] and PEGMA [236], which might have potential as tissue engineering scaffolds. The same group has also developed side chain amino acid-based cross-linked superabsorbent organogels and hydrogels with pH and salt responsiveness that have great potential for implants and tissue engineering scaffolds [237, 238]. The pH reversible amino acid-based dynamic gels [239] and other cross-linked gels have also been developed, which have potential in bio-applications due the presence of amino acid units [240, 241].

7.5.4 Nanostructures for imaging and therapy

Surface modification with synthetic polymers via the RAFT technique is an important method in biotechnological applications such as tissue engineering, biosensors, implants manufacturing and is also used to regulate protein, microbial, and cell adhesion. Controlled PDMAEMA chains grown from a cellulose surface via RAFT and subsequently quaternized with alkyl bromides of various chain lengths exhibit high biocidal activity against *E. coli* [242]. A self-assembled hydrogels scaffold prepared from PHPMA copolymer grafted from β-sheet peptide (RDG and Beta11A) can assist bone tissue regeneration [243]. pH responsive, endosomolytic micellar nanoparticle prepared from DMAEMA-*b*-(PAA-*co*-BMA-*co*-DMAEMA) can protect and encapsulate siRNA. This NP was further incorporated into nontoxic, biodegradable polyurethane tissue scaffolds for potential injectable delivery devices for siRNA delivery [244]. Novel DNA-biosensors have also been developed using SI-RAFT polymerization. *Graft*-P[DMA-*b*-(DMA-*co*-*N*-acryloyl succinamide)] (DMA-*b*-(DMA-*co*-NAS) diblock copolymers were grown from glass surface and these surface-tethered polymers were exploited to immobilize DNA [245].

Polyethersulfone (PES) membrane was prepared by mixing of PES and PDMAEMA grafted SiO_2 NP brushes via SI-RAFT followed by quaternization. This PES/SiO_2-*g*-PDMAEMA showed strong anti-fouling and anti-bacterial activity against *E. coli* and *S. aureus* [246]. P(DMAEA-*co*-PEGMA) decorated mesoporous silica nanoparticles (MSN) prepared via combination of RAFT and *grafting-from* strategy showed higher efficacy of DOX delivery and suppressed the side effect of drug [247]. Wide varieties of RAFT-made polymers have also been exploited for surface modification for biomedical applications [248]. Also, AuNPs, silica particles, and more were chosen as substrate to graft RAFT polymers for their biomedical application [249].

7.6 Conclusions

In this chapter, synthesis of polymers via RDRP techniques for biomedical applications has been discussed. The potential for designing novel polymeric architectures via RDRP methods for biomedical applications is clearly vast, and in this chapter we have reviewed applications of RDRP-made polymers in glycopolymers, bioconjugates, drug delivery, gene therapy, and as biomaterials. The field is very promising, and there are several exciting and novel opportunities to explore. Most of the biomedical applications were investigated for in vitro conditions. Therefore, one big challenge is their application in vivo, including in humans. However, it is expected that this next big step might not be as far away as it appears.

Acknowledgements: SGR acknowledges Indian Institute of Science Education and Research Kolkata (IISER-Kolkata) for postdoctoral research fellowship.

References

[1] R. Duncan, Nature Reviews Drug Discovery, 2003, 2, 347.
[2] A.S. Hoffman, Advanced Drug Delivery Reviews, 2013, 65, 10.
[3] K. Matyjaszewski and J. Spanswick, Materials Today, 2005, 3, 26.
[4] D.A. Shipp, Polymer Reviews, 2011, 51, 99.
[5] A.D. Jenkins, R.G. Jones and G. Moad, Pure and Applied Chemistry, 2010, 82, 483.
[6] K. Matyjaszewski, Macromolecules, 1993, 26, 1787.
[7] K. Matyjaszewski, Journal of Physical Organic Chemistry, 1995, 8, 197.
[8] E. Rizzardo and D.H. Solomon, Polymer Bulletin, 1979, 1, 529.
[9] G. Moad, E. Rizzardo and D.H. Solomon, Macromolecules, 1982, 15, 909.
[10] M.K. Georges, R.P.N. Veregin, P.M. Kazmaier and G.K. Hamer, Macromolecules, 1993, 26, 2987.
[11] M.K. Georges, R.P.N. Veregin, P.M. Kazmaier and G.K. Hamer, Trends in Polymer Science, 1993, 2, 66.
[12] D. Benoit, E. Harth, P. Fox, R.M. Waymouth and C.J. Hawker, Macromolecules, 2000, 33, 363.
[13] E. Harth, B. Van Horn and C.J. Hawker, Chemical Communications, 2001, 823.
[14] D. Benoit, S. Grimaldi, S. Robin, J.P. Finet, P. Tordo and Y. Gnanou, Journal of the American Chemical Society, 2000, 122, 5929.
[15] P. Lacroix-Desmazes, J.F. Lutz and B. Boutevin, Macromolecular Chemistry and Physics, 2000, 201, 662.
[16] D. Benoit, V. Chaplinski, R. Braslau and C.J. Hawker, Journal of the American Chemical Society, 1999, 121, 3904.
[17] E. Drockenmuller, J.P. Lamps and J.M. Catala, Macromolecules, 2004, 37, 2076.
[18] L. Couvreur, C. Lefay, J. Belleney, B. Charleux, O. Guerret and S. Magnet, Macromolecules, 2003, 36, 8260.
[19] T. Diaz, A. Fischer, A. Jonquieres, A. Brembilla and P. Lochon, Macromolecules, 2003, 36, 2235.
[20] K. Schierholz, M. Givehchi, P. Fabre, F. Nallet, E. Papon, O. Guerret and Y. Gnanou, Macromolecules, 2003, 36, 5995.

[21] F. Chauvin, P.E. Dufils, D. Gigmes, Y. Guillaneuf, S.R.A. Marque, P. Tordo and D. Bertin, Macromolecules, 2006, 39, 5238.
[22] B. Charleux, J. Nicolas and O. Guerret, Macromolecules, 2005, 38, 5485.
[23] J. Vinas, N. Chagneux, D. Gigmes, T. Trimaille, A. Favier and D. Bertin, Polymer, 2008, 49, 3639.
[24] V. Delplace, S. Harrisson, H.T. Ho, A. Tardy, Y. Guillaneuf, S. Pascual, L. Fontaine and J. Nicolas, Macromolecules, 2015, 48, 2087.
[25] D. Gigmes, J. Vinas, N. Chagneux, C. Lefay, T.N.T. Phan, T. Trimaille, P.E. Dufils, Y. Guillaneuf, G. Carrot, F. Boue and D. Bertin, ACS Symposium Series, 2009, 1024, 245.
[26] P. Brémond, K. Kabytaev and S.R.A. Marque, Tetrahedron Letters, 2012, 53, 4543.
[27] M. Chenal, C. Boursier, Y. Guillaneuf, M. Taverna, P. Couvreur and J. Nicolas, Polymer Chemistry, 2011, 2, 1523.
[28] B.L. Droumaguet and J. Nicolas, Polymer Chemistry, 2010, 43, 4835.
[29] A. Narumi, T. Matsuda, H. Kaga, T. Satoh and T. Kakuchi, Polymer, 2002, 43, 4835.
[30] K. Ohno, O. Izu, S. Yamamoto, T. Miyamoto and T. Fukuda, Macromolecular Chemistry and Physics, 1999, 200, 1619.
[31] S.R.S. Ting, E.H. Min, P. Escalé, M. Save, L. Billon and M.H. Stenzel, Macromolecules, 2009, 42, 9422.
[32] K. Ohno, Y. Tsujii, T. Miyamoto, T. Fukuda, M. Goto, K. Kobayashi and T. Akaike, Macromolecules, 1998, 31, 1064.
[33] K. Ohno, T. Fukuda and H. Kitano, Macromolecular Chemistry and Physics, 1998, 199, 2193.
[34] H. Kaga, A. Narumi, T. Satoh, T. Matsuda, M. Sharfuddin and T. Kakuchi, Macromolecular Symposia, 2002, 181, 95.
[35] A. Narumi, T. Satoh, H. Kaga and T. Kakuchi, Macromolecules, 2002, 35, 699.
[36] Y. Miura, D. Koketsu and K. Kobayashi, Polymers for Advanced Technologies, 2007, 18, 647.
[37] S. Chen, M.H. Alves, M. Save and L. Billon, Polymer Chemistry, 2014, 5, 5310.
[38] M.L. Becker, J. Liu and K.L. Wooley, Biomacromolecules, 2005, 6, 220.
[39] M.L. Becker, J. Liu and K.L. Wooley, Chemical Communications, 2003, 180.
[40] T. Trimaille, K. Mabrouk, V. Monnier, L. Charles, D. Bertin and D. Gigmes, Macromolecules, 2010, 43, 4864.
[41] K. Molawi and A. Studer, Chemical Communications, 2007, 5173.
[42] M. Moller, C. Hentschel, L. Chi and A. Studer, Organic & Biomolecular Chemistry, 2011, 9, 2403.
[43] J. Nicolas, Y. Guillaneuf, C. Lefay, D. Bertin, D. Gigmes and B. Charleux, Progress in Polymer Science, 2013, 38, 63.
[44] R.J.I. Knoop, G.J.M. Habraken, N. Gogibus, S. Steig, H. Menzel, C.E. Koning and A. Heise, Journal of Polymer Science Part A: Polymer Chemistry, 2008, 46, 3068.
[45] K.E. Sapsford, W.R. Algar, L. Berti, K.B. Gemmill, B.J. Casey, E. Oh, M.H. Stewart and I.L. Medintz, Chemical Reviews, 2013, 113, 1904.
[46] S. Harrisson, J. Nicolas, A. Maksimenko, D.T. Bui, J. Mougin and P. Couvreur, Angewandte Chemie International Edition, 2013, 52, 1678.
[47] V. Delplace, A. Tardy, S. Harrisson, S. Mura, D. Gigmes, Y. Guillaneuf and J. Nicolas, Biomacromolecules, 2013, 14, 3769.
[48] E. Guégain, V. Delplace, T. Trimaille, D. Gigmes, D. Siri, S.R.A. Marque, Y. Guillaneuf and J. Nicolas, Polymer Chemistry, 2015, 6, 5693.
[49] M. Ouchi, T. Terashima and M. Sawamoto, Chemical Reviews, 2009, 109, 4963.
[50] W.A. Braunecker and K. Matyjaszewski, Progress in Polymer Science, 2007, 32, 93.
[51] T. Patel, J. Zhou, J.M. Piepmeier and W.M. Saltzman, Advanced Drug Delivery Reviews, 2012, 64, 701.

[52] C. Li and S. Wallace, Advanced Drug Delivery Reviews, 2008, 60, 886.
[53] M.A. Gauthier and H.A. Klok, Chemical Communications, 2008, 2591.
[54] A. Rösler, G.W.M. Vandermeulen and H.A. Klok, Advanced Drug Delivery Reviews, 2012, 64, 270.
[55] R.A. Cordeiro, D. Farinha, N. Rocha, A.C. Serra, H. Faneca and J.F.J. Coelho, Macromolecular Bioscience, 2015, 15, 215.
[56] J.P. Magnusson, S. Bersani, S. Salmaso, C. Alexander and P. Caliceti, Bioconjugate Chemistry, 2010, 21, 671.
[57] R.M. Broyer, G.N. Grover and H.D. Maynard, Chemical Communications, 2011, 47, 2212.
[58] R.M. Broyer, G.M. Quaker and H.D. Maynard, Journal of the American Chemical Society, 2008, 130, 3, 1041.
[59] D. Bontempo, R.C. Li, T. Ly, C.E. Brubaker and H.D. Maynard, Chemical Communications, 2005, 4702.
[60] L. Xu, L. Yuan and S. Liu, RSC Advances, 2014, 4, 140.
[61] H. Murata, C.S. Cummings, R.R. Koepsel, and A.J. Russell, Biomacromolecules, 2014, 15, 7, 2817.
[62] T. Steinbach, F. Wurm and H.-A. Klok, Polymer Chemistry, 2014, 5, 4039.
[63] J. Nicolas, S. Mura, D. Brambilla, N. Mackiewicz and P. Couvreur, Chemical Society Review, 2013, 42, 1147.
[64] H. Rettig, E. Krause and H.C. Börner, Macromolecular Rapid Communications, 2004, 25, 13, 1251.
[65] R. Trzcinska, D. Szweda, S. Rangelov, P. Suder, J. Silberring, A. Dworak and B. Trzebicka, Journal of Polymer Science Part A: Polymer Chemistry, 2012, 50, 15, 3104.
[66] B. Trzebick, B. Robak, R. Trzcinsk, D. Szweda, P. Suder, J. Silberring and A. Dworak, European Polymer Journal, 2013, 49, 499.
[67] D. Bontempo and H.D. Maynard, Journal of the American Chemical Society, 2005, 127, 18, 6508.
[68] W. Gao, W. Liu, J. Mackay, M. Zalutsky, E. Toone and A. Chilkoti, Proceedings of the National Academy of Sciences of the United States of America, 2009, 106, 15231.
[69] W.P. Gao, W.G. Liu, T. Christensen, M.R. Zalutsky and A. Chilkoti, Proceedings of the National Academy of Sciences of the United States of America, 2010, 107, 16432.
[70] J. Hu, W. Zhao, Y. Gao, M. Sun, Y. Wei, H. Deng and W. Gao, Biomaterials, 2015, 47, 13.
[71] J. Peeler, B. Woodman, S. Averick, S. Miyake-Stoner, A. Stokes, K. Kess, K. Matyjaszewski and R. Mehl, Journal of the American Chemical Society, 2010, 132, 13575.
[72] S.E. Averick, C.G. Bazewicz, B.F. Woodman, A. Simakova, R.A. Mehl and K. Matyjaszewski, European Polymer Journal, 2013, 49, 10 2919.
[73] D.J. Siegwart, J.K. Oh, H. Gao, S.A. Bencherif, F. Perineau, A.K. Bohaty, J.O. Hollinger and K. Matyjaszewski, Macromolecular Chemistry and Physics, 2008, 209, 2180.
[74] L. Ayres, M.R.J. Vos, P.J.H.M. Adams, I.O. Shklyarevskiy and J.C.M. van Hest, Macromolecules, 2003, 36, 5967.
[75] A. Abuchowski, J.R. McCoy, N.C. Palczuk, T. Van Es and F.F Davis, The Journal of Biological Chemistry, 1977, 252, 3582.
[76] F. Lecolley, L. Tao, G. Mantovani, I. Durkin, S. Lautru and D.M. Haddleton, Chemical Communications, 2004, 18, 2026.
[77] K. Wang, X. Zhang, X. Zhang, X. Fan, Z. Huang, Y. Chen and Y. Wei, Polymer Chemistry, 2015, 6, 32, 5891.
[78] L.Q. Xu, N.N. Li, B. Zhang, J.C. Chen and E.T. Kang, Polymers, 2015, 7, 2119.
[79] P.C. Li, Y.C. Lin, M. Chen and S.W. Kuo, Soft Matter, 2013, 9, 47, 11257.
[80] S. Mendrek, A. Mendrek, H.J. Adler, A. Dworak and D.J. Kuckling, Journal of Polymer Science Part A: Polymer Chemistry, 2009, 47, 1782.

[81] J. Du, L. Fan and Q. Liu, Macromolecules, 2012, 45, 8275.
[82] X. Zhao, W. Liu, D. Chen, X. Lin and W.W. Lu, Macromolecular Chemistry and Physics, 2007, 208, 1773.
[83] Y. Xu, L. Shi, R. Ma, W. Zhang, Y. An and X.X. Zhu, Polymer, 2007, 48, 1711.
[84] X. Liu, P. Ni, J. He and M. Zhang, Macromolecules, 2010, 43, 4771.
[85] D.M. Henn, R.A.E. Wright, J.W. Woodcock, B. Hu and B. Zhao, Langmuir, 2014, 30, 2541.
[86] S.V. Lale, A. Kumar, F. Naz, A.C. Bharti and V. Koul, Polymer Chemistry, 2015, 6, 2115.
[87] K. Nagase, Y. Hatakeyama, T. Shimizu, K. Matsuura, M. Yamato, N. Takeda and T. Okano, Biomacromolecules, 2015, 16, 532.
[88] S.P. Rwei, Y.Y. Chuang, T.F. Way and W.Y. Chiang, Colloid and Polymer Science, 2015, 293, 493.
[89] Y.J. Shih, Y. Chang, A. Deratani and D. Quemener, Biomacromolecules, 2012, 13, 2849.
[90] J. Xuan, D. Han, H. Xia and Y. Zhao, Langmuir, 2014, 30, 410.
[91] Z. Dong, J. Mao, D. Wang, M. Yang and X. Ji, Langmuir, 2015, 31, 8930.
[92] F. Jiang, S. Chen, Z. Cao and G. Wang, Polymer, 2016, 83, 85.
[93] N.G. Patil, N.B. Basutkar and A.V. Ambade, Chemical Communications, 2015, 51, 17708.
[94] Q. Zhang, S. Aleksanian, S.M. Noh and J.K. Oh, Polymer Chemistry, 2013, 4, 351.
[95] N. Chan, S.Y. An and J.K. Oh, Polymer Chemistry, 2014, 5, 1637.
[96] N. Chan, S.Y. An, N. Yee and J. K. Oh, Macromolecular Rapid Communications, 2014, 35, 752.
[97] W. Chen, L.A. Shah, L. Yuan, M. Siddiq, J. Hua and D. Yang, RSC Advances, 2015, 5, 7559.
[98] C.R. Becer, M.I. Gibson, J. Geng, R. Ilyas, R. Wallis, D.A. Mitchell and D.M. Haddleton, Journal of the American Chemical Society, 2010, 132, 15130.
[99] I. Otsuka, K. Fuchise, S. Halila, S. Fort, K. Aissou, I. Pignot-Paintrand, Y. Chen, A. Narumi, T. Kakuchi and R. Borsali, Langmuir, 2010, 26, 2325.
[100] K.M. Zepon, I. Otsuka, C. Bouilhac, E.C. Muniz, V. Soldi and R. Borsali, Biomacromolecules, 2015, 16, 2012.
[101] C. von der Ehe, J.A. Czaplewska, M. Gottschaldt and U.S. Schubert, European Polymer Journal, 2013, 49, 2660.
[102] A. Muñoz-Bonilla, O. León, M.L. Cerrada, J. Rodríguez-Hernández, M. Sánchez-Chaves and M. Fernández-García, European Polymer Journal, 2015, 62, 167.
[103] V. Vzquez-Dorbatt, J. Lee, E.W. Lin and H.D. Maynard, ChemBioChem, 2012, 13, 2478.
[104] V. Ladmiral, L. Monaghan, G. Mantovani and D.M. Haddleton, Polymer, 2005, 46, 8536.
[105] W. Gao, T. Wang, X. Tang, Q. Zhang, F. Yu and M. Pei, Journal of Polymer Science Part A: Polymer Chemistry, 2014, 52, 2131.
[106] M. Trinadh, K. Govindaraj, T. Rajasekhar, M. Dhayal and A.V.S. Sainath, Polymer International, 2015, 64, 795.
[107] X.H. Dai, Z.M. Wang, Y-fei Huang, J.M. Pan, Y.-sheng Yan, D.M. Liuc and L. Sun, RSC Advances, 2014, 4, 42486.
[108] B.B. Ke, L.S. Wan, W.X. Zhang and Z.K. Xu, Polymer, 2010, 51, 2168.
[109] H. Park, R.R. Rosencrantz, L. Elling and A. Böker, Macromolecular Rapid Communications, 2015, 36, 45.
[110] Z. Lu, L. Mei, X. Zhang, Y. Wang, Y. Zhao and C. Li, Polymer Chemistry, 2013, 4, 5743.
[111] Y. Wang, H. Wang, G. Liu, X. Liu, Q. Jin and J. Ji, Macromolecular Bioscience, 2013, 13, 1084.
[112] Y. Liu, C. Lin, J. Li, Y. Qua and J. Ren, Journal of Materials Chemistry B, 2015, 3, 688.
[113] J. Guo, H. Hong, G. Chen, S. Shi, T.R. Nayak, C.P. Theuer, T.E. Barnhart, W. Cai and S. Gong, ACS Applied Materials & Interfaces, 2014, 6, 21769.
[114] C. Zhao, Q. Chen, K. Patel, L. Li, X. Li, Q. Wang, G. Zhangc and J. Zheng, Soft Matter, 2012, 8, 7848.

[115] S.E. Averick, E. Paredes, A. Irastorza, A.R. Shrivats, A. Srinivasan, D.J. Siegwart, A.J. Magenau, H.Y. Cho, E. Hsu, A.A. Averick, J. Kim, S. Liu, J.O. Hollinger, S.R. Das and K. Matyjaszewski, Biomacromolecules, 2012, 13, 3445.
[116] M. Li, Z. Tang, H. Sun, J. Ding, W. Song and X. Chen, Polymer Chemistry, 2013, 4, 1199.
[117] A.R. Shrivats, Y. Mishina, S. Averick, K. Matyjaszewski and J.O. Hollinger, Bioengineering, 2015, 2, 160.
[118] Y. Li, D. Maciel, J. Rodrigues, X. Shi and H. Tomas, Chemical Reviews, 2015, 115, 8564.
[119] D.J. Siegwart, J.K. Oh, H. Gao, S.A. Bencherif, F. Perineau, A.K. Bohaty, J.O. Hollinger and K. Matyjaszewski, Macromolecular Chemistry and Physics, 2008, 209, 2179.
[120] S.E. Averick, A.J.D. Magenau, A. Simakova, B.F. Woodman, A. Seong, R.A. Mehl and K. Matyjaszewski, Polymer Chemistry, 2011, 2, 1476.
[121] R.H. Harrison, J.A.M. Steele, R. Chapman, A.J. Gormley, L.W. Chow, M.M. Mahat, L. Podhorska, R.G. Palgrave, D.J. Payne, S.P. Hettiaratchy, I.E. Dunlop and M.M. Stevens, Advanced Functional Materials, 2015, 25, 5748.
[122] Z. Yang, S. Yuan, B. Liang, Y. Liu, C. Choong and S.O. Pehkonen, Macromolecular Bioscience, 2014, 14, 1299.
[123] A.E. Rodda, F. Ercole, V. Glattauer, J. Gardiner, D.R. Nisbet, K.E. Healy, J.S. Forsythe and L. Meagher, Biomacromolecules, 2015, 16, 2109.
[124] W. Yuana, Y. Feng, H. Wang, D. Yang, B. An, W. Zhang, M. Khan and J. Guo, Materials Science and Engineering: C, 2013, 33, 3644.
[125] N.Y. Kostina, O. Pop-Georgievski, M. Bachmann, N. Neykova, M. Bruns, J. Michalek, M. Bastmeyer and C. Rodriguez-Emmenegger, Macromolecular Bioscience, 2016, 16, 83.
[126] H.S. Kima and H.S.Yoo, Chemical Communications, 2015, 51, 306.
[127] D.E. Borchmann, T.P. Carberry and M. Weck, Macromolecular Rapid Communications, 2014, 35, 27.
[128] Z. Lin, S. Cao, X. Chen, W. Wu and J. Li, Biomacromolecules, 2013, 14, 2206.
[129] S. Atzet, S. Curtin, P. Trinh, S. Bryant and B. Ratner, Biomacromolecules, 2008, 9, 3370.
[130] Z. Ren, Y. Wang, S. Ma, S. Duan, X. Yang, P. Gao, X. Zhang and Q. Cai, ACS Applied Materials & Interfaces, 2015, 7, 19006.
[131] A. Galperin, T.J. Long, S. Garty and B.D. Ratner, Journal of Biomedical Materials Research Part A, 2013, 101, 775.
[132] Y. Jiang, J. Chen, C. Deng, E.J. Suuronen and Z. Zhong, Biomaterials, 2014, 35, 4969.
[133] J.C. Igwe. P.E. Mikael and S.P. Nukavarapu, Journal of Tissue Engineering and Regenerative Medicine, 2014, 8, 131.
[134] C. Rodriguez-Emmenegger, M. Houska, A.B. Alles and E. Brynda, Macromolecular Bioscience, 2012, 12, 1413.
[135] A.R. Kuzmyn, A. de los Santos Pereira, O. Pop-Georgievski, M. Bruns, E. Brynda and C. Rodriguez-Emmenegger, Polymer Chemistry, 2014, 5, 4124.
[136] Z. Tang, Y. Luan, D. Li, H. Du, D.M. Haddleton and H. Chen, Chemical Communications, 2015, 51, 14263.
[137] W. Jaeger, J. Bohrisch and A. Laschewsky, Progress in Polymer Science, 2010, 35, 511.
[138] Y. Sui, X. Gao, Z. Wang and C. Gao, Journal of Membrane Science, 2012, 394–395, 107.
[139] L.Q. Xu, N.N. Li, J.C. Chen, G.D. Fu and E.-T. Kangd, RSC Advances, 2015, 5, 61742.
[140] S. Yuan, G. Xiong, F. He, W. Jiang, B. Liang and C. Choong, Journal of Materials Chemistry B, 2015, 3, 8088.
[141] J. Qian and X. Gao, ACS Applied Materials & Interfaces, 2013, 5, 2845.
[142] S. Li, Y. Li, C.A. Wisner, L. Jin, N. Leventis and Z. Peng, RSC Advances, 2014, 4, 35823.
[143] N. Erathodiyil and J.Y. Ying, Accounts of Chemical Research, 2011, 44, 925.

[144] C.Y. Chen, C.K. Syu and H.C. Lin, Macromolecular Bioscience, 2016, 2, 188.
[145] S. Wang, F. Yuan, K. Chen, G. Chen, K. Tu, H. Wang and L.Q. Wang, Biomacromolecules, 2015, 16, 2693.
[146] Y. Liu, P. Chen and Z. Li, Macromolecular Rapid Communications, 2012, 33, 287.
[147] H. Zhou, W. Jiang, N. An, Q. Zhang, S. Xiang, L. Wang and J. Tang, RSC Advances, 2015, 5, 42728.
[148] Y. Zhu, G-P. Tang and F.-J. Xu, ACS Applied Materials & Interfaces, 2013, 5, 1840.
[149] Z.H. Wang, W.B. Li, J. Ma, G.P. Tang, W.T. Yang and F.J. Xu, Macromolecules, 2011, 44, 230.
[150] A. Kumar, S.V. Lale, S. Mahajan, V. Choudhary and V. Koul, ACS Applied Materials & Interfaces, 2015, 7, 9211.
[151] L. Zhang, W. Zhao, X. Liu, G. Wang, Y. Wang, D. Li, L. Xie, Y. Gao, H. Deng and W. Gao, Biomaterials, 2015, 64, 2.
[152] R.Q. Li, Y. Hu, B.R. Yu, N.N. Zhao and F.J. Xu, Bioconjugate Chemistry, 2014, 25, 155.
[153] T. Zhao, H. Zhang, B. Newland, A. Aied, D. Zhou and W. Wang, Angewandte Chemie International Edition, 2014, 53, 6095.
[154] Y. Hu, Y. Zhu, W.T. Yang and F.J. Xu, ACS Applied Materials & Interfaces, 2013, 5, 703.
[155] J. Liu, Y. Xu, Q. Yang, C. Li, W.E. Hennink, R. Zhuo and X. Jiang, Acta Biomaterialia, 2013, 9, 7758.
[156] C. Boyer, N.A. Corrigan, K. Jung, D. Nguyen, T.-K. Nguyen, N.N.M. Adnan, S. Oliver, S. Shanmugam and J. Yeow, Chemical Reviews, 2016, 116, 1803.
[157] A. Simakova, M. Mackenzie, S.E. Averick, S. Park and K. Matyjaszewski, Angewandte Chemie International Edition, 2013, 52, 12148.
[158] E.P. Magennis, F. Fernandez-Trillo, C. Sui, S.G. Spain, D.J. Bradshaw, D. Churchley, G. Mantovani, K. Winzer and C. Alexander, Nature Materials, 2014, 13, 748.
[159] M. Li, H. Li, P. De and B.S. Sumerlin, Macromolecular Rapid Communications, 2011, 32, 354.
[160] H. Li, M. Li, X. Yu, A.P. Bapata and B.S. Sumerlin, Polymer Chemistry, 2011, 2, 7, 1531.
[161] S.N.S. Alconcel, S.H. Kim, L. Tao and H.D. Maynard, Macromolecular Rapid Communications, 2013, 34, 12, 983.
[162] Y. Zhao and S. Perrier, Chemical Communications, 2007, 4294.
[163] J. Liu, H. Liu, V. Bulmus, L. Tao, C. Boyer, T.P. Davis, Journal of Polymer Science Part A: Polymer Chemistry, 2010, 48, 6, 1399.
[164] N. Vanparijs, R. De Coen, D. Laplace, B. Louage, S. Maji, L. Lybaert, R. Hoogenboom and B.G. De Geest, Chemical Communications, 2015, 51, 73, 13972.
[165] P. De, M. Li, S.R. Gondi and B.S. Sumerlin, Journal of the American Chemical Society, 2008, 130, 34, 11288.
[166] S.S. Zhang, K. Cui, J. Huang, Q.L. Zhao, S.K. Cao and Z. Ma, RSC Advances, 2015, 5, 44571.
[167] M. Guerre, B. Ameduri and V. Ladmiral, Polymer Chemistry, 2016, 7, 441.
[168] C. Boyer, V. Bulmus, T.P. Davis, V. Ladmiral, J. Liu and S. Perrier, Chemical Reviews, 2009, 109, 5402.
[169] N. Vanparijs, S. Maji, B. Louage, L. Voorhaar, D. Laplace, Q. Zhang, Y. Shi, W.E. Hennink, R. Hoogenboom and B.G. De Gees, Polymer Chemistry, 2015, 6, 5602.
[170] G.N. Grover, J. Lee, N.M. Matsumoto and H.D. Maynard, Macromolecules, 2012, 45, 4958.
[171] D.J. Phillips and M.I. Gibson, Chemical Communications, 2012, 48, 1054.
[172] K.L. Heredia, G.N. Grover, L. Tao and H.D. Maynard, Macromolecules, 2009, 42, 2360.
[173] K.L. Heredia, T.H. Nguyen, C.W. Chang, V. Bulmus, T.P. Davis and H.D. Maynard, Chemical Communications, 2008, 3245.
[174] X. Wang, L. Liu, Y. Luo, H. Shi, J. Li and H. Zhao, Macromolecular Bioscience, 2012, 12, 1575.
[175] M. Li, P. De, H. Li and B.S. Sumerlin, Polymer Chemistry, 2010, 1, 854.

[176] G. Pound, J.M. McKenzie, R.F.M. Lange and B. Klumperman, Chemical Communications, 2008, 3193.
[177] N.Y. Xiao, A.L. Li, H. Liang and J. Lu, Macromolecules, 2008, 41, 2374.
[178] L. Tao, J. Liu, J. Xu and T.P. Davis, Organic & Biomolecular Chemistry, 2009, 7, 3481.
[179] M. Li, P. De, S.R. Gondi and B.S. Sumerlin, Macromolecular Rapid Communications, 2008, 29, 1172.
[180] D. Moatsou, J. Li, A. Ranji, A. Pitto-Barry, I. Ntai, M.C. Jewett and R.K. O'Reilly, Bioconjugate Chemistry, 2015, 26, 1890.
[181] M. Foldvari, D.W. Chen, N. Nafissi, D. Calderon, L. Narsineni and A. Rafiee, Journal of Controlled Release, 2015, doi:10.1016/j.jconrel.2015.12.012.
[182] M. Ahmed and R. Narain, Progress in Polymer Science, 2013, 38, 767.
[183] S. Sun and C. Gao, Biomacromolecules, 2010, 11, 3609.
[184] S.G. Roy and P. De, Journal of Applied Polymer Science, 2014, 41084 (1–12).
[185] K. Bauri, S.G. Roy and P. De, Macromolecular Chemistry and Physics, 2016, 217, 65.
[186] S. Kumar, S.G. Roy and P. De, Polymer Chemistry, 2012, 3, 1239.
[187] S. Kumar, R. Acharya, U. Chatterji and P. De, Langmuir, 2013, 29, 49, 15375.
[188] K. Bauri, S.G. Roy, S. Pant and P. De, Langmuir, 2013, 29, 2764.
[189] K. Bauri, S. Pant, S.G. Roy and P. De, Polymer Chemistry, 2013, 4, 4052.
[190] S.G. Roy, R. Acharya, U. Chatterji and P. De, Polymer Chemistry, 2013, 4, 1141.
[191] S.G. Roy, K. Bauri, S. Pal and P. De, Polymer Chemistry, 2014, 5, 3624.
[192] S. G. Roy and P. De, Polymer Chemistry, 2014, 5, 6365.
[193] A.B. Lowe, B.S. Sumerlin and C.L. McCormick, Polymer, 2003, 44, 6761.
[194] L. Albertin, M.H. Stenzel, C. Barner-Kowollik and T.P. Davis, Polymer, 2006, 47, 1011.
[195] Z. Deng, M. Ahmed and R. Narain, Journal of Polymer Science Part A: Polymer Chemistry, 2009, 47, 614.
196. H. Shi, L. Liu, X. Wang and J. Li, Polymer Chemistry, 2012, 3, 1182.
[197] S.R.S. Ting, A.M. Gregory and M.H. Stenzel, Biomacromolecules, 2009, 10, 342.
[198] S. Pearson, D. Vitucci, Y.Y. Khine, A. Dag, H. Lu, M. Save, L. Billon and M.H. Stenzel European Polymer Journal, 2015, 69, 616.
[199] S. Kumar, B. Maiti and P. De, Langmuir, 2015, 31, 9422.
[200] M. Obata, T. Kobori, S. Hirohara and M. Tanihara, Polymer Chemistry, 2015, 6, 1793.
[201] E.H. Min, S.R.S. Ting, L. Billon and M.H. Stenzel, Journal of Polymer Science Part A: Polymer Chemistry, 2010, 48, 3440.
[202] M. Toyoshima and Y. Miura, Journal of Polymer Science Part A: Polymer Chemistry, 2009, 47, 1412.
[203] M. Takara, M. Toyoshima, H. Seto, Y. Hoshino and Y. Miura, Polymer Chemistry, 2014, 5, 931.
[204] M. Toyoshima, T. Oura, T. Fukuda, E. Matsumoto and Y. Miura, Polymer Journal, 2010, 42, 172.
[205] J.L. Zhu, H. Cheng, Y. Jin, S.X. Cheng, X.Z. Zhang and R.X. Zhuo, Journal of Materials Chemistry, 2008, 18, 4433.
[206] W. Li, M. Nakayama, J. Akimoto and T. Okano, Polymer, 2011, 52, 3783.
[207] P. Chytil, E. Koziolová, O. Janoušková, L. Kostka, K. Ulbrich, T. Etrych, Macromolecular Bioscience, 2015, 15, 839.
[208] B. Maiti and P. De, RSC Advances, 2013, 3, 24983.
[209] B. Maiti, S. Kumar and P. De, RSC Advances, 2014, 4, 56415.
[210] U. Haldar, M. Nandi, B. Ruidas and P. De, European Polymer Journal, 2015, 67, 274.
[211] K. Bauri, P. De, P.N. Shah, R. Li and R. Faust, Macromolecules, 2013, 46, 5861.
[212] K. Bauri, R. Li, R. Faust and P. De, Macromolecular Symposia, 2015, 349, 65.

[213] E. Gallon, T. Matini, L. Sasso, G. Mantovani, A.A. de Benito, J. Sanchis, P. Caliceti, C. Alexander, M.J. Vicent and S. Salmaso, Biomacromolecules, 2015, 16, 1924.
[214] K. Bauri, A. Narayanan, U. Haldar and P. De, Polymer Chemistry, 2015, 6, 6152.
[215] S. Kumar, V. Bheemireddy and P. De, Macromolecular Bioscience, 2015, 15, 1447.
[216] S. Kumar, R. Acharya, U. Chatterji and P. De, Polymer Chemistry, 2014, 5, 6039.
[217] S. Pal, S.G. Roy and P. De, Polymer Chemistry, 2014, 5, 1275.
[218] Y. Li, K. Xiao, W. Zhu, W. Deng and K.S. Lam, Advanced Drug Delivery Reviews, 2014, 66, 58.
[219] Z. Xu, D. Wang, S. Xu, X. Liu, X. Zhang and H. Zhang, Chemistry – An Asian Journal, 2014, 9, 199.
[220] Y. Zhuang, Y. Su, Y. Peng, D. Wang, H. Deng, X. Xi, X. Zhu and Y. Lu, Biomacromolecules, 2014, 15, 1408.
[221] P. Chytil, E. Koziolová, O. Janoušková, L. Kostka, K. Ulbrich and T. Etrych Macromolecular Bioscience, 2015, 15, 839.
[222] D.D. Lane, F.Y. Su, D.Y. Chiu, S. Srinivasan, J.T. Wilson, D.M. Ratner, P.S. Stayton and A.J. Convertine, Polymer Chemistry, 2015, 6, 1255.
[223] L. Qiu, C.Y. Hong and C.Y. Pan, International Journal of Nanomedicine, 2015, 10, 3623.
[224] C.E. Wang, P.S. Stayton, S.H. Pun and A.J. Convertine, Journal of Controlled Release, 2015, 219, 345.
[225] B.S. Tucker, S.G. Getchell, M.R. Hill and B.S. Sumerlin, Polymer Chemistry, 2015, 6, 4258.
[226] K. Mishra and A. Joy, Polymer, 2015, 66, 110.
[227] P. De, S.R. Gondi and B.S. Sumerlin, Biomacromolecules, 2008, 9, 1064.
[228] D. Das, S. Srinivasan, A.M. Kelly, D.Y. Chiu, B.K. Daugherty, D.M. Ratner, P.S. Stayton and A.J. Convertine, Polymer Chemistry, 2016, 7, 826.
[229] R. Zhang, J. Yang, Y. Zhou, P.J. Shami and J. Kopeček, Macromolecular Bioscience, 2016, 16, 121.
[230] W. Shen, Y. Chang, G. Liu, H. Wang, A. Cao and Z. An, Macromolecules, 2011, 44, 2524.
[231] O. Shimoni, A. Postma, Y. Yan, A.M. Scott, J.K. Heath, E.C. Nice, A.N. Zelikin and F. Caruso, ACS Nano, 2012, 6, 1463.
[232] Y. Dong, Y. Qin, M. Dubaa, J. Killion, Y. Gao, T. Zhao, D. Zhou, D. Duscher, L. Geever, G.C. Gurtner and W. Wang, Polymer Chemistry, 2015, 6, 6182.
[233] S.E. Kirkland, R.M. Hensarling, S.D. McConaughy, Y. Guo, W.L. Jarrett and C.L. McCormick, Biomacromolecules, 2008, 9, 481.
[234] N. Patil, J. Soni, N. Ghosh and P. De, Journal of Physical Chemistry B, 2012, 116, 13913.
[235] N. Patil, S.G. Roy, U. Haldar and P. De, Journal of Physical Chemistry B, 2013, 117, 16292.
[236] U. Haldar, M. Nandi, B. Maiti and P. De, Polymer Chemistry, 2015, 6, 5077.
[237] S.G. Roy, U. Haldar and P. De, ACS Applied Materials & Interfaces, 2014, 6, 4233.
[238] S.G. Roy and P. De, Polymer, 2014, 55, 5425.
[239] U. Haldar, K. Bauri, R. Li, R. Faust and P. De, ACS Applied Materials & Interfaces, 2015, 7, 8779.
[240] A. Vaish, S.G. Roy and P. De, Polymer, 2015, 58, 1.
[241] B. Maiti, B. Ruidas and P. De, Reactive and Functional Polymers, 2015, 93, 148.
[242] D. Roy, J.S. Knapp, J.T. Guthrie and S. Perrier, Biomacromolecules, 2008, 9, 91.
[243] L.C. Wu, J. Yang and J. Kopeček, Biomaterials, 2011, 32, 5341.
[244] C.E. Nelson, M.K. Gupta, E.J. Adolph, J.M. Shannon, S.A. Guelcher and C.L. Duvall, Biomaterials, 2012, 33, 1154.
[245] G.D. Carlo, F. Damin, L. Armelao, C. Maccato, S. Unlud, P.S. Spuhler and M. Chiari, Applied Surface Science, 2012, 258, 3750.

[246] L.-J. Zhu, L.-P. Zhu, Y.-F. Zhao, B.-K. Zhu and Y.-Y. Xu, Journal of Materials Chemistry A, 2014, 2, 15566.
[247] M. Ma, S. Zheng, H. Chen, M. Yao, K. Zhang, X. Jia, J. Mou, H. Xu, R. Wu and J. Shi Journal of Materials Chemistry B, 2014, 2, 5828.
[248] J.S. Basuki, A. Jacquemin, L. Esser, Y. Li, C. Boyer and T.P. Davis, Polymer Chemistry, 2014, 5, 2611.
[249] H. Deng, F. Dai, G. Ma and X. Zhang, Advanced Materials, 2015, 27, 3645.

Abbreviations

AGET	Activator generated by electron transfer
AgNP	Silver nanoparticles
AIBN	Azobisisobutyronitrile
ARGET	Activator regenerated by electron transfer
AROP	Anionic ring-opening polymerization
ATRA	Atom transfer radical addition
ATRC	Atom transfer radical coupling
ATRP	Atom transfer radical polymerization
AuNP	Gold nanoparticles
BA_6TREN	Tris(2-bis(3-butoxy-3-oxopropyl)aminoethyl)amine
BnSH	Benzyl mercaptan
bpy	2, 2′-Bipyridine
BnMA	Benzyl methacrylate
CD	Circular dichroism
CDDS	Controlled drug delivery system
CMC	Critical micellar concentration
CMS	p-chloromethyl styrene
CNT	Carbon nanotube
CRP	Controlled Radical Polymerization
CTA	Chain transfer agent
CuAAC	Copper catalysed alkyne-azide click
DA	Diels-Alder reaction
DBF	Diblock fluoropolymers
DBU	1,8-Diazabicyclo[5.4.0]undec-7-ene
DEA	N,N-Diethylacrylamide
DEGMA	Di(ethylene glycol)methyl ether methacrylate
DMA	N,N-dimethylacrylamide
DMAc	Dimethylacetamide
DMF	N,N-dimethylformamide
DMPA	2,2-Dimethoxy-2-phenylacetophenone
DMPP	Dimethylphenylphosphine
DMSO	Dimethylsulfoxide
dNbpy	4,4′-Di-5-nonyl-2,2′-bipyridine
DOX	Doxorubicin
DP	Degree of polymerization
DSC	Differential scanning Calorimetry
FRP	Free radical polymerization
FTIR	Fourier transform infrared
GA	Glycidyl acrylate
GMA	Glycidyl methacrylate
GTP	Group transfer polymerization
HDA	Hetero Diels-Alder reaction
HEA	2-Hydroxy ethyl acrylate
HEAm	N-(2-hydroxyethyl)acrylamide
HEMA	2-Hydroxy ethyl methacrylate
HMTETA	1, 1, 4, 7, 10, 10-hexamethyltriethylenetetramine
ITP	Iodine transfer polymerization

https://doi.org/10.1515/9783110643695-008

LAMs	Less-activated monomers
LCST	Lower critical solution temperature
LRP	Living radical polymerization
MAA	Methacrylic acid
MA	Methyl acrylate
MALDI-TOF MS	Matrix-assisted laser desorption/ionization time of flight mass spectrometry
MAMs	More-activated monomers
Me_6TREN	Tris(2-dimethylaminoethyl)amine
MMA	Methyl methacrylate
MMT	Montmorillonite
Mn	Number average molecular weight
MNP	Magnetite nanoparticles
MSN	Mesoporous silica nanoparticle
MWD	Molecular weight distribution
NG	Nanogel
NI	Nano-imprinting
NIPAM	N-isopropylacrylamide
NMP	Nitroxide-mediated polymerization
NMR	Nuclear magnetic resonance
NP	Nano-particle
P3HT	Poly(3-hexylthiophene)
PAN	Poly(acrylonitrile)
PC	Polycarbonate
PCL	Poly(ε-caprolactone)
PDEAEMA	Poly(N,N-diethylaminoethyl methacrylate)
PDEAM	Poly(N,N-diethylacrylamide)
PDEGMA	Poly(diethylene glycol methacrylate)
PDMAEMA	Poly(N,N-dimethylaminoethyl methacrylate)
PDMAEMA	Poly(N,N-dimthylaminoethyl methacrylate)
PDMAEMA	Polydimethylaminoethyl methacrylate
PEGA	Polyethyleneglycol acrylate
PEGMA	Poly(ethylene glycol methacrylate)
PEGMA	Poly(ethylene glycol) monomethacrylate
PFPMA	Pentafluorophenyl methacrylate
PFPS	Pentafluorophenylstyrene
PHEMA	Polyhydroxyethyl methacrylate
PHFMA	Polyhexafluromethyl acrylate
PLMA	Polylauryl methacrylate
PMA	Polymethyl acrylate
PMDETA	N, N, N',N",N"-pentamethyldiethylenetriamine
PMMA	Poly(methyl methacrylate)
PNIPAM	Poly(N-isopropylacrylamide)
POEGMA	poly[oligo(ethylene glycol) methyl ether methacrylate]
POSS	Polyhedral oligomeric silsesquioxane
PPMA	3-phenylpropyl methacrylate
PR	Primary radical
PRE	Persistent radical effect
PSt	Polystyrene

PtBA	Poly(t-butyl acrylate)
PTFE	Polytetrafluoroethylene
PVP	Polyvinyl pyrrolidone
RAFT	Reversible addition-fragmentation chain transfer
rDA	Retro Diels-Alder reaction
RDRP	Reversible deactivation radical polymerization
RITP	Reverse iodine transfer polymerization
ROMP	Ring opening metathesis polymerization
ROP	Ring opening polymerization
SAM	Self-assembled monolayer
SARA	Supplemental activator and reducing agent
SBS	Styrene-butadiene-styrene
SEC	Size exclusion chromatography
SEM	Scanning electron microscopy
SET-LRP	Single electron transfer-living radical polymerization
SET-RAFT	Single electron transfer-RAFT
SI-ATRP	Surface-initiated atom transfer radical polymerization
SiNP	Silica nanoparticles
SIP	Surface initiated polymerization
SI-RAFT	Surface-initiated reversible addition-fragmentation chain transfer
SPAAC	Stress-promoted azide-alkyne coupling
TAD	Triazolinedione
TBF	Triblock fluoropolymer
tBMA	t-butyl methacrylate
TCEP	Tris(2-carboxyethyl)phosphine
TEM	Transmission electron microscopy
TEMPO	2,2,6,6-tetramethylpiperidine-1-oxyl
TEOS	Tetraethoxysilane
TERP	Tellurium-mediated radical polymerization
TFA	Trifluoroacetic acid
THF	Tetrahydrofuran
THPMA	2-tetrahydropyranyl methacrylate
TIPS	Triisopropyl silyl
TMS	Trimethyl silyl
TMSMA	Trimethylsilyl methacrylate
UV-Radiations	Ultra-violet radiations
VDF	Vinylidene fluoride
WCA	Water contact angle

Index

https://doi.org/10.1515/9783110643695-009

Bei Fragen zur Produktsicherheit wenden Sie sich bitte an:
If you have any questions regarding product safety,
please contact:

Walter de Gruyter GmbH
Genthiner Straße 13
10785 Berlin
productsafety@degruyterbrill.com